Flexible Flyer and other Great Sleds for Collectors

Joan Palicia

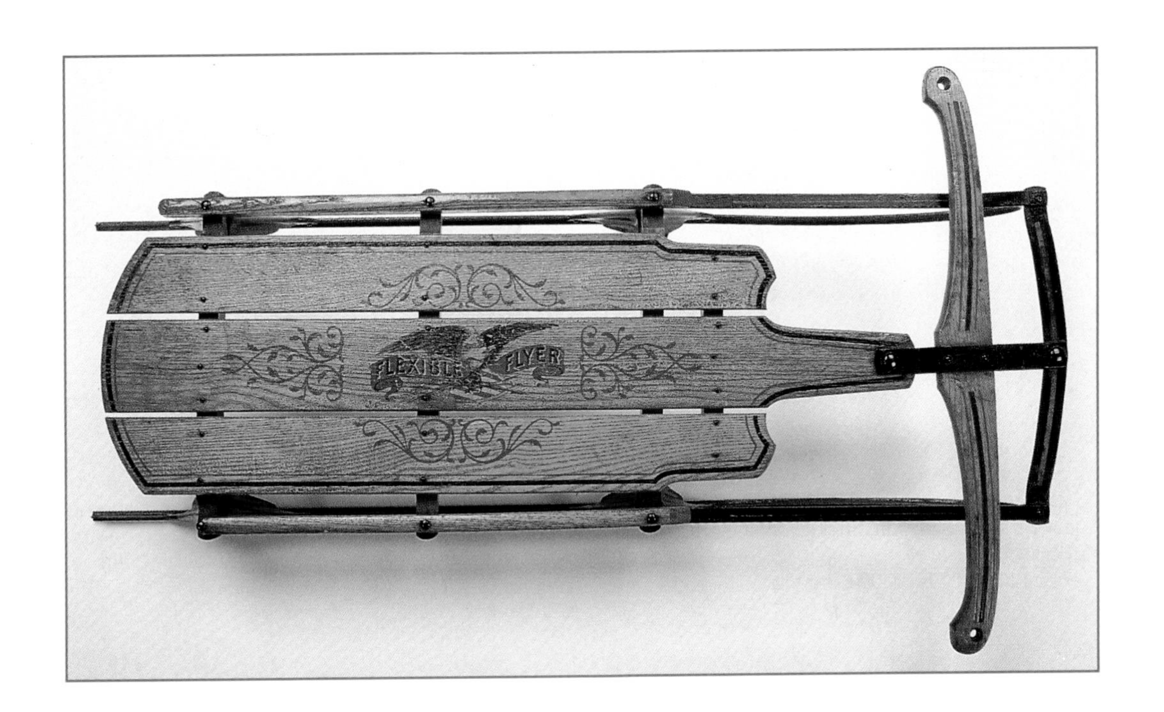

4880 Lower Valley Road, Atglen, PA 19310

Dedication

To my husband Mel, a friend and confidante without whose encouragement, support, great patience, and enthusiasm, this book would still be a dream. Thanks for making my dream a reality, but most of all, thank you for your love.

To my children Glen and Deborah for allowing me to remain a child for more years than I care to remember, for exceeding my expectations and reminding me it is better to have tried and failed than not to have tried at all. Thanks for your encouragement and support. This book is for you. I'm honored to be your mother.

In Memoriam

In Memory of Bessie A. Reise, my mother, and Dr. Gerald F. Reise D.D.S., my brother.

Library of Congress Catalog Card Number: 97-65989

Design by Blair R.C. Loughrey

ISBN: 0-7643-0103-9
Printed in Hong Kong

Published by Schiffer Publishing Ltd.
4880 Lower Valley Road
Atglen, PA 19310
Phone: (610) 593-1777; Fax: (610) 593-2002
E-mail: Schifferbk@aol.com
Please write for a free catalog.
This book may be purchased from the publisher.
Please include $2.95 for shipping.
Try your bookstore first.

We are interested in hearing from authors
with book ideas on related subjects.

Contents

Group of children near the Lightning Guider sled factory, Ducannon, Pennsylvania, c.1920. *Courtesy of The Sledworks.*

Acknowledgments

No one writes a book alone. The completion of this book would have been impossible without the generous assistance of many individuals, collectors, historians, and institutions around the country.

I owe a debt of gratitude to the collectors who shared their collections with me and to the historians for access to historical ephemera, trade literature, and other research material.

I would like to thank the following individuals for contributing to making this book a success. Let me first begin with thanking the original Flexible Flyer® Company, S.L. Allen, Philadelphia, Pennsylvania, for a lifetime of wonderful memories and to Roadmaster-Flexible Flyer® for perpetuating the memories with the release of their new collector's edition, Fall, 1996.

New books would be difficult to write without the assistance of many libraries. Many thanks to the following for their contributions: Cedar Falls Public Library, Iowa (Wagner Auto Coaster), Fletcher Free Library in Burlington, Vermont (Emerson-Johnson), Kalamazoo Public Library, Michigan (Champion), Mead Public Library in Sheboygan, Wisconsin (Garton), and North Tonawanda Public Library and Historical Society, New York (Auto Wheel Coaster).

A very special thanks to Selma Davidson for supplying the Hunt, Helm, and Ferris catalog and to Don Schumacher of Cannon Ball Industries for supplying patent numbers for Hunt, Helm, and Ferris patents. To Barbara Brandage the administrator of the Agriculture and Industrial Museum of York County, Pennsylvania for taking her time and opening the museum to share information on the American Acme Company. To Tom Johnson of Hopkinton, New Hampshire for stumbling onto the son of the inventor of the Icecycle and putting me in touch with John Everett Bean. Thanks to John, son of the inventor, we have the only known photo of the adult Icecycle.

To former strangers and now new friends, I would like to thank Lou and Carol Scudillo and Jimmy Rosen for allowing us to photograph their wonderful sled collections and to Henry R. Morton for his hospitality and generosity for allowing us the use of his Paris Catalogs.

A very special thanks to Sotheby's, New York, for lending their catalog photo and granting exclusive rights to reproduce "Rosebud." No book on sleds would be complete without it.

And to my family and special friends Muriel, Helen, and Marie for sharing in my excitement and accompanying me on those journeys to "where?" "Let's call AAA!!!" Many thanks.

Last but not least, I would like to thank the entire staff at Schiffer Publishing, but most of all to Peter Schiffer for seeing the value in this type of book and for taking a chance with me. To Douglas Congdon-Martin, editor, for calming me down and keeping me focused when everything looked "blurred." Those who know me understand that this was a monumental task in itself.

As the author, I hope you enjoy reading this book as much as I enjoyed collecting, researching, and writing on my favorite subject, sleds.

I welcome any additional information and/or corrections for future editions. You may write to Joan Palicia, 15 Canton Road, Wayne, NJ 07470. Please send self-addressed stamped envelope for reply.

Clark Gable's childhood sled. It recently sold at Christies for a reported $8,050.00. It is marked 1899 and is scripted "Clark" on the deck. 45" in length.
Courtesy of Christie's Inc. Used with Permission.

Introduction

It is said that imitation is the sincerest form of flattery and imitated Flexible Flyers® were. As the patent rights expired virtually every company that manufactured sleds "copied" the revolutionary new design of the Flexible Flyer®, in some cases right down to the advertising.

The companies mentioned in this book are the ones I have been able to research. As with any research, it is painstakingly slow. Most of the companies have been plagued with fire or floods making it a miracle that any information was retrieved at all.

Sled collecting as well as research is an adventure. For example, if you have two sleds with the same name you would think they were made by the same company, but this is not necessarily so. Names were interchanged from one company to another. So where do you go from here?

First you should look for any company names, marks, or numbers which are almost always found on the back side of the center deck board unless they are worn off. In that case, you must rely on visual differences. Look at the design and style of the frame, steering bar, and deck. Thanks to the expert photography of Douglas Congdon-Martin you will be able to see these differences.

Original catalogs, magazine advertisements, and patent numbers have been used to date and identify sled companies.

All sleds pictured are in original condition none have been repaired or restored. Because I consider myself a purist, the sleds are washed with a mild soap and dried well. At times varnish has been applied over the trademark to keep it from flaking. The words "skillfully restored in original colors" make me cringe! How skillful is "skillful" and what are original colors? You will have to make your own decisions as to repair or restoration. In my opinion, value is subtracted once a sled has been "restored."

A word about value. Value is determined by condition, availability, desirability, rarity, and potential for resale. Values quoted in this book are based on my own experiences. I do not claim or intend to set a market price. Price is determined solely by the buyer and the seller, and many factors go into determining a fair price. Some people will automatically double book value, assuming a book value is low! Remember, buy it because you love it and for no other reason. Let the price guide be just that: a guide. Do not assume you have a "rare" item if it is not included in the book. Newer items have intentionally been left out. Buyer beware...nostalgia has no price.

As you leaf through these pages take special notice of the small details. Remember a slight change in style, design, or color can put your sled in a different decade, increasing or decreasing its value.

When sleds in mint condition were not available for photography, I have included many company catalogs along with printed ephemera and magazine advertisements to help with identification.

Whatever your interest, sledding, advertising, or simple pleasure in old, nostalgic things, I hope you are able to find something of value in these pages that, if only for a short time, allows you to revisit the past.

The Anatomy of a Sled

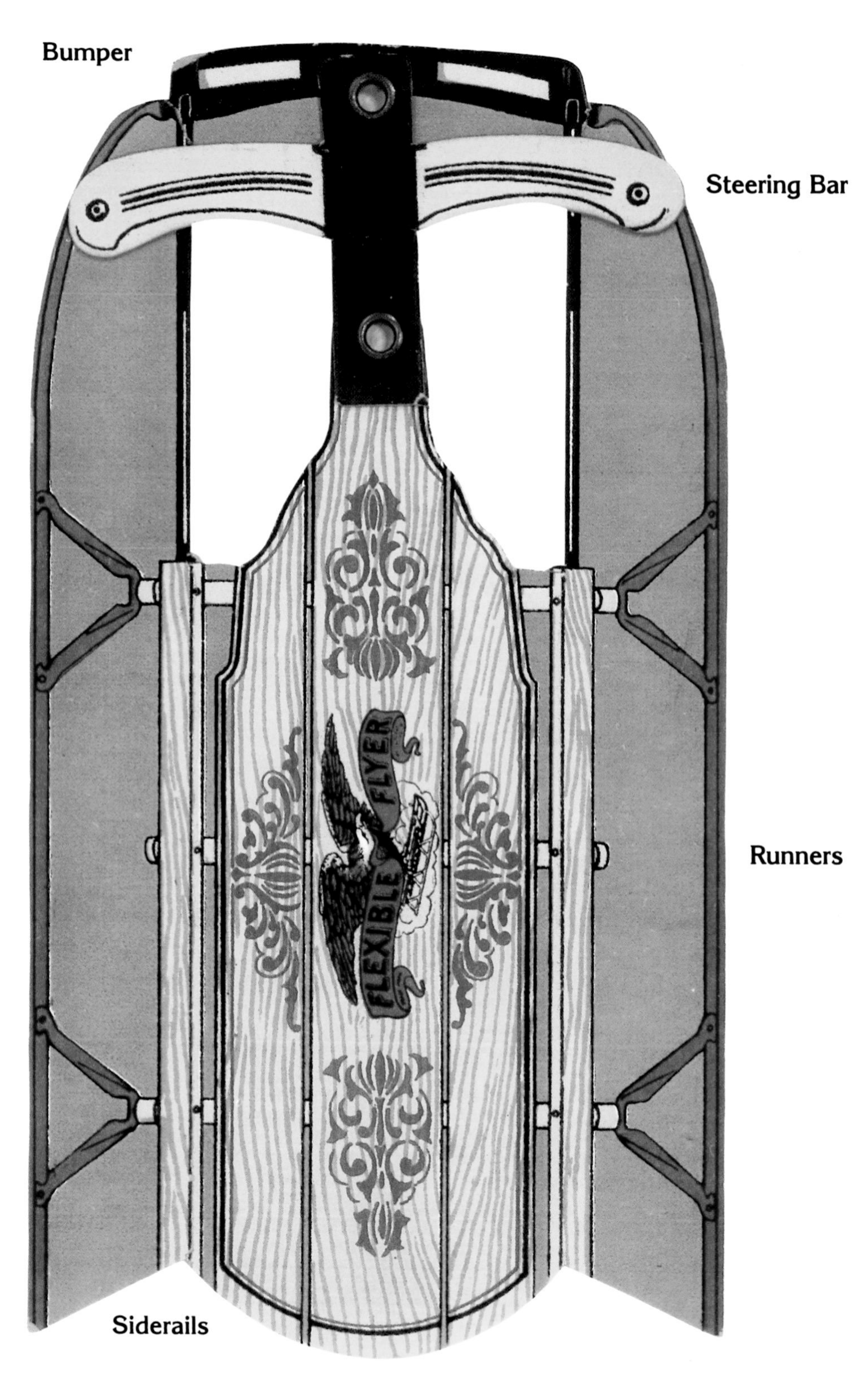

Common Sledding Terms

Sledge. A vehicle with low runners that is used for transporting loads over snow and ice.

Sled. Used by children for coasting down snow-covered hills.

Sledding. Conditions under which one may use a sled.

Sleigh. A vehicle on runners used for transporting persons or goods on snow or ice.

Clipper. Long and low slung sled with the deck mounted directly onto low, "squatty" wooden runners. Designed for boys and perfect for belly flops. The rider throws himself on the deck and speeds down the hill head first. Clipper runners thrust to a point. Speed was most important for boys.

Cutters. More refined and sedate, the deck set high on an open framework above wooden runners. Designed for girls or younger children to ride sitting up. Decorated (painted) with flowers or dainty motifs with rounded or bow runners curled elegantly upward in front. Beauty, not speed, was most important for girls.

Toboggan. A long, flat-bottomed sled made of thin boards curved up at the front end to form a "hood," usually four to nine feet long, about 18 inches wide with low hand rails at the sides, cushioned seat to accommodate one to five persons, depending on the length of the sled.

Hitching. A sport and means of locomotion, by catching the rope of your sled around the rear strut of a sleigh runner while in motion (horse-drawn, that is), flop on your sled and hang on.

Tailing: Hitching a number of sleds in pairs behind a horse drawn sleigh. The rider of the first sled hooks his feet into the front of the steering bar on the second sled.

Bob Sled: Two short sleds, coupled together in tandem having a steering wheel and brake, used in racing down a steeply banked, open chute.

I. Flexible Flyers®

"What puts color in the cheeks and sparkle in the eyes and sets the whole body tingling with health and happiness?" Is this the ranting of a Snake Oil medicine man on the virtues of a new patent medicine? Actually not—"It's coasting!"

Coasting or sledding has existed in America since colonial times. Racing was what it was all about. Early sleds bore names like Comet, Reliance, Thunder, or Flying Cloud. These names reflected the hopes of their owners to descend the hill as fast as the wind to a victory!

Early American sleds were made of hard wood with runners that frequently turned up in "bow fashion." Sleds also had "gender," and sledding by the 19th Century had strict rules.

Clippers were for boys, sleds with runners that thrust to a point. This was most important for speed as the boys would "belly whopper" head first down the hill, going as fast as they could. **Cutters** were for girls, being more sedate and refined. Their runners curled elegantly upward, in bow fashion. Beauty, not speed, was most important for girls, who "sat straight up" as not to show their petticoats as they descended the hill.

In 1861, one of the nation's first sled makers, Henry F. Morton, opened the Paris Manufacturing Corporation of South Paris, Maine, where he built non-steerable sleds that his wife hand-painted with various motifs. Today these sleds are highly collectible and bring hundreds of dollars at auction.

Sledding drastically changed in the late 1800s with the advent of the "Flexible Flyer®," the nation's first steerable and most recognizable sled. The Flexible Flyer® was the invention of a grown-up boy, Samuel Leeds Allen. Mr. Allen was born on May 5, 1841, in Philadelphia into a prominent Quaker family. At the age of eleven he was sent to Westtown Boarding School in Chester County, Pennsylvania. Later this would prove to be the testing ground for all the sleds he designed, preceding and including the Flexible Flyer®.

After graduation, he moved to the family farm in 1861 near Westfield, New Jersey, half-way between Moorestown and Riverton. There he married (November 22, 1867) and settled into farm life. To make life easier he conceived an idea to construct a device that could be rolled over the earth and had the means of fertilizing and planting seeds. This device resembled the Planet Saturn with its rings. Before long the soon-to-be world-renowned "S.L. Allen Planet Jr. Company," makers of farm and garden equipment, was born.

Being a seasonal business, Mr. Allen needed to come up with a product he could manufacture in the summer and sell in the winter. His good workers would leave for summer employment elsewhere and not return. Mr. Allen searched the dictionary for an idea when he came upon the word "sled."

An avid outdoorsman, he committed himself to extensive research. His marketing took him to all the hot spots for coasting, Orange, New Jersey, Albany and Rochester, New York, and Burlington, Vermont.

The first sled on the market, under the name "Fairy Coaster," was a double runner bob sled that held three or four adults. Runners and supports were made of steel and the seats were plush. It was light weight and folded for easy travel. At $50.00, however, the retail price was too high for the sled to sell in quantity. This and "The Phantom," a tricky little canvas three-seater as described in *Intimate Recollections and Letters*, a book by Elizabeth Allen, the daughter of S.L. Allen, was tested at Westtown School. In 1884 "The Fleetwing" was given to the Westtown girls. The Fleetwing would carry six comfortably, it was light weight and could easily be pulled by five or six persons. It was painted a gray-blue and had a silver nameplate and a gong. It steered with handles and the front sled was fastened by chains; thus the front sled guided the rear one.

Mr. Allen continued to experiment, developing a sled that would carry eight men. This sled, also a bob, was called "The Ariel." For years Mr. Allen worked on the bob sled.

It was not until he came up with the ideas for a T-shaped runner and slatted seat, both new concepts at this time, that he made any progress. After it was proven Mr. Allen called his sled the Flexible Flyer®. It was an appropriate name because the sled was fast considering its weight and size, and it was the only steerable sled made at this time. Mr. Allen applied for his patent on February 14, 1889, and the patent was granted six months later, in August.

For many years the name Flexible Flyer® was not put on the sleds. The center boards were decorated with flowers and a sundry of other things. The waste of good advertising space was seen by Mr. Allen and a trademark was adopted made up of an Eagle, Shield, and Ribbon with "Flexible Flyer" on it.

The popularity of the sled was not instantaneous. Vacations were cut short trying to market the sleds. A salesman asked Mr. Allen to sell the patent but he would

The Sled of the Nation

Flexible Flyer® poster, 1929. Donald Harrison Corp., New York City. Signed by David Paige. 4'6" x 5'. $1500-2500.

not. As with his farm and garden equipment, it took years to find success.

For many years profits were absorbed by advertising. But with the introduction of golf in the United States and the revival of other sports like tennis, skating, and tobogganing, Mr. Allen finally succeeded in convincing two great department stores, John Wanamaker in Philadelphia and R.H. Macy in New York, of the merits of the Flexible Flyer®. He began to spend money liberally on advertising, and it paid off. By 1915, as Elizabeth Allen wrote, "We are selling sleds at a rate of 2000 per day and there seems little doubt that we will sell out clean, 120,000 this season." It may have been the extensive advertising that enticed the founding fathers of the exclusive Tuxedo Club in Tuxedo Park, New York to invest in the Flexible Flyer®. Tobogganing was already in vogue at Tuxedo Park when the Tuxedo Club purchased the Flexible Flyer® sleds. Coasting at Tuxedo became a major competitive sport, and Mr. Allen produced a sled worthy of its name, "The Tuxedo Racer."

By the 1920s the ability to steer was dominating the market and others followed Flexible Flyers'® lead and began advertising the virtues of their steerable sleds. The Sears catalog offered the "Flying Arrow." The advertisement read, "Watch it Fly Around the Corner, these Sleds Really Steer. Runners are highly tempered spring steel. They curve up abruptly in front, which leaves more of the long flat runner on the ground, making it easier to steer and giving a better coasting surface." The price: $2.87.

During the 1930s, Sears offered "The Snow Bird," almost exactly the same as the Flying Arrow, now newly improved to steer the entire length of its runners.

In 1936, Flexible Flyer® was given the rights by Walt Disney Enterprises to produce "The Mickey Mouse Sled." The center decal shows Mickey riding a Flexible Flyer® and Minnie standing behind in the snow waving. Two years later another Disney Flexible Flyer® was produced, the "Donald Duck." Its decal shows Donald Duck being chased by Huey, Louie, and Dewey on a Flexible Flyer®.

The year 1939 marked Flexible Flyer's Golden Anniversary, and a special gift was offered with each No. 44 Airline Patrol shipped from the factory: "A pair of Running Lights." The brochure reads, "This is our special gift to you in celebration of our Golden Anniversary; they can be attached to any Flexible Flyer® Sled."

The thirties saw Flexible Flyer® introduce a new line of sleds, "The Airline Series," the Yankee Clipper, and Disney Sleds; along with the introduction of the Ski Racer and Flexible Flyer® Splitkein Skis. The Flexy Racer #300 was also introduced, much improved over its predecessors, the Flexy #100 and #200. The Firefly line of sleds was also improved.

Never has another sled been so revered as the Flexible Flyer®, except perhaps "Rosebud," immortalized by Hollywood in the 1941 movie "Citizen Kane." The Flexible Flyer® was mentioned in Hollywood's original classic film, "Miracle on 34th Street." Because he received a Flexible Flyer® for Christmas, a young boy states with assurance that "Santa Claus does exist."

The love of the Flexible Flyer® continued to grow. In a vignette, Bing Crosby wrote for Ed Sullivan in 1959; he called his Flexible Flyer® "one of the sweetest memories of all."

S.L. Allen continued to manufacture the Flexible Flyer® until 1968, when the family owned company was sold to "Leisure Group Inc." of Los Angeles. The Leisure Group also bought Blazon Inc., an Ohio based swing set maker with a plant in Mississippi.

The Leisure Group had among its affiliates Ben Pearson Archery, Thompson Sprinklers, Hayes Spray Guns, and Black Magic for plants, to name a few. With garden and outdoor sports products around them, Flexible Flyer® was amidst familiar associates; however, tough sledding was ahead. The Leisure Group continued to manufacture the Flexible Flyer® under the name "Winter Products, Medina, Ohio."

In 1973 a group of employees and investors bought the swing and sled business and formed Blazon-Flexible Flyer® Inc. It was former TV anchorman Harry Reasoner who alerted the nation in December, 1975 of Blazon's financial problems. He ended his broadcast with, "If society cannot afford a Flexible Flyer, it is in trouble."

The following year, and four million dollars in debt, Blazon sought Chapter 11 bankruptcy protection, consolidating its operations in Mississippi, and closing its sled-making plant in Medina, Ohio. Mr. C. Garland Dempsey, President of Blazon-Flexible Flyer, attributed the decline in sled sales to the popular demand of the less expensive plastic tubes and saucers.

The 1980s proved to be exciting. The exclusive "Smithsonian Edition" was issued to the nation for all to view in 1986, followed in 1989 by the "Centennial Racer," marking a century of "Joy Riding."

Roadmaster Corporation purchased Blazon-Flexible Flyer® in 1993 and continues to manufacture the world's most recognizable sled, with plans to release a collector's edition for Christmas 1996.

Yes, Flexible Flyer® is alive and well and living in the hearts of everyone over the age of 50, who can remember the joy and excitement of being "King of the Hill."

References

The World with a Fence Around It. George M. Rushmore. New York: Pageant Press: 1957

The Book of Winter Sports. J.C. Duer, Editor. New York: MacMillan Co., 1912.

Smithsonian Magazine, December 1987. Peter Stock.

Signature Magazine (Diner's Club). 1976

Chronology of Flexible Flyer

1841: Samuel Leeds Allen, inventor of "Flexible Flyer" born in Philadelphia, Pennsylvania.

1861: S.L. Allen begins farming on the family farm in New Jersey.

1867: S.L. Allen marries and invents his first farm equipment. The machine resembles the plant Saturn, so he calls his company "Planet Junior."

1881: S.L. Allen Company of Philadelphia is a leading manufacturer of farm equipment, marketing products throughout the U.S. and Europe.

1884-1887: Samuel Allen experiments with various types of sleds. The "Phantom," "Fleetwing," and "Ariel" were never marketed. All sleds were tested at Westtown School, Westtown, Pennsylvania.

1888: "Fairy Coaster," a double runner bobsled and predecessor to the Flexible Flyer® was unsuccessfully marketed.

Feb. 14, 1889: S.L. Allen applies for "Flexible Flyer" patent.

Aug. 1889: Patent granted for Flexible Flyer® and production begins.

1890-1899: No trademark. All sleds have wood frames. Center deck decorated with flowers. Back of sled marked only with the letters FF and numbers 1 through 5 with the letter A.

Fire Fly Coasters: named after Planet Jr. Seeder. Deck decorated with a Holly wreath. Back of sled marked with numbers 9 through 12 and a letter B.

1900: S.L. Allen actively markets the Flexible Flyer®. Wanamakers in Philadelphia and R.H. Macy's in New York stock sleds for Christmas. Trademark of Eagle, Shield, and Ribbon adopted. All models 1 through 6 were decorated with fancy red scroll work encircling trademark. All sleds from this time on are marked on the back center deck board with the model number, letter, and "Made in U.S.A. by SL Allen and Co. Inc. Philadelphia, Pa. Manufacturers of Planet Jr. Farm and Garden Tools."

1905: Cardboard model and colorful booklet offered for free from company.

1913: Flexible Flyer® 56 inch Racer introduced. No scroll work on deck. Trademark with large red arrow appears.

1915: New Construction. "All Steel Frame" introduced along with the 48 inch Junior Racer.

1921: Trademark change. Eagle, Sled and Ribbon. The introduction of Flexy Racer, numbers 100 and 200. New diamond pattern introduced on model numbers 1 through 6 with the letter C.

1924: Expert Coaster Button, dark blue, offered, *Saturday Evening Post,* December 13, 1924.

1925: Expert Sledder Button, light blue, offered *Youth's Companion,* December, 1925. Buttons manufactured by Whitehead and Hoag, Newark, New Jersey.

1927: The "Wee" Racer introduced, the only Flexible Flyer® to have an airplane on the center deck above the trademark. Plucky Lindy button introduced. The advertisement read "out front among Lindbergh's." *Youth's Companion,* December, 1927

1928: Flexible Flyer® No. 5G, Admiral Byrd at the South Pole. Advertisement read "You, too, can buy from your dealer a Flexible Flyer® exactly the same as Commander Byrd is using. *Youth's Companion,* 1928.

1935: Airline Flexible Flyer® sleds introduced Safety Runners, plain steering bar with no decorations. Large arrow added to sled deck with Trademark. The name, "Airline," and model name also on sled deck. Back of sled deck had name, model, and the words "Series A." Splitkein Laminated Skis and Ski Racers introduced. Fire Fly Coasters reintroduced with metal Frame. Back of center deck had numbers 8 through 12 and the letter C. Also introduced Fire Fly Racer and Firefly Special.

1936: Mickey Mouse and Yankee Clipper Sleds introduced, model Numbers 8 through 14. Airline Sled's steering bar is decorated with wings. Series A letter dropped.

1938: Donald Duck Sled introduced.

1939: Golden Anniversary. Special gift of Flexible Flyer® Running Lights. Flexible Flyer® Membership Fan Club Card and "Sure Sledders" membership pin offered.

1940: All straight runner models phased out.

1950: Airline Sleds had model names removed from deck. Only trademark with arrow remains. Model name appears on back of center deck only with number and letter H, etc.

1951: Flexible Flyer® Puzzle offered. *American Boy.*

1955: New model No. 551 with new air flow deck and rails. Two-tone like Dad's new car, new chrome plated bumper, natural deck, side rails and runners gleaming red or black. *Boy's Life.*

1968: Flexible Flyer® sold to the Leisure Group of Los Angeles.

1969: The Leisure Group buys Blazon Inc. and moves to Medina, Ohio where it manufactures the famous Flexible Flyer® Sled, returning to the original style trademark with eagle, shield, and ribbon. Back of deck reads "Winter Products Medina Ohio."

1973: Private Investors buy the Leisure Groups toy division and consolidates all operations under the name of Blazon-Flexible Flyer® Inc., West Point, Mississippi.

1986: Smithsonian Edition exclusively issued for the nation to view in Washington, D.C.

1989: Centennial Racer issued marking a century of "Joy Riding."

1993: Roadmaster Corporation purchased Flexible Flyer® and continues to manufacture the world's most recognizable sled.

1996: Collector's Edition released.

Advertising

Post Cards

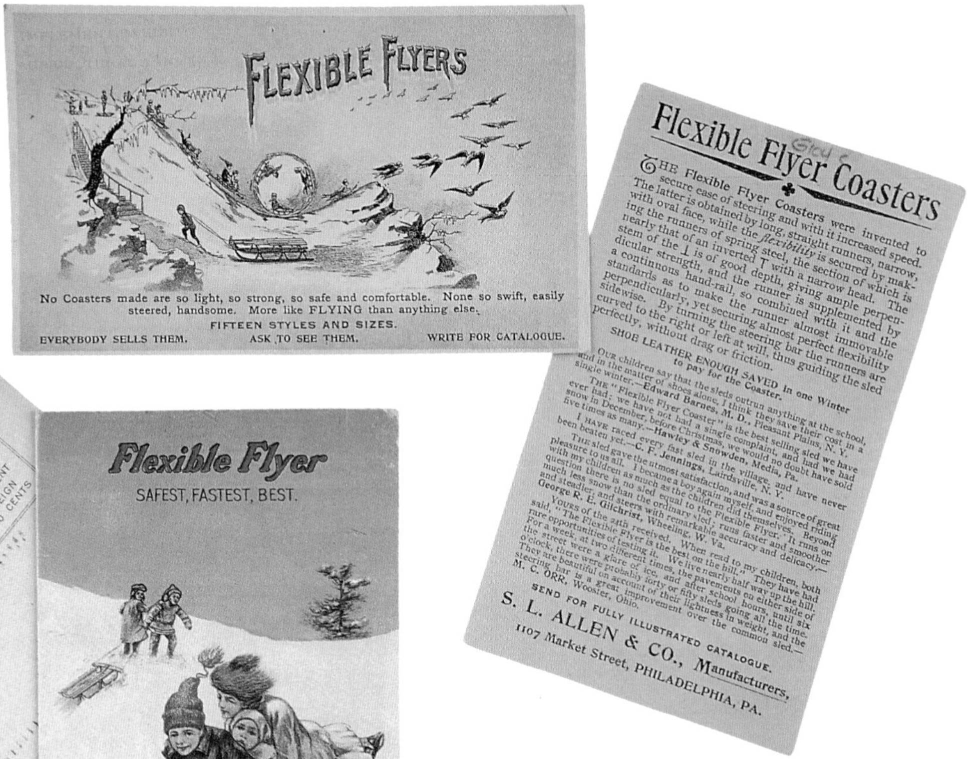

FLEXIBLE FLYERS

No Coasters made are so light, so strong, so safe and comfortable. None so swift, easily steered, handsome. More like FLYING than anything else.
FIFTEEN STYLES AND SIZES.
EVERYBODY SELLS THEM. ASK TO SEE THEM. WRITE FOR CATALOGUE.

Flexible Flyer Coasters

THE Flexible Flyer Coasters were invented to secure ease of steering and with it increased speed. The latter is obtained by long, straight runners, narrow, with oval face, while the *flexibility* is secured by making the runners of spring steel, the section of which is nearly that of an inverted T with a narrow head. The stem of the ⊥ is of good depth, giving ample perpendicular strength, and the runner is supplemented by a continuous hand-rail, so combined with it and the standards as to make the runner almost immovable perpendicularly, yet securing almost perfect flexibility sidewise. By turning the steering bar the runners are curved to the right or left at will, thus guiding the sled perfectly, without drag or friction.

SHOE LEATHER ENOUGH SAVED in one Winter to pay for the Coaster.

OUR children say that the sleds outrun anything at the school, and in the matter of shoes alone, I think they save their cost in a single winter.—**Edward Barnes, M. D.**, Pleasant Plains, N. Y.

THE "Flexible Flyer Coaster" is the best selling sled we have ever had; we have not had a single complaint, and had we had snow in December, before Christmas, we would no doubt have sold five times as many.—**Hawley & Snowden**, Media, Pa.

I HAVE raced every fast sled in the village, and have never been beaten yet.—**C. F. Jennings**, Lairdsville, N. Y.

THE sled gave the utmost satisfaction, and was a source of great pleasure to us all. I became a boy again myself, and enjoyed riding with my children as much as the children did themselves. Beyond question there is no sled equal to the Flexible Flyer. It runs on much less snow than the ordinary sled; runs faster and smoother and steadier, and steers with remarkable accuracy and delicacy.—**George R. E. Gilchrist**, Wheeling, W. Va.

YOURS of the 24th received. When read to my children, both said, "The Flexible Flyer is the best on the hill." They have had rare opportunities of testing it. We live nearly half way up the hill. For a week, at two different times, the pavements on either side of the street were a glare of ice, and after school hours, until six o'clock, there were probably forty or fifty sleds going all the time. They are beautiful on account of their lightness in weight, and the steering bar is a great improvement over the common sled.—M. C. ORR, Wooster, Ohio.

SEND FOR FULLY ILLUSTRATED CATALOGUE.
S. L. ALLEN & CO., Manufacturers,
1107 Market Street, PHILADELPHIA, PA.

The earliest post card, c. 1892. 3.5" x 5.5". $25-50.

Post card, 1914. R.H. Macy's. 5.5" x 3.5". $50-75.

Post card, 1907. 3.5" x 5.5". $25-50.

Old Flexible Flyer brochure.®

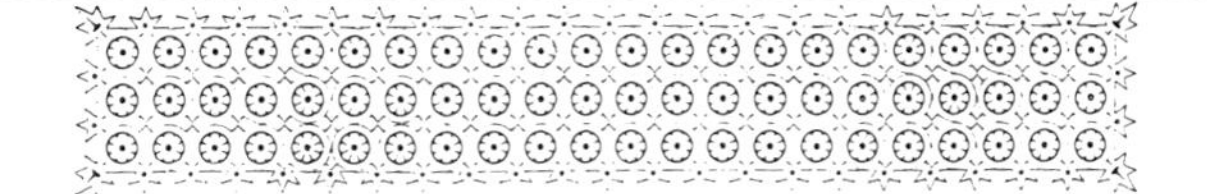

UNTIL quite recently there has been very little improvement in the construction of sleds. Except that they are lighter, the sleds of the present day are much the same as those of our grandfathers. No good device for steering single sleds has been introduced; the old fashion of digging the heels in the snow, or dragging the foot behind, being still practised. Everybody knows what it is to have his trousers' legs filled with flying snow, and every boy's father knows what it is to supply new shoes for those worn and cut out on the coasting hill.

Not only is the old way of steering destructive of comfort and shoes, but also of *speed*. It is of course impossible that a sled should travel at its best speed when the feet are dug in the snow at front and sides, or dragged in it behind.

The "Flexible Flyer" Coasters

were invented to secure ease of steering, and with it increased speed. The latter is obtained by long, straight runners, narrow, with oval face, while the flexibility is secured by making the runners of spring steel, the section of which is nearly that of an inverted tee with a narrow head. The stem of the tee is of good depth, giving ample perpendicular strength, and the runner is supplemented by a continuous hand rail, so combined with it and the standards as to make the runner almost immovable perpendicularly, yet increasing its flexibility sidewise. Then, by turning the steering-bar, *the runners are curved to right or left at will*, by this means guiding the sled perfectly, without drag or friction. The steering, thus made simple and easy, becomes *the greatest charm* of the coaster, and it is easy to understand why *no other style can compete in speed*.

When two or more ride, the steering is most conveniently done by the second person, though the first can steer if preferred. In either position, it is done by using the feet and the drawing-ropes also, the latter being attached to the outer ends of the steering-bar both for drawing the coaster and for steering. When there is but one rider, the steering can be done either sitting (when hands and feet are both used), or lying down, or on the hip (when the hands only need be used).

These features of the "FLEXIBLE FLYER" COASTERS make them especially desirable, and being made in the most thorough and workman-like manner of a combination of spring steel and hard woods, we feel certain of not saying too much when we claim them to be the

LIGHTEST, FASTEST,
NEATEST, CHEAPEST,
STRONGEST, MOST COMFORTABLE,
AND MOST EASILY STEERED

coasters in the world. These advantages make the pastime of coasting almost a new sensation, delightful for old and young. The steering device

Saves Shoe Leather by the Ton,

while exemption from the discomfort of flying snow, and greater freedom from accidents, much increase the pleasure of their use.

We desire to express our appreciation of the favor with which our goods were received by the trade last season, and regret that it was impossible for us to fill all orders promptly. Although we anticipated a largely increased demand and prepared for it, we were not prepared for the flood of orders that poured in on us for some sizes.

Our facilities for making these goods are at least equal to any in the country, and we hope the coming season to supply any possible demand. We have exercised great care in the selection of our materials, and shall pay special attention to finish and appearance of the goods.

TERMS.--All sled bills due January 1st next, subject to 1 per cent. per month discount for cash before that time, after which no cash discount will be allowed.

2

Flexible Flyer No. 3.

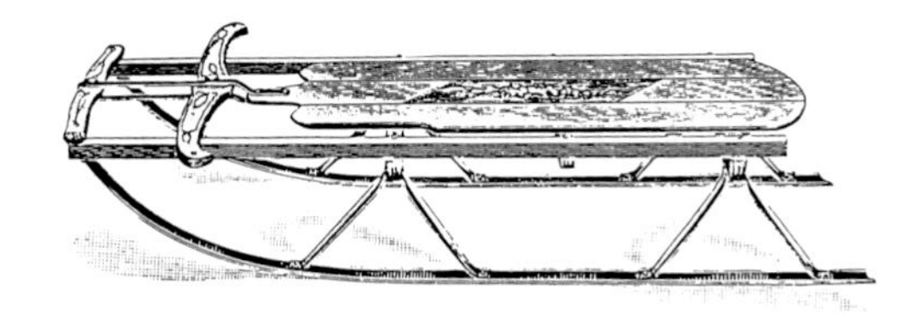

Length, 44 in. Height, 7 in. Width, 13 in. Weight, 11¼ lbs.

Price, $45 per dozen.

This sled carries two large children with entire comfort, and is amply strong for two adults. The side rails are carried forward to the front end of the runner and riveted there, forming a truss which gives firmness and strength to the sled, yet adds flexibility and ease to the steering. The metal work is spring steel; the woodwork, white ash, handsomely finished in natural color. This sled is a great favorite.

Flyer Sleigh No. 12.

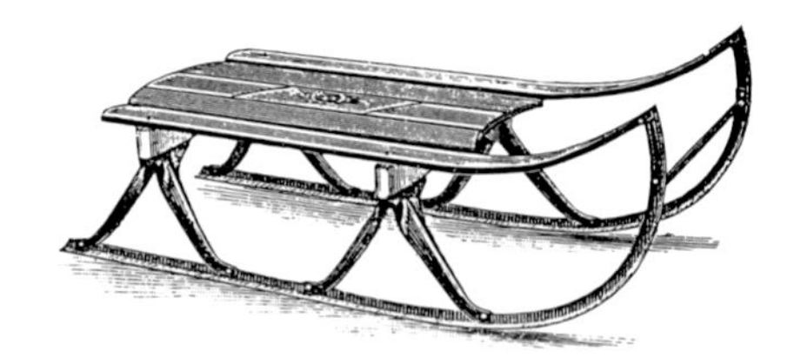

Length, 36 in. Height, 8 in. Width, 13 in. Weight, 8 lbs.

Price, $30 per dozen.

This is the proper size for large children, the extra height giving increased comfort. Very light, strong and handsome.

Flyer Sleigh No. 13.

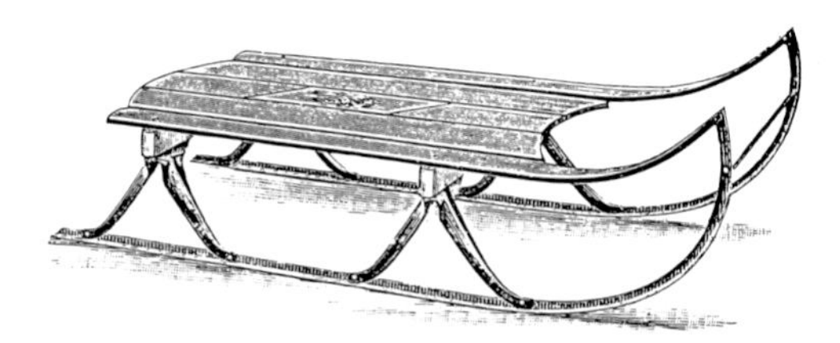

Length, 42 in. Height, 8 in. Width, 15 in. Weight, 9¼ lbs.

Price, $36 per dozen.

This number is the most comfortable Flyer Sleigh we make, being extra long and broad, and very high; yet it is very light, strong, and prettily finished.

8

Flyer Coaster No. 23.

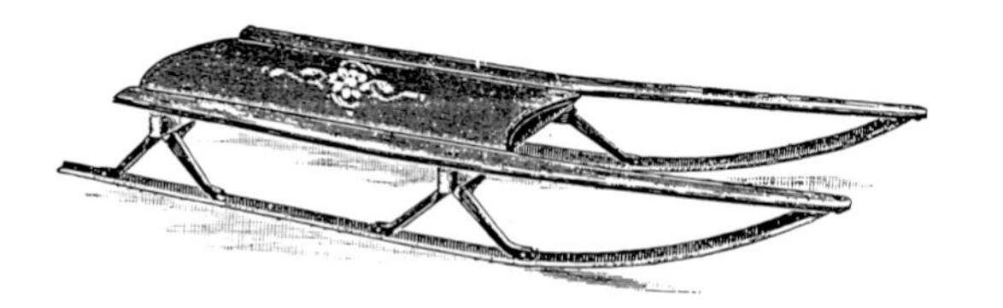

Length, 48 in. Height, 4¾ in. Width, 13 in. Weight, 8½ lbs.

Price, $36 per dozen.

This size is a beautiful Coaster for large boys and grown persons. Very strong and very fast.

Flyer Coaster No. 24.

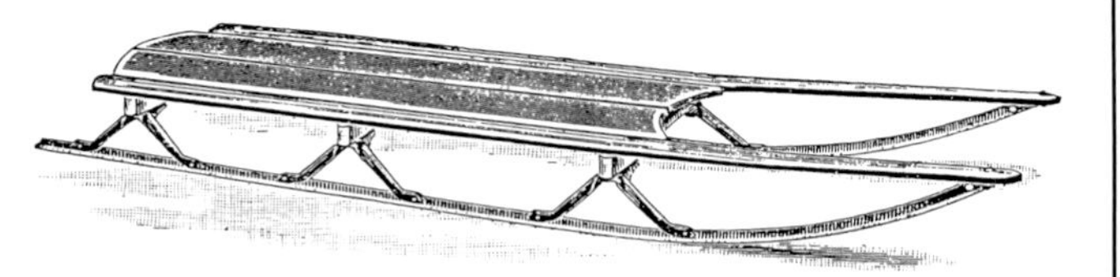

Length, 60 in. Height, 4¾ in. Width, 13 in. Weight, 11½ lbs.

Price, $48 per dozen.

This is probably the very best Coaster ever made. Extra long and strong, yet light. The handsomest, fastest and best of this class.

10

Old Flexible Flyer brochure.®

TESTIMONIALS.

It is impossible in our limited space to give the opinion of many of our customers on these goods, but we have hundreds more like those published below. Note the representative houses that our list covers, also the wide section of country and that they embrace the toy, hardware, house-furnishing, implement and department store trade.

THE "Flexible Flyers" proved satisfactory, and proved good sellers.—THOMAS L. MARSHALL, Toy Department, John Wanamaker, Philadelphia, Pa.

STRONG, durable, speedy and handsome. The "Flexible Flyers" are the best sleds in the market.—G. A. SCHWARZ, Philadelphia, Pa.

WE sold all that we had last year, and could have disposed of more if we could have gotten them when wanted.—CONWAY BROS., Philadelphia, Pa.

THE only objection we have ever found to the "Flexible Flyer" was that we could not get enough of them in the season. Our own experience has been that we have more than doubled our sales on them every year.—BUEHLER, BONBRIGHT & CO., Philadelphia, Pa.

WE were very well pleased with our trade on these goods —SUPPLEE HARDWARE Co., Philadelphia, Pa.

WE have no fault to find with the "Flexible Flyers," they were good sellers with us; the only complaint we have to make is, you could not furnish them when we wanted them. We trust you will be in better shape this year.—SELTZER KLAHR HARDWARE Co., Philadelphia, Pa.

OUR experience last year with the "Flexible Flyers" was quite satisfactory.—BIDDLE HARDWARE Co., Philadelphia, Pa.

WE had a very fair trade on the "Flexible Flyers" last year, more than doubling our sales of the year before.—JAMES M. VANCE & Co., Philadelphia, Pa.

THE "Flexible Flyers" that we purchased from you were very satisfactory, and proved to be good sellers with us; we sold about double what we did the season before last, and expect to increase our sales this season.—J. B. SHANNON & SONS, Philadelphia, Pa.

REPLYING to your recent favor, we will take pleasure in placing your complete line of sleds in our next catalogue, which we will publish about September 1st. We have found them the most desirable novelty for winter sports that we have ever advertised. In fact, they sell themselves. We do not see how you can improve on those of last year's manufacture.—PECK & SNYDER, New York City.

THE sleds received from you last season were satisfactory in every respect, and I have no doubt that in course of time we shall sell still larger quantities.—F. A. O. SCHWARZ, New York City.

WE are pleased to say that your *Sleds*, "Flexible Flyers," which we have handled these past few seasons, have proved both to ourselves and our customers a most satisfactory article, and we have had no sort of trouble or fault to find in their workmanship, finish and fit. We have, particularly during the season of 1893 and 1894, found them to be a decided favorite as a sled.—SCHARLES BROS., New York City.

SOLD them all very readily and expect to handle them again.—GEO. W. TRAVERS, New York City.

THE sleds we had from you were very satisfactory, and gave our customers equal satisfaction.—SLAZENGER & SONS, New York City.

YOUR goods proved very satisfactory last year, and the only suggestion we have to make is, keep the quality and bright colors the same and we will be able to use our share of them this season.—HILTON, HUGHES & Co., New York City. Successors to A. T. Stewart & Co.

THE "Flexible Flyers" proved most satisfactory in every case.—HENRY C. SQUIRES & SON, New York City.

WE sold double the quantity of "Flexible Flyers" last year that we expected and our dealers were all well pleased with them —E. B. COLBY & Co., New York City.

YOURS is the only sled in the market which you don't have to break your lungs to "sell it."—J. A. CRANDALL, Brooklyn, N. Y.

I CONSIDER them by far the strongest and best sled made in every respect and predict for you a large increase every year in the sale of your sleds.—D. C. HENRY, Auburn, N. Y.

THE "Flexible Flyers" which you shipped us last winter were entirely satisfactory and were rapidly sold. We hope to be able to dispose of a great many this coming season.—E. F. SCHWARZ & BRO., Chicago, Ill.

THE "Flexible Flyers" sold well with us, and gave good satisfaction to our customers.—MARQUA CARRIAGE AND TOY CO., Cincinnati, Ohio.

YOUR sleds have given the best of satisfaction. We have never seen anything in the way of a boy's sled which equalled it in points of merit.—CLARK, QUIEN & MORSE, Peoria, Ill.

YOUR sleds have given good satisfaction.—SCHWERDTMANN TOY CO., St. Louis, Mo.

THE "Flexible Flyers" have proven the best hand sleds that we have ever carried in stock. We find that they are not only rapid coasters, but they are rapid sellers. Will require at least double the quantity this season.—SICKELS, PRESTON & NUTTING Co., Davenport, Iowa.

THE "Flexible Flyers" we received from you last year and the year before have given the best of satisfaction, and are very good sellers in this section of country.—KNAPP, BURRELL & Co., Spokane, Wash.

YOUR sled is good enough.—BELCHER & LOOMIS, Providence, R. I.

WHILE last year was a very bad season to sell sleds here, the sleds sold some and we think will sell better next winter, as they are very much liked.—R. D. HAWLEY & Co., Hartford, Conn.

WE sold quite a few of your sleds last year, and they gave good satisfaction.—B. L. BRAGG Co., Springfield, Mass.

THE "Flexible Flyers" bought of you last season met with ready sale, and we were obliged to order more from your agency in Boston. Shall want to handle them this season.—W. H. WILLARD & Co., Worcester, Mass.

THE "Flexible Flyers" pleased us very much.—R. S. REED & Co., Fall River, Mass.

WE were pleased with your "Flexible Flyers," they proved to be good sellers.—C. J. CONOLLY, Rochester, N. Y.

WE believe we sold all of these goods which we received from you, and inasmuch as we heard no complaint in regard to them, we conclude that they must have been satisfactory.—SWEET & JOHONNOT, Buffalo, N. Y.

YOUR "Flyer Sleighs" Nos. 10, 11, 12, etc., are giving excellent satisfaction to our customers, and we consider them the best and neatest sleds manufactured, and expect to handle more of them the coming season than last.—THE W. K. BOONE Co., Lima, Ohio.

NO sled on the market attracts half the attention that the "Flexible Flyer" does.—CLEVELAND CYCLE DEPOT, Cleveland, Ohio.

THE "Flexible Flyers" we bought of you last season have been very satisfactory.—THE SPRINGFIELD HARDWARE Co., Springfield, Ohio.

THE sleds you sent us last year gave the best of satisfaction.—THE BRYAN HARDWARE Co., Bryan, Ohio.

THE sleds gave the utmost satisfaction. They are the best sleds ever manufactured.—WISE & Co., Butler, Ohio.

THE "Flexible Flyers" are certainly all you claim for them and lead everything we ever heard of for a sled.—SLOSSER & KAHLO, Defiance, Ohio.

ALMOST impossible to sell the ordinary sleds and coasters when they were placed beside the "Flyers" upon our floor.—PARMELEE BROS., Burton, Ohio.

THE No. 3 "Flexible Flyers" that we bought of you last year were the wonder of the coasting hill, the object of the envy of every man and boy and the desire of every lassie.—BIESE & BLAIR, Chattanooga, Tenn.

THE "Flexible Flyers" gave good satisfaction and were very much liked by the purchasers.—FRANK OWENS HARDWARE Co., Maysville, Ky.

WE think the "Flexible Flyer" the best sled on the market, and the only one in our experience that will stand the work required of a sled by the average American boy. Can suggest no improvement.—DITZLER & LINSLEY, Hinsdale, Ill.

THEY attracted a good deal of attention and gave good satisfaction when sold.—H. R. CLEARS & Co., Kewanee, Ill.

WE have several of your "Flexible Flyers" in use in our town, and consider them the best we have ever used or seen. They work easily and cannot be beat and are not liable to get out of repair. We cheerfully commend them to any one wanting a good article.—BREWER & IRELAND, Ashton, Ill.

I THINK they will be good sellers if season is all right. I know they will give good satisfaction.—E. HOLBROOK, Batavia, Ill.

WE are well pleased with the goods and carry a good assortment. We desire the exclusive agency and will keep well stocked.—THE LANE IMPLEMENT Co., Red Oak, Iowa.

THE "Flexible Flyers" are all right and good sellers.—CHRISTMAN & HEALEY, Dubuque, Iowa.

Old Flexible Flyer brochure.®

WE beg to report our experience with your system of sleds as eminently satisfactory, as well as profitable, for the past three winters. We could have easily doubled our sale for holiday trade if we had bought twice as many sleds. There being nothing like them on the market, we had a monopoly on the sled trade as long as our stock held out, and at full prices. We will take all the sled trade of this vicinity.—M. BENHAM & Co., Muscatine, Iowa.

THE sleighs you sent us take first-rate. Sold all out.—ROGERS & Co., Washta, Iowa.

YOUR "Flexible Flyers" gave good satisfaction, and I sold all I had within ten days.—W. P. HOHENSCHUH, Iowa City, Iowa.

WE found your sleds all you claimed for them. They are the best sleds on the market and are good sellers.—ROBERTS BROS., Independence, Mo.

YOUR "Flexible Flyers" are immense and more than claimed for them.—D. H. MILLS, Litchfield, Mich.

THEY are all tip-top goods and give satisfaction.—L. W. SPRAGUE, Greenville, Mich.

YOUR sleds are the finest we ever saw. Best on earth, we think.—A. A. ALDRICH & SON, Hickory Corners, Mich.

THE sleds we bought of you were the best sleds we ever bought, and were the best sellers.—F. F. PALMER & SON, Hudson, Mich.

THE "Flexible Flyers" proved to be A1 sleds in every way and splendid sellers.—RECHLIN & FRANK, Bay City, Mich.

THE medium sizes sold the best with us. The sled is a "hummer."—J. S. WHITE & Co., Marshall, Mich.

THE "Flexible Flyers" sent us last year sold.—L. S. PLAUT & Co., Newark, N. J.

THE "Flexible Flyers" sent us last season gave splendid satisfaction.—HILL & HOWELL, Newton, N. J.

WE find "Flexible Flyers" good sellers; we sold three of your make to one of any other. The boys tell us they out-coast all other kinds. You have a *good thing* in them.—J. P. BODINE & SONS, Flemington, N. J.

THE "Flyers" bought of you last winter gave perfect satisfaction to my customers.—G. E. VOORHEES, Morristown, N. J.

IT is impossible to sell any other than the "Flexible Flyer." All the trouble was I could not get them last season as soon as I would have liked to.—A. L. COOK, Marksboro, N. J.

WE liked the "Flexible Flyers" very much.—J. P. LAINE & Co., Plainfield, N. J.

THE "Flexible Flyers" were all we could desire, and gave splendid satisfaction. They were light, strong and swift, just what the boys and girls want.—G. W. CONKLIN, Huntington, N. J.

THE sled proved a good seller and gave entire satisfaction.—HENRY WEILER, Dunkirk, N. Y.

I TAKE pleasure in recommending your line of sleds.—J. W. FEETER, Highland, N. Y.

I WAS very much pleased with the "Flexible Flyers" had from you last winter. They were good sellers and gave satisfaction.—J. E. NEWKIRK, Roxbury, N. Y.

THE "Flexible Flyers" sent us last year were all sold, proved good sellers and gave good satisfaction.—H. F. HERRICK, Southampton, N. Y.

THEY are the best made sled for the price we ever saw.—H. J. ALLEN & SON, Clinton, N. Y.

YOUR "Flexible Flyers" were good sellers, and we expect to handle them again this Fall.—REILLY BROS. & RAUB, Lancaster, Pa.

THE "Flexible Flyers" sent us last winter pleased us very well and have proven to be good sellers.—W. D. SPRECHER, SON & Co., Lancaster, Pa.

OUR sales are increasing from year to year. The boys and girls who have them are envied by every other boy and girl in town.—REES PALMER, West Chester, Pa.

WE shall want more another season. They proved a very good seller.—BEARDSLEY & MCKEAN, Troy, Pa.

I CAN say for the "Flexible Flyers" that they are the *best* coasters made, and as I am a professional steersman in the coasting season, I have used and know the qualities of all the coasters made. I have repeatedly carried a load of half a ton or more down a hill a mile long and very steep on the start. I will warrant a No. 5 to carry one thousand pounds every time, and am perfectly safe in so doing. I dare take any risk with a "Flexible Flyer" and challenge any one on any other sled made to follow me over a hard snow crust, over fences, on ice, and turning short corners. You are always certain where a "Flexible Flyer" is going to and can trust it to do its part every time. I am an enthusiastic coaster myself and know a coaster that is constructed properly when I see it. When you find some one who wants to accept my challenge, let me know and I will pay his expenses up here, and show him some hills as well as some coasting.—LUCIEN I. YOEMANS, Walworth, N. Y.

Old Flexible Flyer brochure.®

Next Two Pages:
Brochure for Flexible Flyers® used by a distributor, Peabody-Whitney Co. Boston, Massachusetts. c. 1890. It includes advertisements for the Flyer Coasters, which, in a later company catalog were omitted. 6" x 6". $25-50.

A Sled that ..Steers..

FLEXIBLE FLYERS
FLYER SLEIGHS
FLYER COASTERS

PEABODY-WHITNEY CO.
Boston, Mass.

Tell the boys
that the self-steering FLEXIBLE FLYER beats any other sled on the hill!

Tell the girls
that other sleds never run into a FLEXIBLE FLYER, for no other sled can catch it!

Tell the mothers
that it is the strongest and safest sled made.

Tell the fathers
that it pays for itself in one winter in shoes saved.

Don't

HINTS FOR COASTERS.

1. Don't Kick. Steer.
Get a real self-steering sled, the famous FLEXIBLE FLYER, and save your shoes.

2. Don't check your sled.
You can't help it with a clumsy old-style sled, scraping and digging in the snow.

3. Don't get beaten.
You certainly will if others have FLEXIBLE FLYERS and you haven't.

4. Don't get wet feet.
With a FLEXIBLE FLYER you keep your shoes out of the snow and the snow out of your shoes.

COASTING is by common consent one of the best of all sports. It is hard to find better fun than flying at breathless speed down a smooth hill covered with swift sleds; and he is a happy boy who rides the champion sled, "the fastest on the hill."

The one great drawback of coasting with the ordinary old style sled has always been the steering, or the trying to steer, until just recently when a grown up boy invented the FLEXIBLE FLYER. The old clumsy method of steering by digging your heels in the snow at the sides, or by digging toes into the snow and ice in the rear were the only ways to guide the sled. If all that the digging and dragging did was to cut out one's shoes and soak

FLEXIBLE FLYER No. 1.

Length, 38 in. Height, 6 in. Width, 12 in. Weight, 6½ lbs.

PRICE, $2.50.

his feet and fill his trousers with snow, it wouldn't be so bad, but it does one thing more, which is the worst of all—it stops the sled.

A coaster doesn't have a steam engine to push it. It will go

FLEXIBLE FLYER No. 2.

Length, 40 in. Height, 6¾ in. Width, 13 in. Weight, 7¾ lbs.

PRICE, $3.00.

if you will let it; but must be free and clear and everything as smooth as ice if you want to run fast.

How can a sled make speed down a hill with a pair of heels scrape, scrape, scraping on the snow all the way?

It can't.

Of course not. And when you see a FLEXIBLE FLYER go gliding down the hill shooting past all the other sleds, you will see what it means to get rid of that dragging.

FLEXIBLE FLYER No. 3.

Length, 44 in. Height, 7 in. Width, 14 in. Weight, 11¼ lbs.

PRICE, $3.75.

The FLEXIBLE FLYER has light runners of T-shaped spring steel, making them very strong vertically, but flexible sidewise. This enables one to guide the sled by putting the feet on a steering bar, by which the runners may be curved and the sled steered to the right or left, as desired. The steering thus made perfect and easy, is the great charm of the Coaster; and as there is no drag or friction, the FLEXIBLE FLYERS then easily go ahead of any other coasters made.

FLEXIBLE FLYER No. 4.

(WITH FOOT RESTS.)

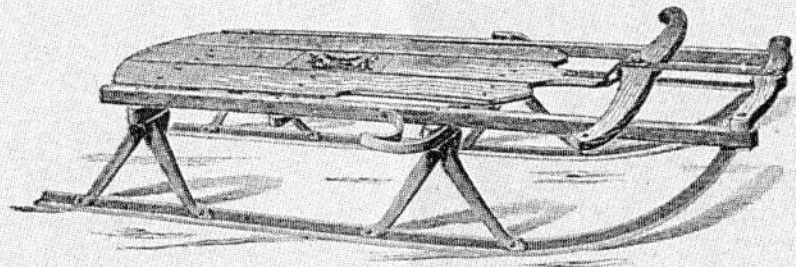

Length, 50 in. Height, 8 in. Width, 16 in. Weight, 13¼ lbs.

PRICE, $4.50.

(WITHOUT FOOT RESTS, $4.00.)

But the speed of the FLEXIBLE FLYER is not its only advantage. It is the most comfortable sled made. Sitting with your feet upon the steering bar, guiding it smoothly and easily past other sleds, avoiding any obstacle, you enjoy coasting as you never did before. Your shoes are dry. Your clothes are not filled with flying snow. You don't run into other sleds, for you can steer past them; other sleds don't run into you, for they can't catch you. No weight will break its runners of spring steel. It is the swiftest, the safest, the strongest sled made.

FLEXIBLE FLYER No. 5.

(WITH FOOT RESTS.)

Length, 62 in. Height, 8 in. Width, 16 in. Weight, 17 lbs.

PRICE, $6.00.

(WITHOUT FOOT RESTS, $5.00.)

And it is the cheapest. To be sure it costs a little more than an old style sled in the first place; but remember that it lasts as long as three old fashion wooden sleds; and don't let the treasurer forget that it saves shoes enough to pay for itself the first season.

THERE'S NO CHRISTMAS GIFT FOR A BOY LIKE A FLEXIBLE FLYER.

FLYER SLEIGHS.

These sleighs are made chiefly for use on the level or on ice, and have no special steering device like the FLEXIBLES. They are, however, superior in speed, construction and durability to any other sleigh made. The metal parts are all of spring steel and the woodwork of white ash, handsomely finished in every part.

FLYER SLEIGH No. 10.

PRICE, $1.50.

Length, 28 in. Width, 11 in.
Height, 6 in. Weight, 5¼ lbs.

FLYER SLEIGH No. 11.

PRICE, $2.00.

Length, 33 in. Width, 12 in.
Height, 6½ in. Weight, 6 lbs.

FLYER SLEIGH No. 12.

PRICE, $2.50.

Length, 36 in. Width, 13 in.
Height, 8 in. Weight, 8 lbs.

FLYER SLEIGH No. 13.

PRICE, $3.00.

Length, 42 in. Width, 15 in.
Height, 8 in. Weight, 9¼ lbs.

FLYER COASTERS.

These are the swiftest, strongest and handsomest coasters made, except the FLEXIBLES. The construction is the same as that of the Flexible Flyers, except that there is no steering arrangement. The metal parts are of spring steel; the woodwork white ash. They are very handsomely finished.

FLYER COASTER No. 21.

PRICE, $1.50.

Length, 36 in. Width, 11 in.
Height, 4½ in. Weight, 5½ lbs.

FLYER COASTER No. 22.

PRICE, $2.00.

Length, 42 in. Width, 12 in.
Height, 4½ in. Weight, 7 lbs.

FLYER COASTER No. 23.

PRICE, $3.00.

Length, 48 in. Width, 13 in.
Height, 4¾ in. Weight, 8½ lbs.

FLYER COASTER No. 24.

PRICE, $4.00.

Length, 60 in. Width, 13 in.
Height, 4¾ in. Weight, 11½ lbs.

A FEW OUT OF MANY.

THE sled gave the utmost satisfaction, and was a source of great pleasure to us all. I became a boy again myself, and enjoyed riding with my children as much as the children did themselves. Beyond question there is no sled equal to the Flexible Flyer. It runs on much less snow than the ordinary sled; runs faster and smoother and steadier, and steers with remarkable accuracy and delicacy.—GEORGE R. E. GILCHRIST, Wheeling, W. Va.

THE handsomest and swiftest sled that was on the road. The guiding device is a most excellent feature in every way. In quality of material, workmanship and beauty of finish, it is beyond my expectations.—E. B. CARTMELL, Lancaster, Ohio.

THEY passed every other sled on the hill, and one young man we had passed said we shot past him so fast he had to feel of the ground to see if he was going at all. While being swift, they are under perfect control.—FRANK KANZ, Cape Girardeau, Mo.

I CONSIDER the sled about perfect, not only for speed and ease of management, but as a great saver of shoe leather. My boys say they can run around any sled in the country, and around the side of the hill, as no other sled can.—JAS. RUSH LINCOLN, Ames, Iowa.

I WAS very agreeably surprised at the "personal appearance" of the Flyer, which from the illustrations in the catalogue I had conceived to be rather plain. Imagine my surprise at seeing a sled very handsomely finished and of the most graceful proportions. Its speed and steering properties are wonderful. The speed which it developed astonished me. It easily outstripped all the other sleds on the first run.—HAROLD W. LATHROP, Livingston Manor, N. Y.

YOURS of the 24th received. When read to my children, both said: "The Flexible Flyer is the best on the hill." They have had rare opportunities of testing it. We live nearly half-way up the hill. For a week, at two different times, the pavements on either side of the street were a glare of ice, and after school hours, until six o'clock, there were probably forty or fifty sleds going all the time. The Flexible Flyer made the best speed. Had we sent and received your circular a little earlier, would have secured several more names. It was but a day or two before Christmas, and some of our neighbors had already purchased for holiday gifts, and were sorry afterwards that they had not waited and secured one of yours. They are beautiful on account of their lightness in weight, and the steering bar is a great improvement over the common sled.—M. C. ORR, Wooster, Ohio.

THE best sled for crust coasting ever made. It is faster than any of the double-runners in town.—JULIAN P. LAUGHLIN, Barnet, Vermont.

I THINK it is great. Everybody calls me the king of the hill. The steering apparatus works admirably.—ELMER HAULENBEEK, 348 Girard Avenue, Baltimore, Md.

OUR children say that the sleds outrun anything at the school, and in the matter of shoes alone, I think that they save their cost in a single winter.—EDWARD BARNES, M. D., Pleasant Plains, N. Y.

THE "Flexible Flyer Coaster" gave entire satisfaction—admired by everybody who saw them "flying." Heard frequent exclamations: "That's a novelty come to stay!" "What a saving of shoe leather for the small boy and a comfort for the large boy!"—F. G. DWIGHT, Reading, Pa.

DAUGHTER thought it a fine thing for girls. It runs at breakneck speed down our hill, as there is no friction.—L. W. STILLWELL, Deadwood, S. D.

Girls don't know what fun is if they have never coasted on the famous FLEXIBLE FLYERS. You steer perfectly with your feet on the cross-bar, never touching the ground. And they go like the wind,—beat any other sled made. No girl should coast on anything but a FLEXIBLE FLYER!

The Allen brochure for sleds used at the Chicago World's Fair of 1892. The brochure pushes the steerable sled and omits the Flyer Coasters seen in the Peabody-Whitney brochure. 6" x 6". $50-75.

Flexible Flyer® catalog, c. 1900. All black-and-white photographs. $75-100.

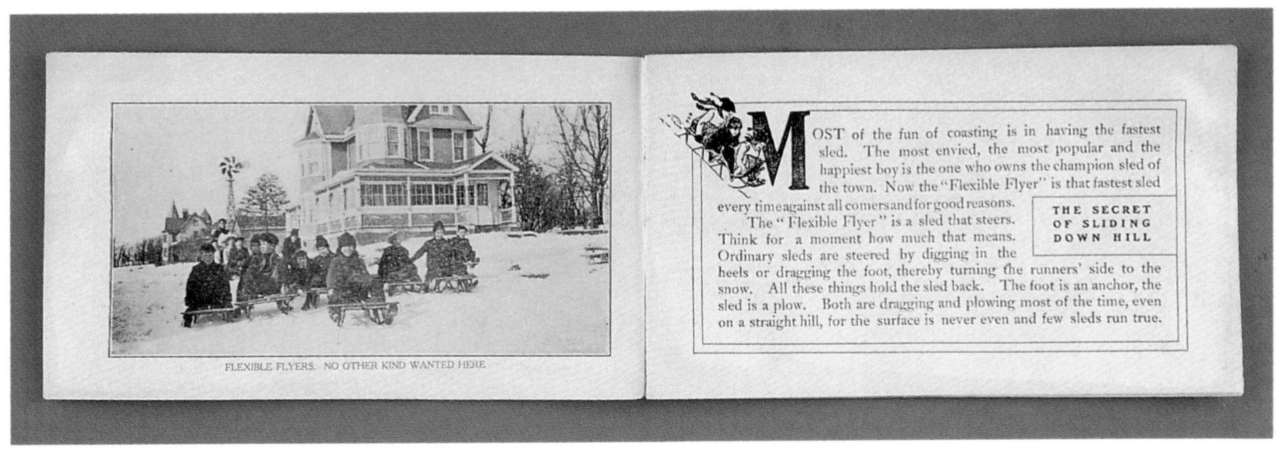

FLEXIBLE FLYERS. NO OTHER KIND WANTED HERE

MOST of the fun of coasting is in having the fastest sled. The most envied, the most popular and the happiest boy is the one who owns the champion sled of the town. Now the "Flexible Flyer" is that fastest sled every time against all comers and for good reasons.

THE SECRET OF SLIDING DOWN HILL

The "Flexible Flyer" is a sled that steers. Think for a moment how much that means. Ordinary sleds are steered by digging in the heels or dragging the foot, thereby turning the runners' side to the snow. All these things hold the sled back. The foot is an anchor, the sled is a plow. Both are dragging and plowing most of the time, even on a straight hill, for the surface is never even and few sleds run true.

SIX GIRLS ON A No. 6. FASTER, SAFER AND LIGHTER THAN A DOUBLE RUNNER

HE Runners of the "Flexible Flyer" are of ⊥ shaped spring steel, very strong and rigid vertically, but free to bend sidewise. They are absolutely true and stay so, making it slide very easily while walking up hill or pulling somebody's sister on level ground. But when you want to steer on the hill, a touch on the cross-bar *curves both runners*, the whole runner going in its own track, without plowing the snow, without dragging the foot. The sled—unhampered—shoots on at full speed, a sure winner. It is FAST—it is not only a sure winner for speed, but also for distance. ¶ You don't run into the others, you steer around them; they can't run into you, for they are not able to catch you. That also makes it SAFE.

WHIZ-Z-Z-Z-Z. WALK BACK A MILE

HE feet do not drag. That keeps them dry, preventing colds, and saves shoe leather—saves the price of the sled the first snow-storm—and the strong steel runners, and hardwood frame, outlast all other sleds in the bargain. That makes it CHEAP.

Besides all that, it is the only sled that girls can have any fun with. They can steer it with the feet or with the hands (by the rope sitting up; or it can be steered by the hands, bicycle fashion, lying flat, or on the hip.

So the simple device for steering makes it FASTER, SAFER, CHEAPER, the Best for Boys, the ONLY sled for Girls.

FLEXIBLE FLYERS ON A SCHOOL HILL

HE *"Flexible Flyer"* is the invention of a grown-up boy. The device that makes it steer is patented, and no other sled is made that compares with it. Made of the finest materials, in a large and perfectly-equipped factory, with spring steel in the runners, pressed steel supports, straight grained hardwood frame and seat. It is light, yet practically indestructible, and handsomely finished. It is lighter and at the same time much stronger than ordinary sleds, and the strictly parallel runners make it easier to pull and faster in coasting.

FLEXIBLE FLYERS TAKING THE BEND IN THE ROAD

Why does the Flexible Flyer beat all other coasters?

Because it *steers*. You don't kick and paw and scrape, and dig your heels in the ground, and half stop your sled. You put your feet on the cross-bar and steer, and let the FLYER go. And it does go!

The Flexible Flyer

The Flexible Flyer is a sled up-to-date,
In all other makes it hasn't a mate,
It takes from the fathers all signs of the blues,
For the self-steering sled is a saver of shoes.

It steers without digging the feet in the snow,
And the less of such dragging, the faster you go,
Hands or feet on the cross-bar, you glide on at ease,
Yet turn to the right or the left as you please.

With dry shoes and clothing in comfort you fly,
And leave all the others to come "bye and bye,"
And when at the end you have time to look back,
You'll find they have stopped, a block short of your track.

This sled gives the daughter a share of the fun,
She can steer it as safely down hill as the son,
It means health and joy when a 'coasting she'd roam,
And absence of worry for mother at home.

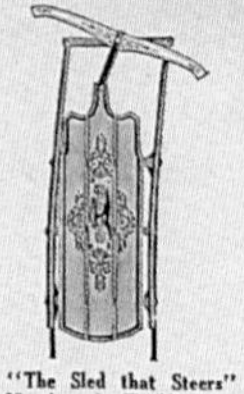

"The Sled that Steers"
"See how the Flexible Flyer" runners can be bent sidewise

STEER IT LYING DOWN, SITTING UP OR ON THE HIP AS YOU PLEASE

Don't

HINTS FOR COASTERS.

1. Don't Kick. Steer.
   Get a real self-steering sled, the famous FLEXIBLE FLYER, and save your shoes.
2. Don't check your sled.
   You can't help it with a clumsy old-style sled, scraping and digging in the snow.
3. Don't get beaten.
   You certainly will if others have FLEXIBLE FLYERS and you haven't.
4. Don't get wet feet.
   With a FLEXIBLE FLYER you keep your shoes out of the snow and the snow out of your shoes.

ALL SIZES OF FLEXIBLE FLYERS. THEY MAKE COASTING A DELIGHT AND NOT A TERROR TO THE FATHER OR MOTHER WHO BUYS THE SHOE LEATHER

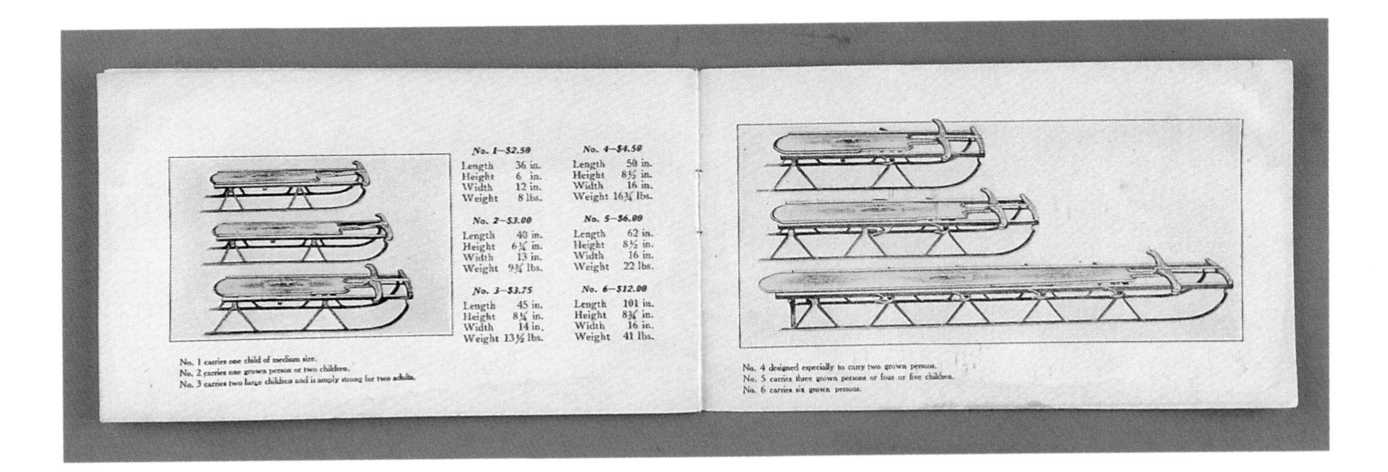

No. 1—$2.50		*No. 4—$4.50*	
Length	36 in.	Length	50 in.
Height	6 in.	Height	8½ in.
Width	12 in.	Width	16 in.
Weight	8 lbs.	Weight	16¾ lbs.
No. 2—$3.00		*No. 5—$6.00*	
Length	40 in.	Length	62 in.
Height	6¾ in.	Height	8½ in.
Width	13 in.	Width	16 in.
Weight	9¾ lbs.	Weight	22 lbs.
No. 3—$3.75		*No. 6—$12.00*	
Length	45 in.	Length	101 in.
Height	8¼ in.	Height	8¾ in.
Width	14 in.	Width	16 in.
Weight	13½ lbs.	Weight	41 lbs.

No. 1 carries one child of medium size.
No. 2 carries one grown person or two children.
No. 3 carries two large children and is amply strong for two adults.

No. 4 designed especially to carry two grown persons.
No. 5 carries three grown persons or four or five children.
No. 6 carries six grown persons.

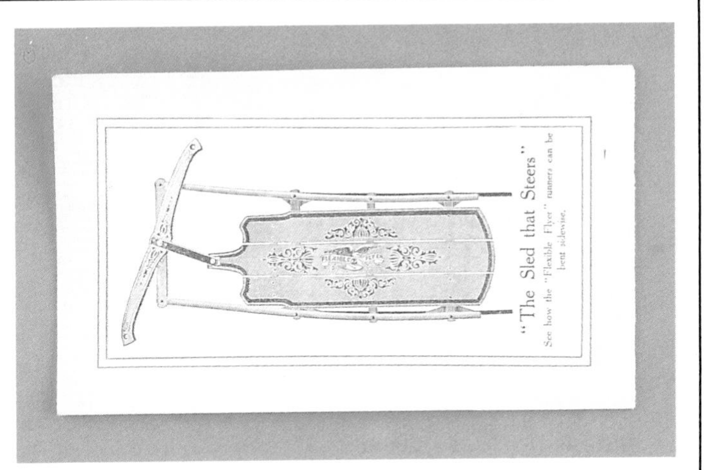

Flexible Flyer® catalog, c. 1900. Differing from the other black-and-white in front and back cover and one page, illustrated here. $75-100.

FLEXIBLE FLYERS TAKING THE BEND IN THE ROAD

Why does the Flexible Flyer. beat all other coasters?

Because it steers. You don't kick and paw and scrape, and dig your heels in the ground, and half stop your sled. You put your feet on the cross-bar and steer, and let the FLYER go. And it does go!

The Flexible Flyer

The Flexible Flyer is a sled up-to-date,
In all other makes it hasn't a mate,
It takes from the fathers all signs of the blues,
For the self-steering sled is a saver of shoes.

It steers without digging the feet in the snow,
And the less of such dragging, the faster you go,
Hands or feet on the cross-bar, you glide on at ease,
Yet turn to the right or the left as you please.

With dry shoes and clothing in comfort you fly,
And leave all the others to come "bye and bye,"
And when at the end you have time to look back,
You'll find they have stopped, a block short of your track.

This sled gives the daughter a share of the fun,
She can steer it as safely down hill as the son,
It means health and joy when a 'coasting she'd roam,
And absence of worry for mother at home.

Flexible Flyer® catalog from 1906, with a sled that steers. 3.75" x 6". $75-100.

THE "Flexible Flyer" has come through the past season with flying colors, the sales beating all previous records to a "frazzle." The "Flexible Flyer" has never been headed and holds the cream of the trade as strongly as it does the hearts of the boys and girls.

THE SECRET OF SLIDING DOWN HILL

The perfect steering is the great feature and is accomplished by a slight movement of the perfected steering bar by the hand or foot or both. This bar bends the runners to the right or left as needed, and they thus follow smoothly in curving tracks when steering, but without plowing up the snow or retarding the progress of the sled.

New patent runners with goose-neck design, quite different in shape from the old ones, are an important improvement, as they greatly lengthen both the bearing and steering surfaces. The extra bearing surface is in front of the first pair of standards, just where it is needed for steering. The consequence is that the latest model "Flexible Flyer" *steers better and quicker* than ever before, and with the lengthened bearing surface *is also faster.*

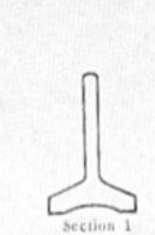

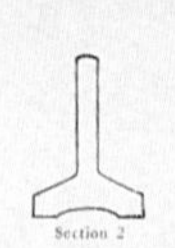

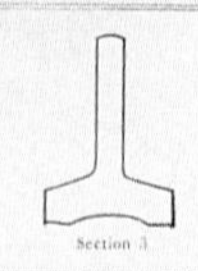

THE various sizes of "Flexible Flyers" are all made with inverted T section spring steel runners, each size being carefully provided with the section best adapted to the load it is to carry. Nos. 1 and 2 "Flexible Flyers" take the No. 1 size, and Nos. 3, 4, 5 and the Racer take No. 2. The coming season No. 6 will take a new section, No. 3, and this being much broader and heavier than heretofore used, will greatly increase the strength, durability and speed of this popular size.

TELL THE BOYS that the "Flexible Flyer" beats any other sled on the hill

WALLACE SHERMAN, Kendallville, Ind., writes:—"With your Flexible Flyer No. 3 I can leave the other boys at the top of the hill."

HAROLD C. HASKY, South Berwick, Me., writes:—"I am very much pleased with your Flexible Flyer No. 2. It will beat any sled on the road."

THE T shaped sections, inverted as shown, are newly patented and, as may easily be seen in the cuts, are of novel design; they are all slightly grooved lengthwise in the center of the bearing surface; they have proven much the best so far discovered, for speed combined with wearing and steering qualities. By the adoption of this patented runner, the danger of skidding on icy hills has been greatly reduced and the speed and safety increased in a marked degree; yet for a snow track the bearing surface is as great as that of a flat runner. That the old style flat or rounded runner is far inferior, is quickly demonstrated upon trial; and a runner grooved its whole width is also greatly inferior for the average track. The combination as used on the "Flexible Flyers" is the great desideratum, as all the boys are finding out; and it is found ONLY upon FLEXIBLE FLYERS, and is thoroughly protected by patent.

TELL THE FATHERS that it pays for itself in one winter in shoes saved

ROBERT CASEY, Boulder, Col., writes:—"I have a No. 2 and No. 5 of your Flexible Flyer sleds. Whenever I take either of them out to coast the other boys speak of it as a 'peach.' The No. 5 goes faster and farther than the big bob-sleds."

E. E. SMITH, Peoria, Ill., writes:—"As to your sled, the writer can give you a pretty strong recommendation for it, if you need it, having bought one for his boy some 13 or 14 years ago, and it is still in good condition and in use when there is opportunity for it."

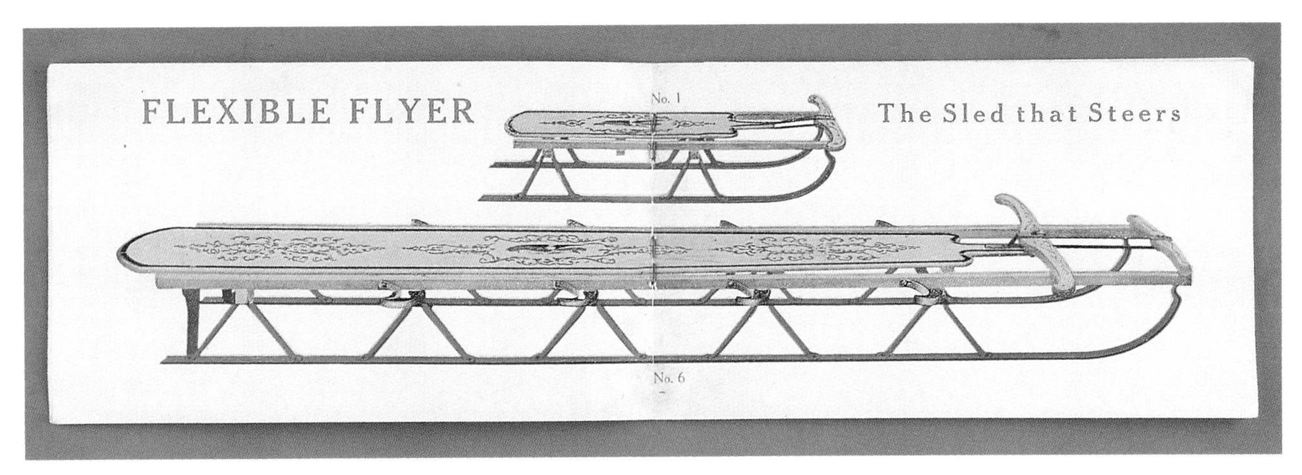

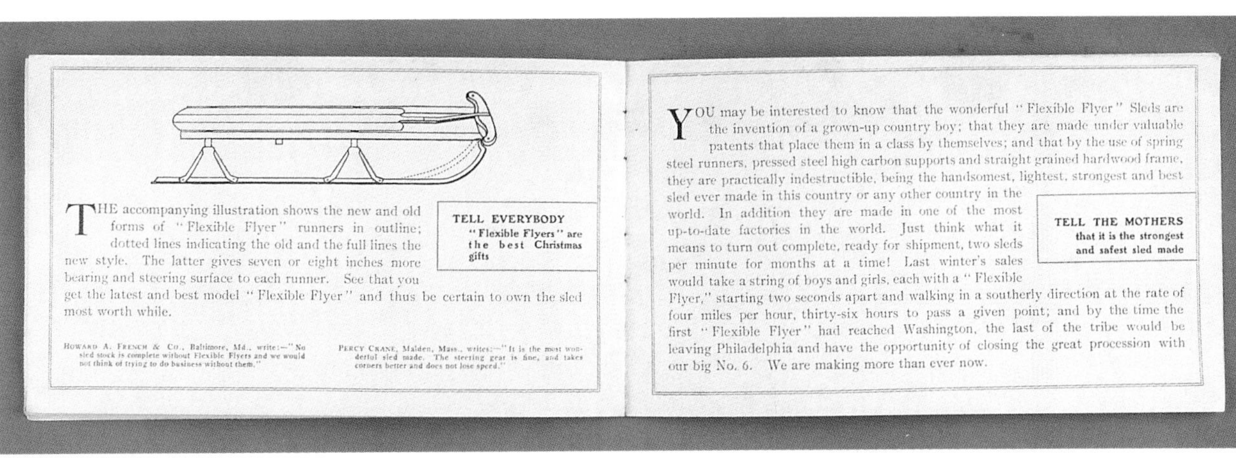

THE accompanying illustration shows the new and old forms of "Flexible Flyer" runners in outline; dotted lines indicating the old and the full lines the new style. The latter gives seven or eight inches more bearing and steering surface to each runner. See that you get the latest and best model "Flexible Flyer" and thus be certain to own the sled most worth while.

TELL EVERYBODY
"Flexible Flyers" are the best Christmas gifts

HOWARD A. FRENCH & CO., Baltimore, Md., write:—"No sled stock is complete without Flexible Flyers and we would not think of trying to do business without them."

PERCY CRANE, Malden, Mass., writes:—"It is the most wonderful sled made. The steering gear is fine, and takes corners better and does not lose speed."

YOU may be interested to know that the wonderful "Flexible Flyer" Sleds are the invention of a grown-up country boy; that they are made under valuable patents that place them in a class by themselves; and that by the use of spring steel runners, pressed steel high carbon supports and straight grained hardwood frame, they are practically indestructible, being the handsomest, lightest, strongest and best sled ever made in this country or any other country in the world. In addition they are made in one of the most up-to-date factories in the world. Just think what it means to turn out complete, ready for shipment, two sleds per minute for months at a time! Last winter's sales would take a string of boys and girls, each with a "Flexible Flyer," starting two seconds apart and walking in a southerly direction at the rate of four miles per hour, thirty-six hours to pass a given point; and by the time the first "Flexible Flyer" had reached Washington, the last of the tribe would be leaving Philadelphia and have the opportunity of closing the great procession with our big No. 6. We are making more than ever now.

TELL THE MOTHERS
that it is the strongest and safest sled made

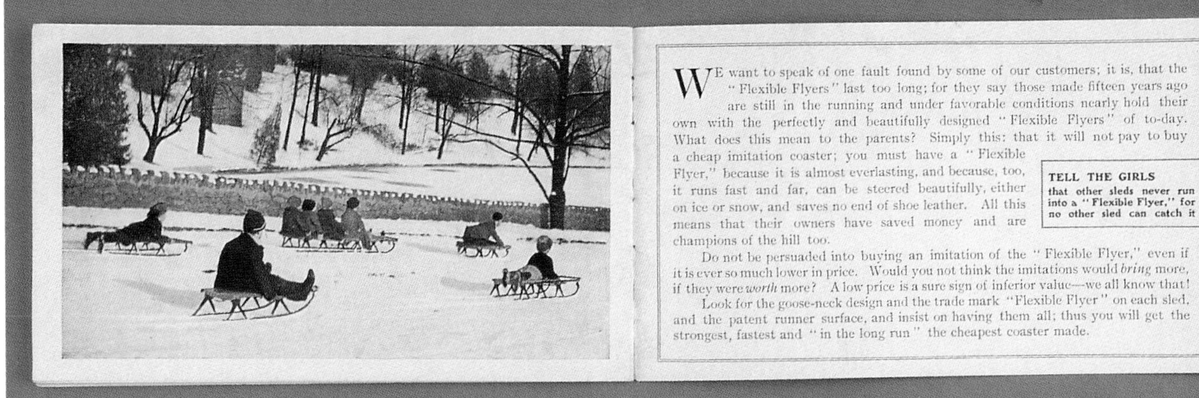

WE want to speak of one fault found by some of our customers; it is, that the "Flexible Flyers" last too long; for they say those made fifteen years ago are still in the running and under favorable conditions nearly hold their own with the perfectly and beautifully designed "Flexible Flyers" of to-day. What does this mean to the parents? Simply this: that it will not pay to buy a cheap imitation coaster; you must have a "Flexible Flyer," because it is almost everlasting, and because, too, it runs fast and far, can be steered beautifully, either on ice or snow, and saves no end of shoe leather. All this means that their owners have saved money and are champions of the hill too.

Do not be persuaded into buying an imitation of the "Flexible Flyer," even if it is ever so much lower in price. Would you not think the imitations would *bring* more, if they were *worth* more? A low price is a sure sign of inferior value—we all know that!

Look for the goose-neck design and the trade mark "Flexible Flyer" on each sled, and the patent runner surface, and insist on having them all; thus you will get the strongest, fastest and "in the long run" the cheapest coaster made.

TELL THE GIRLS
that other sleds never run into a "Flexible Flyer," for no other sled can catch it

Catalog with the John Wanamaker name on the cover. $75-100.

Flexible Flyer® catalog, 1908. $75-100.

There is nothing like coasting to bring a healthy ruddy glow to boy, or girl or grown-up

3

IT IS with pardonable pride that we point to the influence the "Flexible Flyer" has had in the development of coasting. To this most delightful and healthful sport it has given new zest by adding greater speed and perfect control to the sled, and removing every element of danger. With the "Flexible Flyer" you can readily steer in any direction and around every obstacle, by the mere pressure of hand or foot on the famous "Flexible Flyer" steering-bar. And you can do this without plowing up the snow or in any way retarding the speed of the sled.

It is hardly necessary to point out how superior this is to the ordinary way of trying to guide the sled by dragging or digging the feet in the snow, which means wet feet, colds, doctors' bills, besides wearing out boots and shoes. You can readily see, therefore, how a "Flexible Flyer" *saves its cost many times in a single season.*

You can also understand from this why the "Flexible Flyer" has practically superseded the old style sled and completely revolutionized coasting. Enthusiastic owners in this and other countries testify to its great speed, perfect control and safety.

Naturally this phenomenal success has stimulated imitations but like all imitations they lack the perfection of the original. Important features of the "Flexible Flyer" are protected by patents and it is the purpose of this booklet to explain in detail each of these points of superiority.

Oh the fun after school time!

5

THE FAMOUS FLEXIBLE FLYER STEERING-BAR

THIS bar, as we have already explained, bends the spring-steel runners to the right or left as desired and the sled thus follows smoothly in the direction the rider wishes to go, without checking in any way the speed of the sled. The simplicity of this is apparent at a glance. The control is perfect and the smallest child can operate the steering-bar just as easily as a grown-up.

We wish next to call your attention to what we term the "goose neck" design of the fore part of our runners. This is an evidence of how we are constantly on the alert to improve still further the efficiency of the "Flexible Flyer." This "goose neck" is formed where the runners join the front bar and it enables us to *increase* the length and steering surface of each runner *seven or eight inches*, (see illustration). This feature is patented and found exclusively on the "Flexible Flyer." *Be sure to look for it.* It adds greatly to the steering control of the sled and is a very important feature.

Dots show lines of ordinary sled.

We're off!

7

PATENTED RUNNERS

Section 2
Cross section of runners showing grooved or hollow bottom.

EACH sized "Flexible Flyer" is equipped with the runner of the size best suited to the load it has to carry. These runners are made of special process spring steel shaped like the illustration, with a groove along the bottom somewhat like ice skates. This groove prevents "skidding" or slipping on icy hills or pavements, yet on snowy surfaces it is far superior both in point of speed and steering control to the old fashioned *flat or rounded* runners used on all other sleds. This grooved runner is patented and is *used only on* the "Flexible Flyer." It alone places the "Flexible Flyer" far ahead of any other sled on the market.

THEODORE NEWCOMB, New London, Conn., writes:—"I have a No. 4 Flexible Flyer which is certainly O. K. in every detail. I could not ask for a better sled."

R. V. HALL, Baltimore, Md., writes:—"I bought a Flexible Flyer No. 4. I never saw anything like it for speed. I am the winner of all the races. Since my friends have seen my Flexible Flyer they are all buying one. I am thinking of buying a larger one this winter."

WM. WAUGH, Alton, Ill., writes:—"I have your Flexible Flyer No. 5 and we beat every sled on the hill in a run of 800 yards."

AMBROSE H. RITTER, Lafayette Hill, Pa., writes:—"I bought a Flexible Flyer No. 1 and have used it two winters and it is as good as new. The Flexible Flyers beat all the other sleds in speed and go much farther. I think the Flexible Flyer is the only sled."

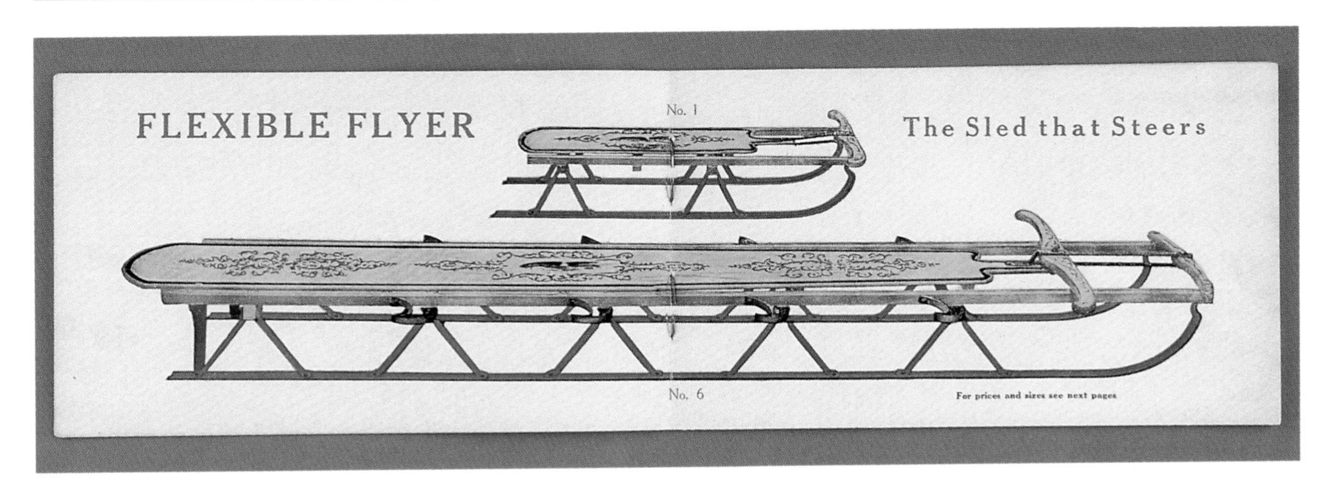

10

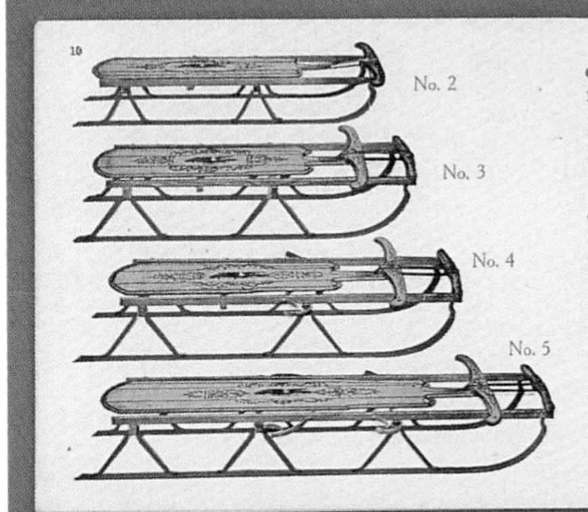

No. 1—$2.50		No. 4—$4.50	
Carries one child of medium size. For illustration see preceding page		Designed especially to carry two grown persons	
Length	36 in.	Length	50 in.
Height	6 in.	Height	8½ in.
Width	12 in.	Width	14 in.
Weight	8½ lbs.	Weight	15½ lbs.
No. 2—$3.00		**No. 5—$6.00**	
Carries one grown person or two children		Carries three grown persons or four or five children	
Length	40 in.	Length	62 in.
Height	6¼ in.	Height	8½ in.
Width	13 in.	Width	16 in.
Weight	10 lbs.	Weight	22 lbs.
No. 3—$3.75		**No. 6—$12.00**	
Carries two large children and is amply strong for two adults		Carries six grown persons. For illustration see preceding page	
Length	45 in.	Length	101 in.
Height	8¼ in.	Height	8¾ in.
Width	14 in.	Width	16 in.
Weight	14 lbs.	Weight	41 lbs.

11

THE FLEXIBLE FLYER RACER

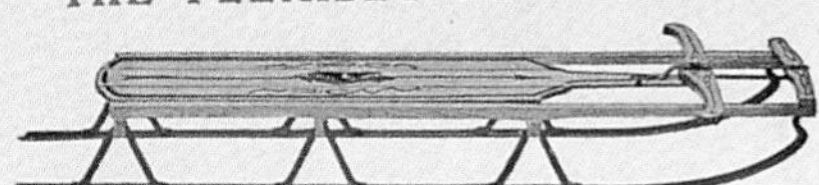

A SPECIAL SLED
for the ambitious, well grown boy

Will stand more hardship than any of the sizes we make. THE BEST VALUE FOR YOUR INVESTMENT

PRICE
$4.25

DIMENSIONS
Length, 50 inches
Height, 6¼ inches
Width, 13 inches

WEIGHT
16 pounds

ENTHUSIASM everywhere greets this latest member of the "Flexible Flyer" family. It is designed for speed and has proven its purpose. Built rather narrow with unusual length of runner extending well forward and to the rear, all of which insures the greatest speed and most efficient control.

Its long, racy lines, lightness, strength, poise and beauty have taken coasting circles by storm. Examine it at your nearest dealer's and see if you can help but grow enthusiastic over it.

Notice particularly the three uprights or supports on each side which make it strong enough for the heaviest load. Yet it is light enough to be easily pulled up the steepest hill. The extreme length of runner in front of the first upright gives exceptional steering control. It is built low, as a Racer should be.

12

CONSTRUCTION

IN the construction of the "Flexible Flyer" we have combined not only the most careful and expert workmanship, but by selecting for every part the very best materials we have placed them in a class by themselves.

For the wood work we use only straight grained second growth ash; the runners are special process spring steel, the uprights or supports are made of pressed high carbon steel, insuring in every way the greatest strength, yet, at the same time, the "Flexible Flyer" is the lightest and easiest to pull up hill.

The runners are japanned, not merely painted. The wood work is handsomely decorated, and finished with *three* coats of the best grade varnish.

This careful, scientific construction explains why the "Flexible Flyer" retains its newness year after year and *outlasts three ordinary sleds.*

DON'T BUY AN UNDERSIZED SLED

Experience has proven to us that a steering sled smaller than our No. 1 (which is 36 inches long) cannot be operated with efficiency or safety. In a smaller sled there is not sufficient bearing or steering surface. We cannot impress this fact too strongly on the minds of parents, who consider above all else the safety of their children. In fact purely from the standpoint of economy, the advantage of buying our larger sized sleds, larger even than would seem necessary, will be apparent because children soon outgrow the smaller sizes. The "Flexible Flyer" lasts season after season, and by selecting a larger size, this likelihood of outgrowing the sled is overcome. All our models, from the largest to the smallest, can be easily operated by the smallest child, so the economy of buying a larger sled will be apparent to everyone.

13

Ice yachting on a No. 2 Flexible Flyer

DEVELOPMENT AND GROWTH

THE "Flexible Flyer" was invented many years ago by a grown up country boy. Realizing the need of just such a sled and being of an inventive turn of mind he set about to see if he could not accomplish what everyone else up to that time had failed to do. After months of experimenting he finally hit upon the idea of a *flexible* runner and from that idea sprang the "Flexible Flyer." It took years to bring it to its present high state of perfection but today our factory, one of the most up-to-date in the world, turns out ready for shipment several sleds per *minute* for months at a time. Just think what that means! Doesn't this prove what the public think of the "Flexible Flyer"? People everywhere have awakened to the merits of the "Flexible Flyer" and no longer buy mere "sleds." They *insist* on the "Flexible Flyer."

C. E. MCKENZIE, So. Essex, Mass., writes:—"Your Flexible Flyer No. 2 beats anything else in the shape of sleds. Have got up a Flexible Flyer Club of seven members."

HENRY YOUNG, Beesleys Point, N. J., writes:—"Please count me as one of the boys that will sing the praises of the Flexible Flyer. It is the best sled for me."

Flexible Flyer® catalog identical to the earlier catalog except that it includes Lord Sempill and ice yachting. 1908. 3.75" x 6". $75-100.

Some of the German Nobility Coasting on a No. 6 Flexible Flyer.

3

THE "Flexible Flyer" was invented many years ago by a grown-up country boy. Realizing the need of just such a sled and being of an inventive turn of mind, he set about to see if he could not accomplish what everybody else up to that time had failed to do. After months of experimenting he finally hit upon the idea of a *flexible* runner and from that idea sprang the "Flexible Flyer." It took years to bring it to its present high state of perfection, but today our factory, one of the most up-to-date in the world, turns out, ready for shipment, an average of over three sleds per minute for months at a time! Just think what that means! Doesn't it prove what the public think of the "Flexible Flyer"? People everywhere have awakened to the merits of the "Flexible Flyer" and no longer buy mere "sleds." If you want the *best, insist on the "Flexible Flyer."*

To fully appreciate its superior quality you should see it. Ask your dealer to show you one and **be sure to look for the name on the sled.**

Naturally, our phenomenal success has stimulated imitations, but, like all imitations, they lack the perfection of the original. Important features of the "Flexible Flyer" are protected by patents and it is the purpose of this booklet to explain in detail each of these points of superiority.

There is nothing like coasting to bring a healthy, ruddy glow to boy, or girl, or grown-up

THE FAMOUS FLEXIBLE FLYER STEERING-BAR

THIS bar, as we have already explained, bends the spring-steel runners to the right or left as desired, and the sled thus follows smoothly in the direction the rider wishes to go, without checking in any way the speed of the sled. The simplicity of this is apparent at a glance. The control is perfect and the smallest child can operate the steering-bar just as easily as a grown-up.

We wish next to call your attention to what we term the "goose-neck" design of the forepart of our runners. This is an evidence of how we are constantly on the alert to improve still further the efficiency of the "Flexible Flyer." This "goose neck" is formed where the runners join the front bar, and it enables us to *increase* the length and steering surface of each runner *several inches* (see illustration). This feature is patented and found exclusively on the "Flexible Flyer." *Be sure to look for it.* It adds greatly to the steering control of the sled and is a very important feature.

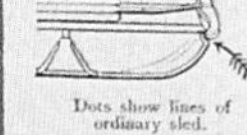

Dots show lines of ordinary sled.

Oh, the fun after school time!

7

PATENTED RUNNERS

Section 2
Cross section of runners showing grooved or hollow bottom.

EACH sized "Flexible Flyer" is equipped with the runner of the size best suited to the load it has to carry. These runners are made of special process spring steel shaped like the illustration, with a groove along the bottom somewhat like ice skates. This groove prevents "skidding" or slipping on icy hills or pavements, yet on snowy surfaces it is far superior, both in point of speed and steering control, to the old-fashioned *flat or rounded* runners used on all other sleds. This grooved runner is patented and is *used only on* the "Flexible Flyer." It alone places the "Flexible Flyer" far ahead of any other sled on the market.

ROBERT HALE GARRISON, Lexington, Mass., writes: "I wish to tell you how satisfied I am with my Flexible Flyers. The first one I used five winters, and, with the exception of the paint, is as good as new. I have *never* had to replace any part. When the first one, which is a No. 2, got too small I got a Racer. I got it the beginning of last winter. For a while it beat every other sled on the coast, but it did not enjoy that distinction long, because as soon as the other boys saw mine they all wanted Racers. So within a month six other boys bought Racers. They are all in perfect condition.
I think the Flexible Flyer is the only *perfect* sled.

F. M. BARTON, Cleveland, Ohio, wires 1-3-11: "Express collect $3.00 Flexible Flyer. Boy won't have any other. Your agent out for 10 days."

SYLVAN KAHN, 624 W. Jackson St., Muncie, Ind., writes: "I have just received one of your Flexible Flyer Racers and I am very much pleased with it. Another boy friend of mine has one. I can say that the Flexible Flyer is a fast and comfortable sled, and every American boy should have one."

WONDER DEPARTMENT STORE, Spokane, Wash.: "We have handled Flexible Flyer Sleds for many years, and found them to give perfect satisfaction."

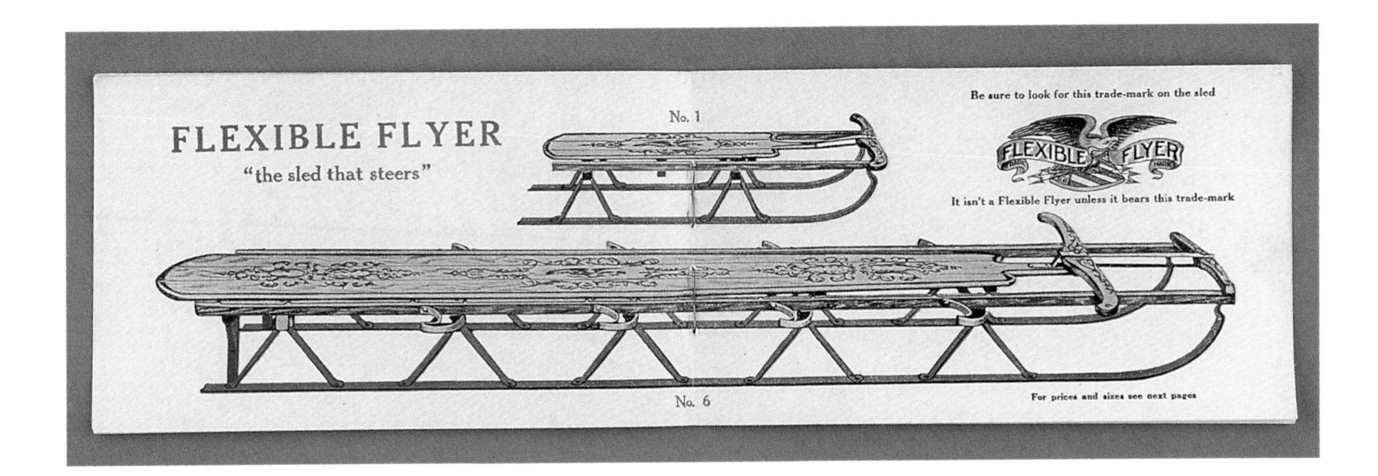

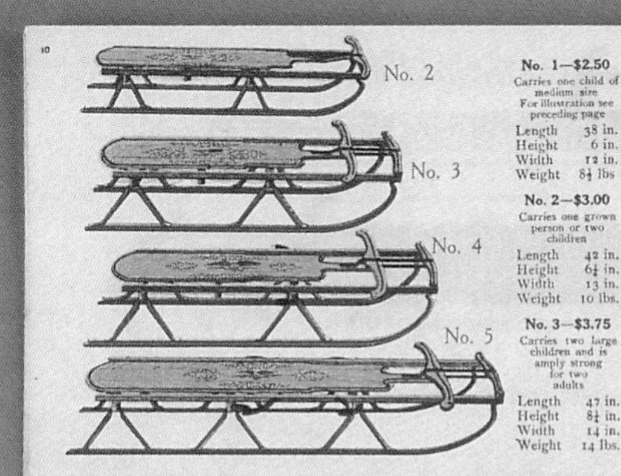

No. 1—$2.50
Carries one child of medium size
For illustration see preceding page

Length	38 in.
Height	6 in.
Width	12 in.
Weight	8½ lbs

No. 2—$3.00
Carries one grown person or two children

Length	42 in.
Height	6½ in.
Width	13 in.
Weight	10 lbs.

No. 3—$3.75
Carries two large children and is amply strong for two adults

Length	47 in.
Height	8½ in.
Width	14 in.
Weight	14 lbs.

No. 4—$4.50
Designed especially to carry two grown persons

Length	52 in.
Height	8½ in.
Width	14 in.
Weight	16 lbs.

No. 5—$6.00
Carries three grown persons or four or five children

Length	63 in.
Height	8½ in.
Width	16 in.
Weight	22 lbs.

No. 6—$12.00
Carries six grown persons
For illustration see preceding page.

Length	101 in.
Height	8½ in.
Width	16 in.
Weight	41 lbs.

THE *Flexible Flyer* RACER

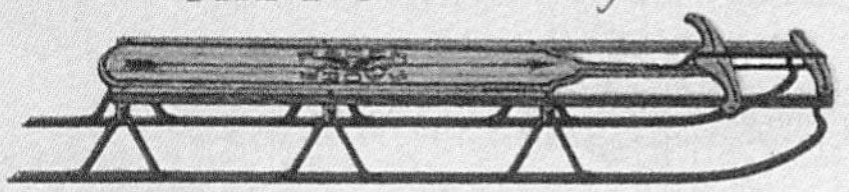

PRICE, $4.25

DIMENSIONS
Length, 56 inches
Height, 6¾ inches
Width, 13 inches
Weight, 14½ pounds

ENTHUSIASM everywhere greets this latest member of the "Flexible Flyer" family. It is designed for speed and has proved its purpose. Built rather narrow, with unusual length of runner extending well forward and to the rear, all of which insures the greatest speed and most efficient control.

Its long, racy lines, lightness, strength, poise and beauty have taken coasting circles by storm. Examine it at your nearest dealer's and see if you can help but grow enthusiastic over it.

Notice particularly the three steel uprights or supports on each side which make it strong enough for the heaviest load. Yet it is light enough to be easily pulled up the steepest hill. The extreme length of runner in front of the first upright gives exceptional steering control. It is built low, as a Racer should be, and is noted for its wonderful speed.

A special sled for the ambitious, well-grown boy. Will stand more hardship than any of the other sizes we make. THE BEST VALUE FOR YOUR INVESTMENT.

CONSTRUCTION

IN the construction of the "Flexible Flyer" we have combined not only the most careful and expert workmanship, but by selecting for every part the very best materials we have placed them in a class by themselves.

For the wood work we use only straight-grained second-growth ash; the runners are special process spring steel; the uprights or supports are made of pressed high carbon steel, insuring in every way the greatest strength, yet, at the same time, the "Flexible Flyer" is the lightest and easiest to pull up hill.

The runners are japanned, not merely painted. The wood work is handsomely decorated, and finished with *three* coats of the best grade varnish.

This careful, scientific construction explains why the "Flexible Flyer" retains its newness year after year and *outlasts three ordinary sleds.*

DON'T BUY AN UNDERSIZED SLED

Experience has proven to us that a steering sled smaller than we make cannot be operated with efficiency or safety. In a smaller sled there is not sufficient bearing or steering surface. We cannot impress this fact too strongly on the minds of parents, who consider above all else the safety of their children. In fact, purely from the standpoint of economy, the advantage of buying our larger sized sleds, larger even than would seem necessary, will be apparent, because children soon outgrow the smaller sizes. The "Flexible Flyer" lasts season after season, and by selecting a larger size, this likelihood of outgrowing the sled is overcome. All our models, from the largest to the smallest, can be easily operated by the smallest child, so the economy of buying a larger sled will be apparent to everyone.

Ice yachting on a No. 2 Flexible Flyer.

WE feel a pardonable pride in the influence our "Flexible Flyer" has had in the development of "coasting." To this most delightful and healthful sport it has given new zest. It has added greater speed and greater control to the sled, and removed every element of danger. With the "Flexible Flyer" you can steer easily in any direction and around every obstacle by the mere pressure of the hand or foot on our famous "Flexible Flyer" steering-bar. And you can do this without plowing up the snow or in any way retarding the speed of the sled.

The way of trying to guide the ordinary sled is by dragging or digging the feet in the snow, which means wet feet, colds, and doctors' bills, besides wearing out boots and shoes. You can readily see, therefore, how superior is the "Flexible Flyer" and how it *saves many times its cost in a single season.*

You can also understand from this why the "Flexible Flyer" has practically superseded the old style sled and completely revolutionized coasting. Enthusiastic owners in this and other countries testify to its great speed, perfect control and absolute safety.

Lord Sempill and daughter, the Hon. Margaret Forbes-Sempill, enjoying a ride on the slopes of St. Moritz. Lord Sempill, formerly captain in the Black Watch, has seen active service in the Soudan and South-African War.—*London Tatler.*

A LAST WORD

WE HAVE endeavored to explain in the preceding pages the various features of the "Flexible Flyer" that make it the best sled on the market. If we have succeeded we need say nothing more. However, to appreciate the many advantages of the "Flexible Flyer" you must see it. Ask your dealer to show you one. Don't be persuaded into buying a cheap imitation coaster. *Insist* on a "Flexible Flyer" and look for the *"goose neck" design*, the *patent grooved runners*, etc., **and be sure** to look for the *name* and *trade-mark* on the sled. *It isn't a "Flexible Flyer" unless it bears this name and trade-mark.*

HARVEY BROS., Hillsdale, N. Y., write: "The Flexible Flyer is the most satisfactory sled we handle and our sales are increasing every year."

"DAN NICHOLS, our former Flexible Flyer Agent at Beloit, Wis., gave his son a No. 2 Flexible Flyer about 20 years ago. The hills are steep and stony. The No. 2 has shot over these rocks every year, and is still in good condition."

HOWARD A. FRENCH & CO., Baltimore, Md., write: A gentleman came into our store Saturday evening and wanted four No. 3 'Flexible Flyers.' Said he had to have them, as he had bought four ——— ——— for his boys and they would not have them; did not have the Eagle on, so he gave the ——— ——— to an orphan asylum. Can you beat it?"

W. H. SMITH & SON, Bath, Maine, write: "We have a stock of your sleds. Don't blame you for being persistent, as you have a *good* sled."

Flexible Flyer® complete dated catalog, 1908. $75-100.

O get the most fun out of coasting, you must have a fast sled. The most envied, most popular and the happiest boy in winter is the one who owns the champion sled of the town. Now, the "Flexible Flyer" is the fastest coaster every time, and for mighty good reasons; for the "Flexible Flyer" has spring steel flexible runners, controlled by a capital steering bar convenient for hands or feet. Think for a moment how much that means. Ordinary sleds are steered by digging in the heels or dragging the foot, or in other words, by checking the speed of the sled, first on one side then on the other. These things hold the sled back. The foot is like a dragging anchor, the sled is like a plow. Both are dragging and plowing most of the time, even on a straight hill, for the surface is never even, and few sleds run exactly true.

THE SECRET OF SLIDING DOWN HILL

JAMES GRADY, Asbury Park, N. J., writes:—"I got your No. 3 Flexible Flyer and am very well pleased with it. Went coasting to-day and left every boy behind. They all wished they had my sled."

WILLIAM VAN DE BOGART, New York City:—"I bought your No. 5 Flexible Flyer and think there is nothing else in the line of sleds that can beat them."

HE Runners of the "Flexible Flyer" are of ⊥ shaped spring steel, very strong and rigid vertically, but free to bend sidewise. They are made true and stay so, making the "Flexible Flyer" run easily and fast down hill and pull very easily while walking up hill, or when pulling loaded on the level. But when you want to steer on the hill, the moving of the steering bar curves both runners, which thus follow smoothly in the curving tracks without plowing the snow and without dragging the foot. The sled thus unhampered, shoots along at full speed, a sure winner;—it is FAST. It is not only a sure winner for speed, but also for distance; and you don't run into the others—you steer around them; they can't run into you, for they are not able to catch you. These are grand elements of both speed and safety, therefore see that you get a "Flexible Flyer," the only coaster that combines these two elements in the highest degree.

TELL THE BOYS that the "Flexible Flyer" beats any other sled on the hill.

W. J. DEARTH, So. Omaha, Neb.:—"I must say the Flexible Flyer is the slickest thing in the sled line I have ever seen and very reasonable in price."

HOWARD HICKEY, Athens, Pa.:—"I have your No. 3 Flexible Flyer. It's great. I hold the record on four hills with it."

HE feet do not drag. This keeps them dry, preventing colds, and saves shoe leather; saves the price of the sleds in shoe leather alone the first snow storm, perhaps; and the strong steel runners, and a hardwood frame with perfection of manufacture, makes them outlast all other sleds into the bargain. That makes them *cheap.*

Besides all that, it is the only suitable sled for girls. They can steer it with the feet or with the hands and rope sitting up; or it can be steered by the hands, bicycle fashion, lying flat, or on the hip.

So you see why the "Flexible Flyer" is FASTER, SAFER and CHEAPER than any other in the world; the Best for Boys, the Only Sled for Girls.

TELL THE FATHERS that it pays for itself in one winter in shoes saved.

R. J. HEMMICK, Port Hope, Ont., writes:—"I bought my son one of your Flexible Flyers in Toronto. To use his own favorite expression, 'It is a jimdandy' and you do a world of good to the young folks the more you sell. Hurrah for the Flexible Flyer! The others are not in it with the boys and girls."

P. W. JACKSON, Merriam Park, Minn., writes:—"The Flexible Flyer certainly beats my old-fashioned sled of forty years ago and the children are now happy; and as the saying is, 'A thing of beauty is a joy forever,' which your Flexible Flyer certainly is."

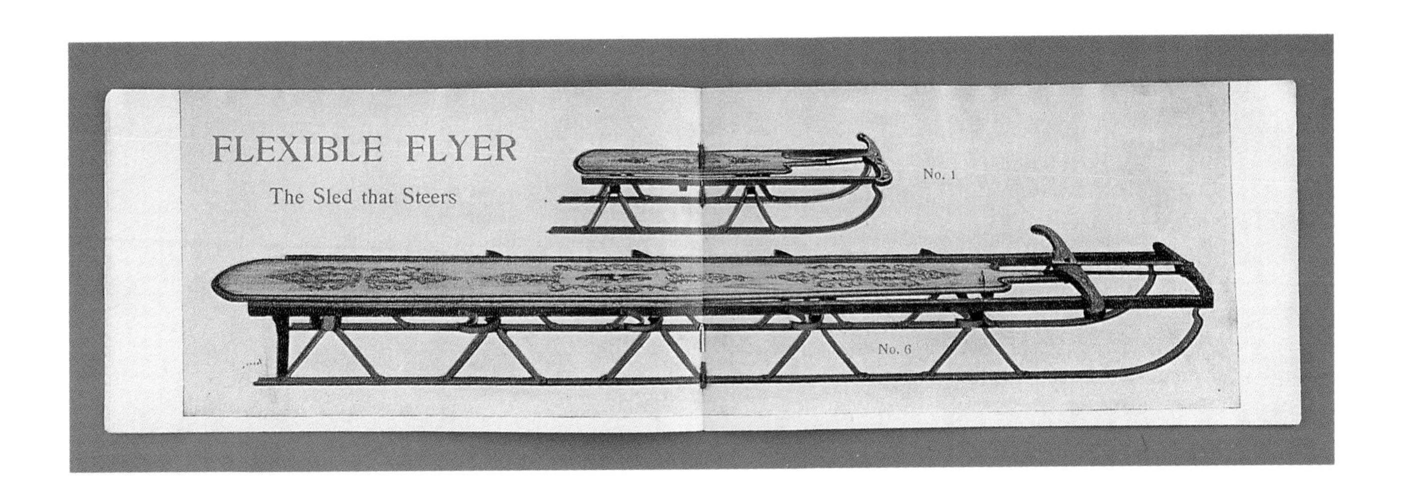

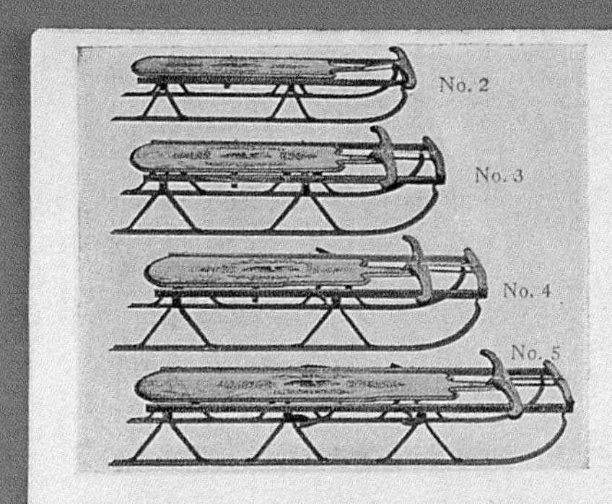

No. 1—$2.50

Carries one child of medium size.

Length 36 in.
Height 6 in.
Width 12 in.
Weight 8½ lbs.

No. 2—$3.00

Carries one grown person or two children.

Length 40 in.
Height 6¼ in.
Width 13 in.
Weight 10 lbs.

No. 3—$3.75

Carries two large children and is amply strong for two adults.

Length 45 in.
Height 8¼ in.
Width 14 in.
Weight 14 lbs.

No. 4—$4.50

Designed especially to carry two grown persons.

Length 50 in.
Height 8½ in.
Width 16 in.
Weight 17 lbs.

No. 5—$6.00

Carries three grown persons or four or five children.

Length 62 in.
Height 8½ in.
Width 16 in.
Weight 22 lbs.

No. 6—$12.00

Carries six grown persons.

Length 101 in.
Height 8¾ in.
Width 16 in.
Weight 41 lbs.

THE "FLEXIBLE FLYER" is the invention of a grown-up boy. It is covered by valuable patents and no other sled compares with it. The "Flexible Flyer" is built of the finest materials, in a large and perfectly equipped steel-working factory. Has spring steel in the runners, pressed steel supports and straight-grained hardwood frame and seat. The "Flexible Flyer" is light in weight, yet practically indestructible, and handsomely finished. The "Flexible Flyer" is lighter and at the same time much stronger and faster than other sleds. In a word, the "Flexible Flyer" is absolutely the best sled ever made in this country, or any other country in the world.

TELL THE MOTHERS
that it is the strongest and safest sled made.

The boys and girls now using "Flexible Flyers" number hundreds of thousands, and the pictures in this booklet, which are taken from life, show how "Flexible Flyers" have displaced the old style sled. Great improvements have been made for 1908 which are fully covered by patents.

THE new patent runners are different in shape from the old ones and enable us to greatly lengthen the bearing and steering surface on the snow. This extra bearing surface is in front of the first pair of standards, just where it is needed for steering. The consequence is that the 1908 model "Flexible Flyer" steers even better and quicker than ever before and is considerably faster. This new runner is easily distinguished by the goose-neck design, and with the trade mark "Flexible Flyer" will enable you to be sure you are buying the latest and best model.

TELL THE GIRLS
that other sleds never run into a "Flexible Flyer," for no other sled can catch it!

All the larger size "Flexible Flyers" for 1908 are also made with a runner slightly guttered in the center. This does not interfere with the speed of the sled on snow hills, but adds materially to the control of the sled on icy hills.

ARTHUR H. ELLIOT, Harbor, Mich.:—"I have a No. 4 Flexible Flyer and am beating all the boys around, even those having 'Bobs.' The Flexible Flyer is A No. 1; you may count on me to sing its praises."

CLEVE McCABE, Chicago, Ill.:—"Have purchased one of your 'Flexible Flyers' and am well pleased with it. Would not take $50 for it if I could not get another."

THESE two improvements mark a new era in Coaster building and are unknown in any other make. Do not be persuaded into buying an inferior imitation of our "Flexible Flyer," even if it is a little lower in price. Look for the trade mark "Flexible Flyer" on the sled, and insist on having it.

The name "Flexible Flyer" as well as the sled itself is patented in order to prevent unscrupulous manufacturers from palming off on the public poor imitations of our famous "Flexible Flyers."

TELL EVERYBODY
"Flexible Flyers" are the best Christmas gifts.

W. S. LETHERBURY, Middletown, Del., writes:—"I purchased some of your Flexible Flyers and would say they have proven perfectly satisfactory."

CLINTON SPRAGUE, Milton, Mass., writes:—"The Flexible Flyer will beat any sled in Milton. There are five Flexibles on my street."

IRA DEVINE, Tyler, Texas, writes:—"I have purchased a Flexible Flyer from my local dealer. It is fine. Would not take $100 for it if I could not get another one."

H. R. CLEVELAND, Danville, Quebec, writes regarding No. 4 Flexible Flyer:—"It's a cracker jack; can run away from all of them."

1908 catalog for S.L. Allen. On the first page is a photograph of the new building being constructed to handle the manufacture of the sled business. 8.75" x 6". $15-25.

The Season of 1908

WE HAVE been behind orders for years past, but hope in 1908 to be equal to all demands, for the great event of the year with us is the occupation of the new five-story addition shown on the opposite page. It was finished in time for this year's trade and adds one and a half acres of floor space. The walls, floors and roof are all of reinforced concrete, making the structure practically fireproof.

The three-story building in the rear (214 feet long) includes our present warehouse and assembling buildings; hidden behind these are the power house, boiler nouse, paint and varnish buildings, large storehouse and the forging shop, the latter nearly four hundred feet long by sixty wide.

The new building is used almost exclusively for manufacturing, the old being newly fitted up for warerooms and the shipping department.

This enlargement comes at an auspicious moment, in view of the alarming scarcity of farm laborers. We shall help to fill up this gap so far as we are able by furnishing first-class labor saving tools for the Farmer and Gardener, and we guarantee our goods equal to or better than our own previous product, and we believe they have no equal in the world's market to-day.

We are always glad to assist our friends in selecting the best tools from our list for their special needs, and we also welcome criticism of our products by practical men—the up-to-date Gardeners and Farmers.

Send in your orders early.

Flexible Flyer Coasters

We make these during the summer let up; they have displaced nearly all other coasters wherever a good article is wanted.

They are now better than ever, with new patented features. Are steered perfectly with a cross bar, either sitting up or lying down; they save shoe leather and prevent wet feet. In a few words they combine lightness with strength and durability, speed with comfort and safety, and economy with satisfaction.

We make six sizes, carrying from one to eight children, or up to six grown persons. We print a special catalogue, which is free for the asking.

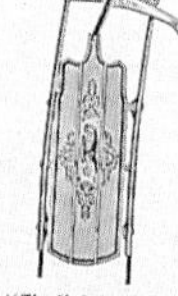

"The Sled that Steers"
"See how the Flexible Flyer" runners can be bent sidewise

S. L. ALLEN & CO.

1107 Market Street, Philadelphia, Penna., U. S. A.

MOST of the fun of coasting is in having the fastest sled. The most envied, the most popular and the happiest boy is the one who owns the champion sled of the town. Now the "Flexible Flyer" is the fastest sled every time against all comers and for good reasons.

THE SECRET OF SLIDING DOWN HILL

The "Flexible Flyer" is a sled that steers. Think for a moment how much that means. Ordinary sleds are steered by digging in the heels or dragging the foot, thereby turning the runner's side to the snow. All these things hold the sled back. The foot is an anchor, the sled is a plow. Both are dragging and plowing most of the time, even on a straight hill, for the surface is never even and few sleds run true.

See how the runners can be bent sidewise.

The Runners of the "Flexible Flyer" are of ⊥ shaped spring steel, very strong and rigid vertically, but free to bend sidewise. They are absolutely true and stay so, making it slide very easily while walking up hill or pulling somebody's sister on level ground. But when you want to steer on the hill, a touch on the cross-bar *curves both runners*, the whole runner going in its own track, without plowing the snow, without dragging the foot. The sled—unhampered—shoots on at full speed, a sure winner. It is FAST—it is not only a sure winner for speed, but also for distance. ¶ You don't run into the others, you steer around them; they can't run into you, for they are not able to catch you. That also makes it SAFE.

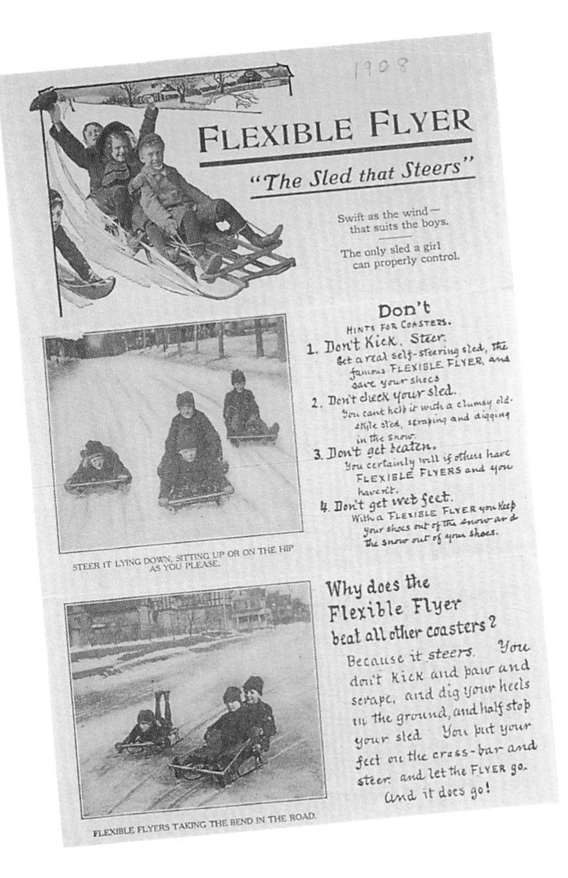

Sepia advertising folder for Flexible Flyer®, 1908. Open size: 9" x 12". $25-50

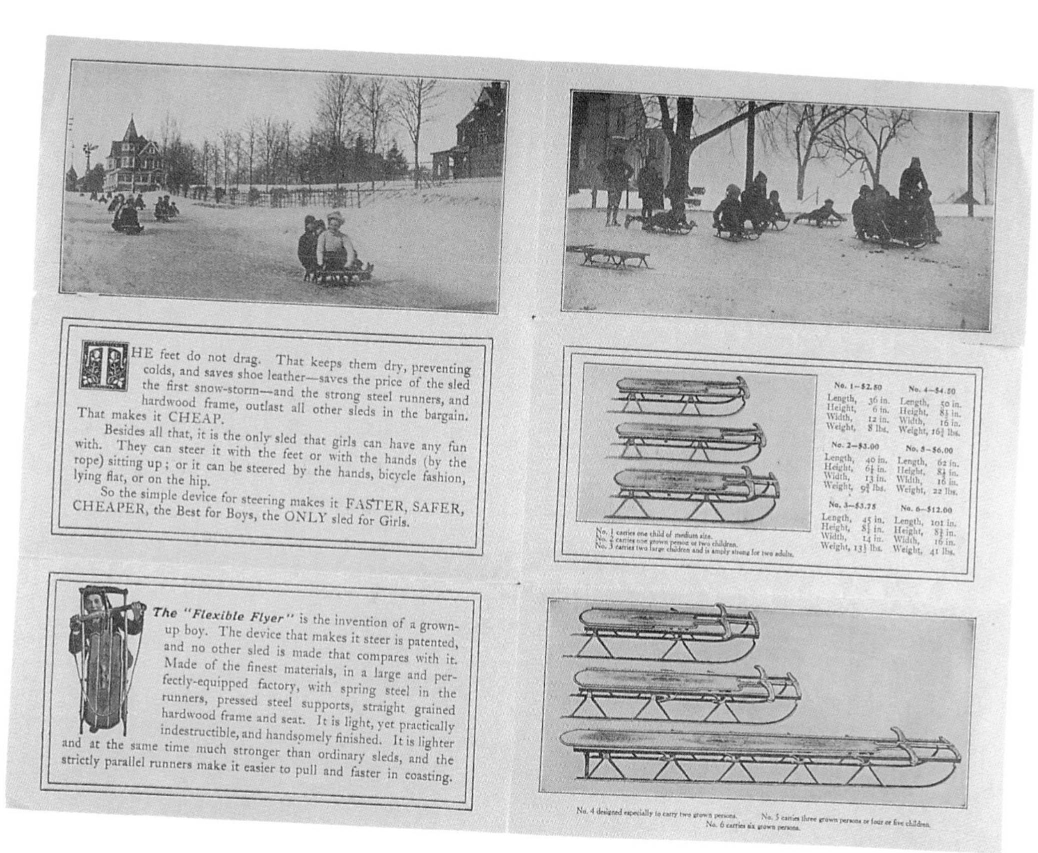

1908 catalog of S.L. Allen & Co.

To get the most fun out of coasting, you must have a fast sled. The most envied, most popular and the happiest boy in winter is the one who owns the champion sled of the town. Now, the "Flexible Flyer" is the fastest coaster every time, and for mighty good reasons: for the "Flexible Flyer" has spring steel flexible runners, controlled by a capital steering bar convenient for hands or feet. Think for a moment how much that means. Ordinary sleds are steered by digging in the heels or dragging the foot, or in other words, by checking the speed of the sled, first on one side then on the other. These things hold the sled back. The foot is like a dragging anchor, the sled is like a plow. Both are dragging and plowing most of the time, even on a straight hill, for the surface is never even, and few sleds run exactly true.

THE SECRET OF SLIDING DOWN HILL

JAMES GRADY,
Asbury Park, N. J., writes.—
"I got your No. 3 Flexible Flyer and am very well pleased with it. Went coasting today and left every boy behind. They all wished they had my sled."

* * *

WILLIAM VANDE BOGART
New York City.
"I bought your No. 5 Flexible Flyer and think there is nothing else in the line of sleds that can beat them."

The runners of the "Flexible Flyer" are of ⊥ shaped spring steel, very strong and rigid vertically, but free to bend sidewise. They are made true and stay so, making the "Flexible Flyer" run easily and fast down hill and pull very easily while walking up hill, or when pulling load on the level. But when you want to steer on the hill, the moving of the steering bar curves both runners, which thus follow smoothly in the curving tracks without plowing the snow and without dragging the foot. The sled thus unhampered, shoots along at full speed, a sure winner;—it is FAST. It is not only a sure winner for speed, but also for distance; and you don't run into the others—you steer around them; they can't run into you, for they are not able to catch you. These are grand elements of both speed and safety, therefore see that you get a "Flexible Flyer," the only coaster that combines these two elements in the highest degree.

TELL THE BOYS
that the "Flexible Flyer" beats any other sled on the hill

W. J. DEARTH,
So. Omaha, Neb.
"I must say that the Flexible Flyer is the slickest thing in the sled line I have ever seen and very reasonable in price."

* * *

HOWARD HICKEY,
Athens, Pa.
"I have your No. 3 Flexible Flyer. It's great. I hold the record on four hills with it."

The feet do not drag. This keeps them dry, preventing colds, and saves shoe leather; saves the price of the sleds in shoe leather alone the first snow-storm, perhaps; and the strong steel runners, and a hardwood frame with perfection of manufacture, makes them outlast all other sleds in the bargain. That makes them *cheap.*

Besides all that, it is the only suitable sled for girls. They can steer it with the feet or with the hands and rope sitting up; or it can be steered by the hands, bicycle fashion, lying flat, or on the hip.

So you see why the "Flexible Flyer" is FASTER, SAFER and CHEAPER than any other in the world; the Best for Boys, the Only Sled for Girls.

TELL THE FATHERS
that it pays for itself in one winter in shoes saved

R. J. HEMMICK,
Port Hope, Ont., writes:—
"I bought my son one of your Flexible Flyers in Toronto. To use his own favorite expression, 'It is a jim-dandy' and you do a world of good to the young folks the more you sell. Hurrah for the Flexible Flyer! The others are not in it with the boys and girls."

* * *

P. W. JACKSON,
Merriam Park, Minn., writes:—
"The Flexible Flyer certainly beats my old-fashioned sled of forty years ago and the children are now happy; and as the saying is, 'A thing of beauty is a joy forever,' which your Flexible Flyer certainly is."

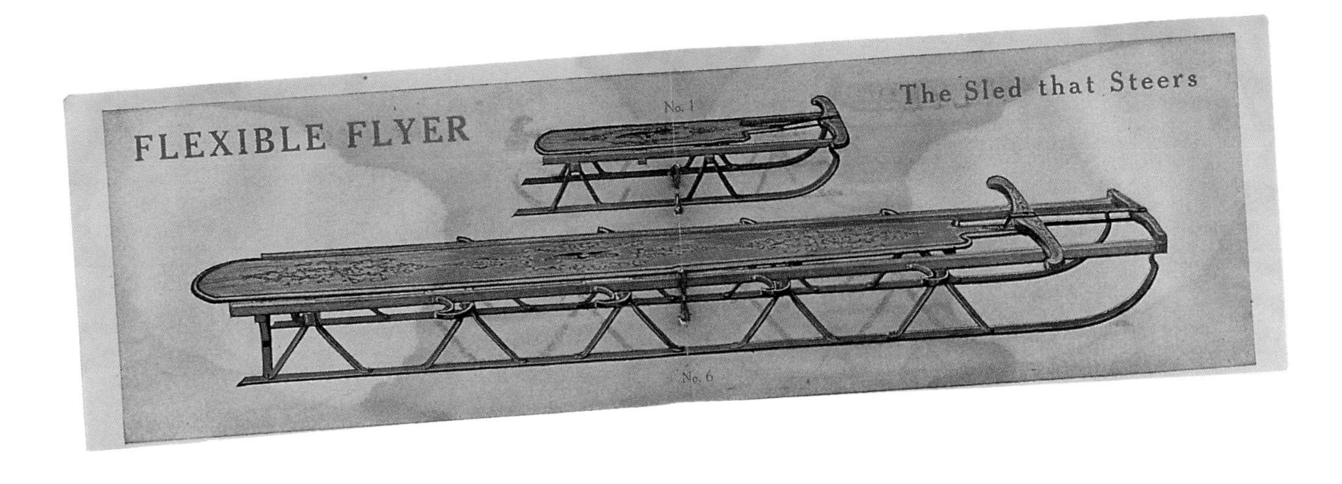

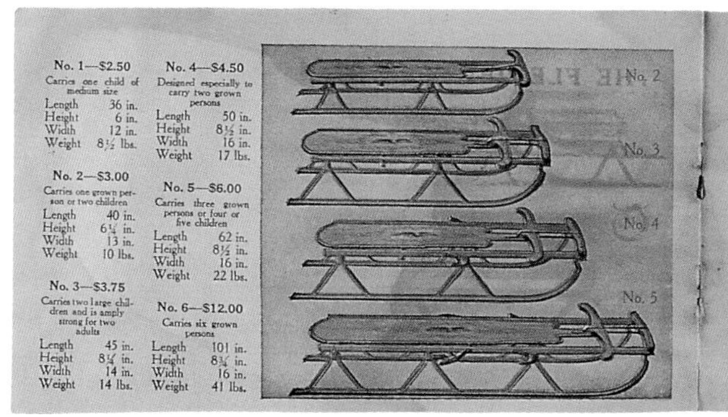

THE FLEXIBLE FLYER RACER

This is the youngest, best and most attractive of the great Flexible Flyer family, if judged by the enthusiasm with which it has been received wherever shown. Its racy lines, its lightness and strength, poise and beauty have taken coasting circles by storm. Though shown first at the tail of the season, its success was immediate, and dealers have already bought thousands for next winter. The Racer is built low and rather narrow, with unusual length of runner extending well forward and to the rear, insuring high speed. With three standards on a side it is extra strong and the springing of the runners by the steering bar is perfect. The runner steel is our patent section insuring successful steering, and the new shape front is pleasing to the eye and correct form for speed. The finish is our best.

A SPECIAL SLED
for the ambitious, well grown boy

Will stand more hardship than any of the sizes we make.
THE BEST VALUE FOR YOUR INVESTMENT

PRICE
$4.25

DIMENSIONS
Length, 56 inches
Height, 6¼ inches
Width, 13 inches

WEIGHT
16 pounds

TELL EVERYBODY
"Flexible Flyers" are the best Christmas gifts

The boys and girls now using "Flexible Flyers" number hundreds of thousands, and the pictures in this booklet, which are taken from life, show how "Flexible Flyers" have displaced the old style sled. Great improvements have been made for this year which are fully covered by patents.

The "FLEXIBLE FLYER" is the invention of a grown-up boy. It is covered by valuable patents and no other sled compares with it. The "Flexible Flyer" is built of the finest materials, in a large and perfectly equipped steel-working factory. Has spring steel in the runners, pressed steel supports and straight-grained hardwood frame and seat. The "Flexible Flyer" is light in weight, yet practically indestructible, and handsomely finished. The "Flexible Flyer" is lighter and at the same time much stronger and faster than other sleds. In a word, the "Flexible Flyer" is absolutely the best sled ever made in this country, or any other country in the world.

The new patent runners are different in shape from the old ones and enable us to greatly lengthen the bearing and steering surface on the snow. This extra bearing surface is in front of the first pair of standards, just where it is needed for steering. The consequence is that the latest model "Flexible Flyer" steers even better and quicker than ever before and is considerably faster. This new runner is easily distinguished by the *goose-neck design,* and with the trade mark *"Flexible Flyer"* will enable you to be sure you are buying the latest and best model.

All the larger size "Flexible Flyers" for this year are also made with a runner slightly guttered in the center. This does not interfere with the speed of the sled on snow hills, but adds materially to the control of the sled on icy hills.

TELL THE MOTHERS
that it is the strongest and safest sled made

ARTHUR H. ELLIOTT,
Harbor, Mich.
"I have a No. 4 Flexible Flyer and am beating all the boys around, even those having 'Bobs.' The Flexible Flyer is A No. 1; you may count on me to sing its praises."

* * *

CLEVE McCABE,
Chicago, Ill.
"Have purchased one of your 'Flexible Flyers' and am well pleased with it. Would not take $50 for it if I could not get another."

These two improvements mark a new era in Coaster building and are unknown in any other make. Do not be persuaded into buying an *inferior imitation* of our "Flexible Flyer," even if it is a little lower in price. Look for the trade mark "Flexible Flyer" on the sled, and insist on having it.

FLEXIBLE FLYER

The name "Flexible Flyer," as well as the sled itself, is patented in order to prevent unscrupulous manufacturers from palming off on the public poor imitations of our famous "Flexible Flyers."

TELL THE GIRLS
that other sleds never run into a "Flexible Flyer" for no other sled can catch it

W. S. LETHERBURY,
Middletown, Del., writes:—
"I purchased some of your Flexible Flyers and would say that they have proven perfectly satisfactory."

* * *

CLINTON SPRAGUE,
Milton, Mass., writes:—
"The Flexible Flyer will beat any sled in Milton. There are five Flexibles on my street."

* * *

IRA DEVINE,
Tyler, Texas, writes:—
"I have purchased a Flexible Flyer from my local dealer. It is fine. Would not take $100 for it if I could not get another one."

* * *

H. R. CLEVELAND,
Danville, Quebec, writes regarding No. 4 Flexible Flyer:—
"It's a crackerjack; can run away from all of them."

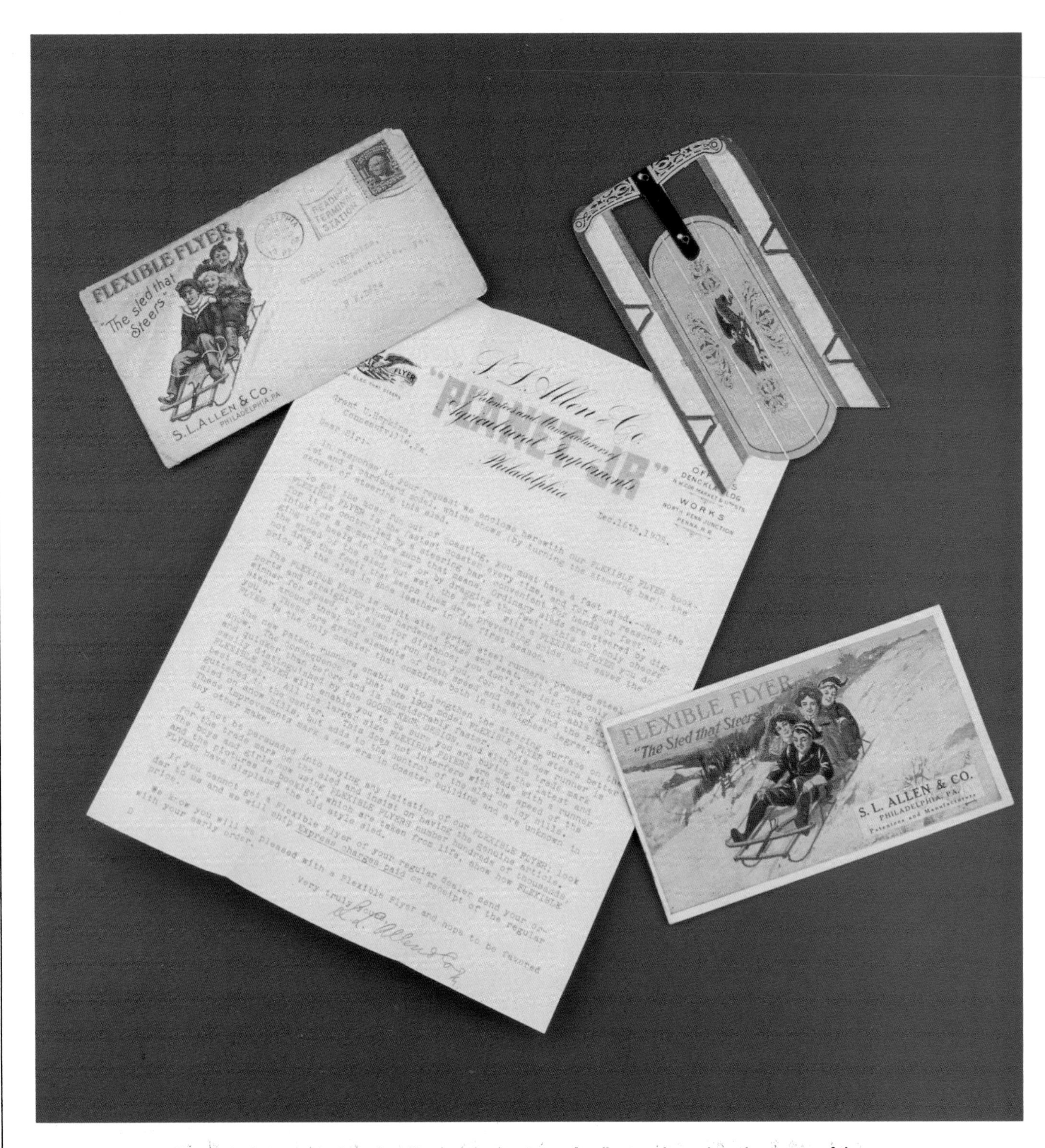

S. L. Allen & Co.
Patentees and Manufacturers of
Agricultural Implements
"PLANET JR"
Philadelphia

OFFICES DENCKLA BLDG N.W. COR. MARKET & 11TH STS
WORKS NORTH PENN JUNCTION PENNA. R.R.

Dec.16th,1908.

Grant V.Hopkins,
Connestville,Pa.

Dear Sir:-

In response to your request we enclose herewith our FLEXIBLE FLYER booklet and a cardboard model, which shows (by turning the steering bar), the secret of steering this sled.

To get the most fun out of coasting, you must have a fast sled.--Now the FLEXIBLE FLYER is the fastest coaster every time, and for good reasons; for it is controlled by a steering bar, convenient for hands or feet. Think for a moment how much that means. Ordinary sleds are steered by digging the heels in the snow or by dragging the feet: this not only checks the speed of the sled, but wets the feet. With a FLEXIBLE FLYER you do not drag the feet; that keeps them dry, preventing colds, and saves the price of the sled in shoe leather in the first season.

The FLEXIBLE FLYER is built with spring steel runners, pressed steel supports and straight grained hardwood frame and seat. It is not only a winner for speed, but also for distance; you don't run into the other... steer around them; they can't run into you, for they are not able to... you. These are grand elements of both speed and safety and the FLEXIBLE FLYER is the only coaster that combines both in the highest degree.

The new Patent runners enable us to lengthen the steering surface on the snow. The consequence is that the 1908 model FLEXIBLE FLYER steers better and quicker than before and is considerably faster. This new runner is easily distinguished by the GOOSE-NECK DESIGN, and with the latest mark FLEXIBLE FLYER will enable you to be sure you are buying the trade mark best model. All the larger size FLEXIBLE FLYERS are made with a runner guttered in the center. This does not interfere with the speed of the sled on snow hills, but adds to the control of the sled on icy hills. These improvements mark a new era in Coaster building and are unknown in any other make.

Do not be persuaded into buying any imitation of our FLEXIBLE FLYER; look for the trade mark on the sled and insist on having the genuine article. The boys and girls now using FLEXIBLE FLYERS number hundreds of thousands, and the pictures in booklet, which are taken from life, show how FLEXIBLE FLYERS have displaced the old style sled.

If you cannot get a Flexible Flyer of your regular dealer send your order to us and we will ship Express charges paid on receipt of the regular price.

We know you will be pleased with a Flexible Flyer and hope to be favored with your early order.

Very truly yours,
S. L. Allen & Co.

D

The letter from the inside of an illustrated cover issue. It tells you about the advantages of the Flexible Flyer® and how to order it. Also included were a cardboard model and a catalog. 1908.

Flexible Flyer® catalog, 1908. 3.75" x 6". $75-100.
Continued on Next Page.

3

THE "Flexible Flyer" was invented many years ago by a grown-up country boy. Realizing the need of just such a sled and being of an inventive turn of mind, he set about to see if he could not accomplish what everybody else up to that time had failed to do. After several winters of experimenting he finally hit upon the idea of a *flexible* runner and from that idea sprang the "Flexible Flyer." It took years to bring it to its present high state of perfection, but today our factory, one of the most up-to-date in the world, turns out, ready for shipment, an average of over one thousand sleds a day for months at a time! Just think what that means! Doesn't it prove what the public think of the "Flexible Flyer"? People everywhere have awakened to the merits of the "Flexible Flyer" and no longer buy mere "sleds." If you want the *best, insist on the "Flexible Flyer."*

To fully appreciate its superior quality you should see it. Ask your dealer to show you one and **be sure to look for the name on the sled.**

Naturally, our phenomenal success has stimulated imitations, but, like all imitations, they lack the perfection of the original. Important features of the "Flexible Flyer" are protected by patents and it is the purpose of this booklet to explain in detail each of the points of superiority.

5

THE FAMOUS *Flexible Flyer* STEERING-BAR

THIS bar bends the springsteel runners to the right or left as desired, and the sled thus follows smoothly in the direction the rider wishes to go, without checking the speed in any way. The simplicity of this is apparent at a glance. The control is perfect and the child can operate the steering-bar just as deftly as a grown-up.

We wish next to call your attention to what we term the "goose-neck" design of the forepart of our runners. This is an evidence of how we are constantly on the alert to improve still further the efficiency of the "Flexible Flyer." This 'goose neck" is formed where the runners join the front bar, and it enables us to *increase* the length and steering surface of each runner *several inches* (see illustration). This feature is patented and found exclusively on the "Flexible Flyer." *Be sure to look for it.* It adds greatly to the steering control of the sled and is a very important feature.

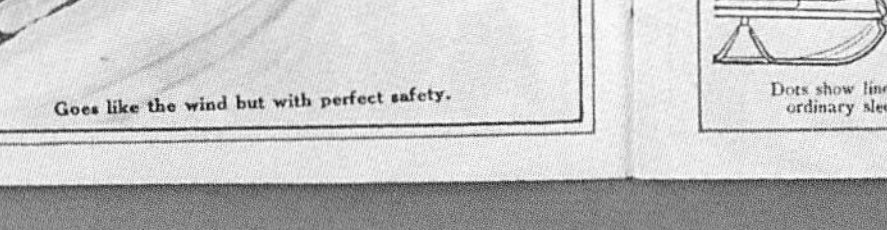

Dots show lines of ordinary sled.

7

PATENTED RUNNERS

Section 2
Cross section of runners showing grooved or hollow bottom.

EACH sized "Flexible Flyer" is equipped with the runner of the size best suited to the load it has to carry. These runners are made of special process spring steel shaped like the illustration, with a groove along the bottom somewhat like ice skates. This groove largely prevents "skidding" or slipping on icy hills or pavements, yet on snowy surfaces it is far superior, *both in point of speed and steering control,* to the old-fashioned *flat or rounded* runners used on all other sleds. This grooved runner is patented and is *used only on* the "Flexible Flyer." It alone places the "Flexible Flyer" far ahead of any other sled on the market.

ROBERT HALE GARRISON, Lexington, Mass., writes: "I wish to tell you how satisfied I am with my Flexible Flyers. The first one I used five winters, and, with the exception of the paint, is as good as new. I have *never* had to replace any part. When the first one, which is a No. 2, got too small, I got a Racer. I got it the beginning of last winter. For a while it beat every other sled on the coast, but it did not enjoy that distinction long, because as soon as the other boys saw mine they all wanted Racers. So within a month six other boys bought Racers. They are all in perfect condition.
I think the Flexible Flyer is the only *perfect* sled.

F. M. BARTON, Cleveland, Ohio, wires 1-3-11: "Express collect $3.00 Flexible Flyer. Boy won't have any other. Your agent out for 10 days."

SYLVAN KAHN, 624 W. Jackson St., Muncie, Ind., writes: "I have just received one of your Flexible Flyer Racers and I am very much pleased with it. Another boy friend of mine has one. I can say that the Flexible Flyer is a fast and comfortable sled, and every American boy should have one."

WONDER DEPARTMENT STORE, Spokane, Wash.: "We have handled Flexible Flyer Sleds for many years, and found them to give perfect satisfaction."

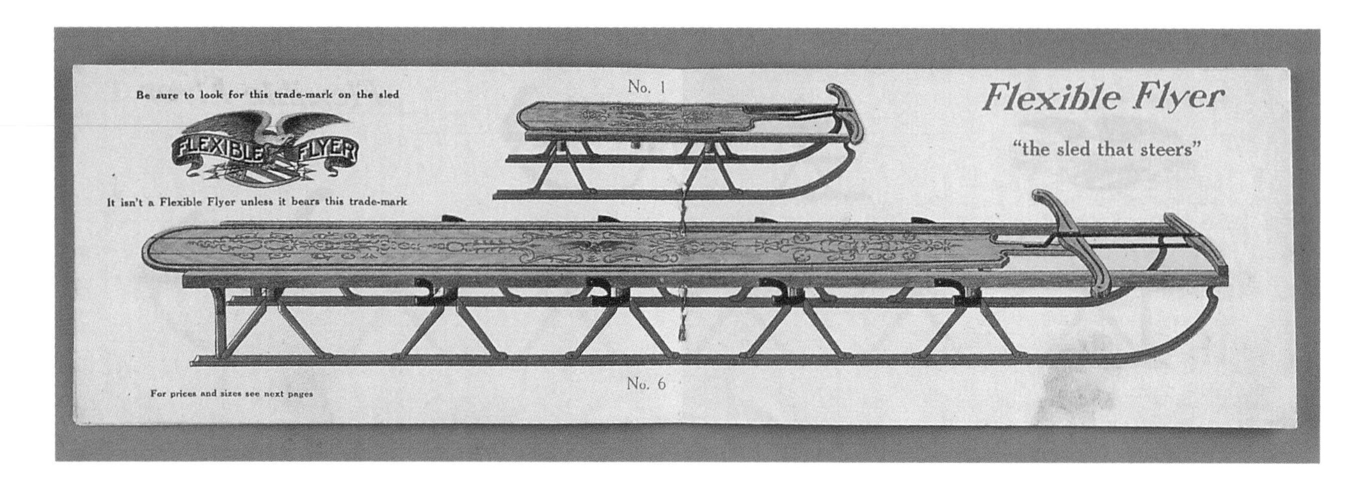

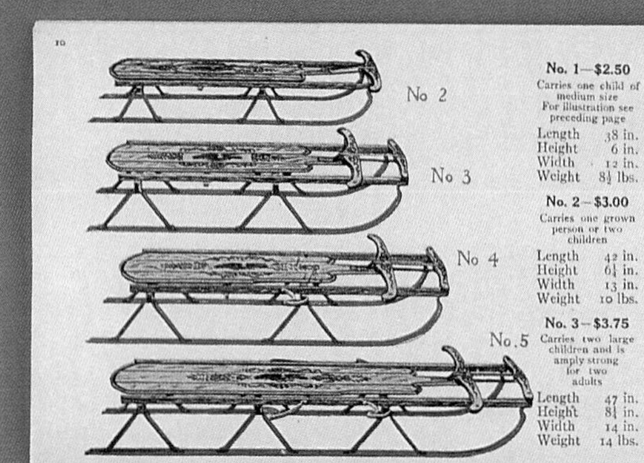

No. 1—$2.50
Carries one child of medium size
For illustration see preceding page

Length	38 in.
Height	6 in.
Width	12 in.
Weight	8½ lbs.

No. 2—$3.00
Carries one grown person or two children

Length	42 in.
Height	6¼ in.
Width	13 in.
Weight	10 lbs.

No. 3—$3.75
Carries two large children and is amply strong for two adults

Length	47 in.
Height	8¼ in.
Width	14 in.
Weight	14 lbs.

No. 4—$4.50
Designed especially to carry two grown persons

Length	52 in.
Height	8½ in.
Width	14 in.
Weight	16 lbs.

No. 5—$6.00
Carries three grown persons or four or five children

Length	63 in.
Height	8½ in.
Width	16 in.
Weight	22 lbs.

No. 6—$12.00
Carries six grown persons
For illustration see preceding page

Length	101 in.
Height	8¾ in.
Width	16 in.
Weight	41 lbs.

THE FLEXIBLE FLYER RACER

PRICE, $4.25

DIMENSIONS
Length, 56 inches
Height, 6¾ inches
Width, 13 inches
Weight, 14½ pounds

ENTHUSIASM everywhere greets this latest member of the "Flexible Flyer" family. It is designed for speed and has proved its purpose. Built rather narrow, with unusual length of runner extending well forward and to the rear, all of which insures the greatest speed and most efficient control.

Its long, racy lines, lightness, strength, poise and beauty have taken coasting circles by storm. Examine it at your nearest dealer's and see if you can help but grow enthusiastic over it.

Notice particularly the three steel uprights or supports on each side which make it strong enough for the heaviest load. Yet it is light enough to be easily pulled up the steepest hill. The extreme length of runner in front of the first upright gives exceptional steering control. It is built low, as a Racer should be, and is noted for its wonderful speed.

A special sled for the ambitious, well-grown boy. Will stand more hardship than any of the sizes we make. THE BEST VALUE FOR YOUR INVESTMENT.

Flexible Flyer TUXEDO RACER.

A SPECIAL size made first for the Tuxedo Club—a fashionable country place near New York—which is famous for its coasting. Every part is designed with particular care for grace, strength, ease of steering and speed. The dimensions are as follows:

Total length	60 inches
Length of seat	40 "
Width	13 "
Height to top of seat (front)	6½ "
Height to top of seat (rear)	5 "
Height of front bar	7½ "
Weight	22½ lbs.

The runners of this sled are of Chrome Nickel steel, which is far superior to ordinary steel, on account of its strength and wear-resisting qualities. They curve upward in front about 1¾ inches more than the regular Racer in order to more easily ride over any sharp bumps in the road. The seat is supported by four steel standards and benches, and in order to facilitate steering, is riveted only to the rear benches. The side rails are made of wood from the rear to the front bench (tapering towards the front), and from the front bench to the front of the sled they are steel. This combination gives ample strength and added flexibility. The sled as a whole slopes about 2½ inches from the front to the rear, thus making it more pleasant for the rider and at the same time giving a racy appearance.

We do not recommend this sled for a child, but we do recommend it for those who want the *strongest, fastest and best* sled that can be had. PRICE $6.00

WE feel a pardonable pride in the influence our "Flexible Flyer" has had in the development of "coasting." To this most delightful and healthful sport it has given new zest. It has added greater speed and greater control to the sled, and removed largely the element of danger.

The method of guiding the ordinary sled, is by dragging or digging the feet in the snow, which means wet feet, colds, and doctors' bills, besides wearing out boots and shoes. You can readily see, therefore, how superior is the "Flexible Flyer" and how it often *saves many times its cost in a single season.*

You can also understand why the "Flexible Flyer" has practically superseded the old style sled and completely revolutionized coasting.

CONSTRUCTION.—In the construction of the "Flexible Flyer" we have combined not only the most careful and expert workmanship, but by selecting for every part the best materials we have placed them in a class by themselves.

For the wood work we use *only straight-grained* second-growth ash; the runners are high carbon spring steel; the uprights or supports are made of pressed steel, insuring by their patented design the greatest strength, yet, at the same time, the "Flexible Flyer" is light and therefore easy to pull up hill.

The runners are japanned, not merely painted. The wood work is handsomely decorated, and finished with *three* coats of the best grade varnish.

This careful and scientific construction explains why the "Flexible Flyer" retains its value year after year and *outlasts all ordinary sleds.*

DON'T BUY AN UNDERSIZED SLED.—Experience has proven that a steering sled smaller than we make cannot be operated with efficiency or safety as there is not sufficient steering surface. We cannot impress this fact too strongly on the minds of parents, who consider above all else the safety of their children. In fact, the advantage of sleds, larger even than would seem necessary, will be apparent, because children soon outgrow their first sled. The "Flexible Flyer" lasts season after season, and by selecting a larger size than necessary at the time, it is not soon outgrown. All our models, from the largest to the smallest, can be easily operated by a child.

They're off!

A LAST WORD

WE HAVE endeavored to exolain in the preceding pages the various features of the "Flexible Flyer" that make it the best sled on the market If we have succeeded we need say nothing more. However, to appreciate the many advantages of the "Flexible Flyer" you must see it. Ask your dealer to show you one. Don't be persuaded into buying a cheap imitation coaster. *Insist* on a "Flexible Flyer" and look for the *"goose neck" design*, the *patent grooved runners*, etc., **and be sure** to look for the *name* and *trade-mark* on the sled. *It isn't a "Flexible Flyer" unless it bears this name and trade-mark.*

JOHN HENSLEY, Pleasant Plains, Ill. writes:—"I would not know how to spend a winter without my Racer."

HARVEY BROS., Hillsdale, N. Y., write: "The Flexible Flyer is the most satisfactory sled we handle and our sales are increasing every year."

"DAN NICHOLS, our former Flexible Flyer Agent at Beloit, Wis., gave his son a No. 2 Flexible Flyer about 20 years ago. The hills are steep and stony. The No. 2 has shot over these rocks every year, and is still in good condition."

HOWARD A. FRENCH & Co., Baltimore, Md., write: A gentleman came into our store Saturday evening and wanted four No 3 "Flexible Flyers." Said he had to have them, as he had bought four ——— ——— for his boys and they would not have them; did not have the Eagle on, so he gave the ——— ——— to an orphan asylum. Can you beat it?"

W. H. SMITH & SON, Bath, Maine, write: "We have a stock of your sleds. Don't blame you for being persistent, as you have a *good* sled."

Flexible Flyer

"THE SLED THAT STEERS"

In response to your request we are sending you herewith our FLEXIBLE FLYER booklet and a cardboard model, which shows (by turning the steering bar) the secret of steering this sled.

On the other side we give the name of our agents from whom you can buy at factory prices. Please call and see the sleds. You will be pleased with them and we hope to be favored with your order. If our agents cannot supply you from their stock,—give us their name and **send your order to us.** We will send you any sled you desire and will prepay the express charges on receipt of regular price.

Read the booklet carefully.

Don't be persuaded to buy an imitation, it is **sure** to be inferior.

Look for the **Flexible Flyer trade mark** and insist on having the genuine article.

S. L. Allen & Co.

PHILADELPHIA

(OVER)

Flexible Flyer® sepia post card, c. 1910. $10-20.

Illustrated Covers

By the mid-1850s envelopes were seen as a novel but useful and inexpensive form of advertising. Beautifully illustrated, they became known as illustrated covers. Mr. Allen was a master at marketing his product and seized every opportunity to advertise, including illustrated covers.

Illustrated cover, 1905.
6.5" x 3.5". $60-75.

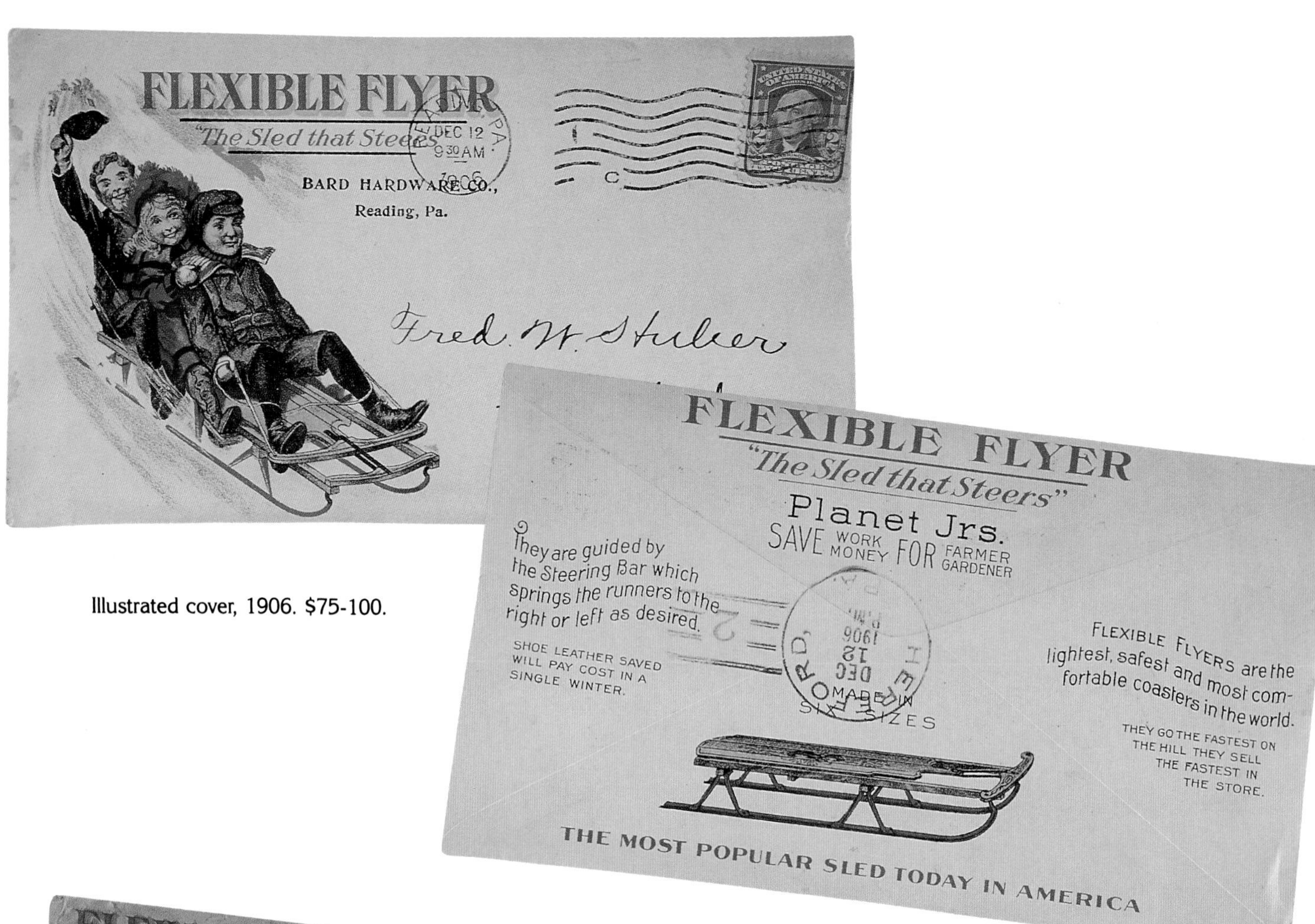

Illustrated cover, 1906. $75-100.

Illustrated cover, 1908.
6.5" x 3.5". $75-100.

Illustrated cover, 1911. Identical to the 1912 version. 6.5" x 3.5". $75-100.

Illustrated cover, 1912. 6.5" x 3.5". $50-75.

Illustrated cover, 1915. 6.5" x 3.5". $60-75.

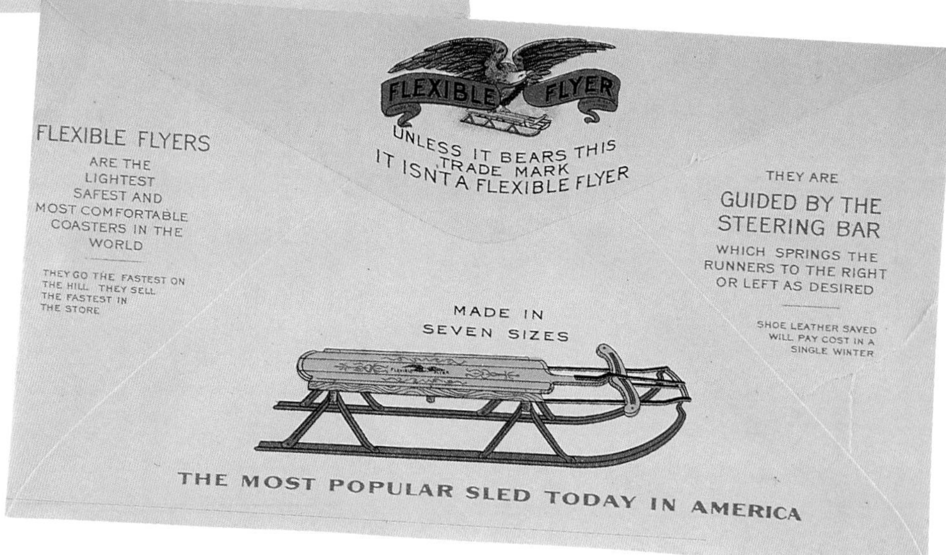

Illustrated cover, 1938. 6.5" x 3.5". $25-50.

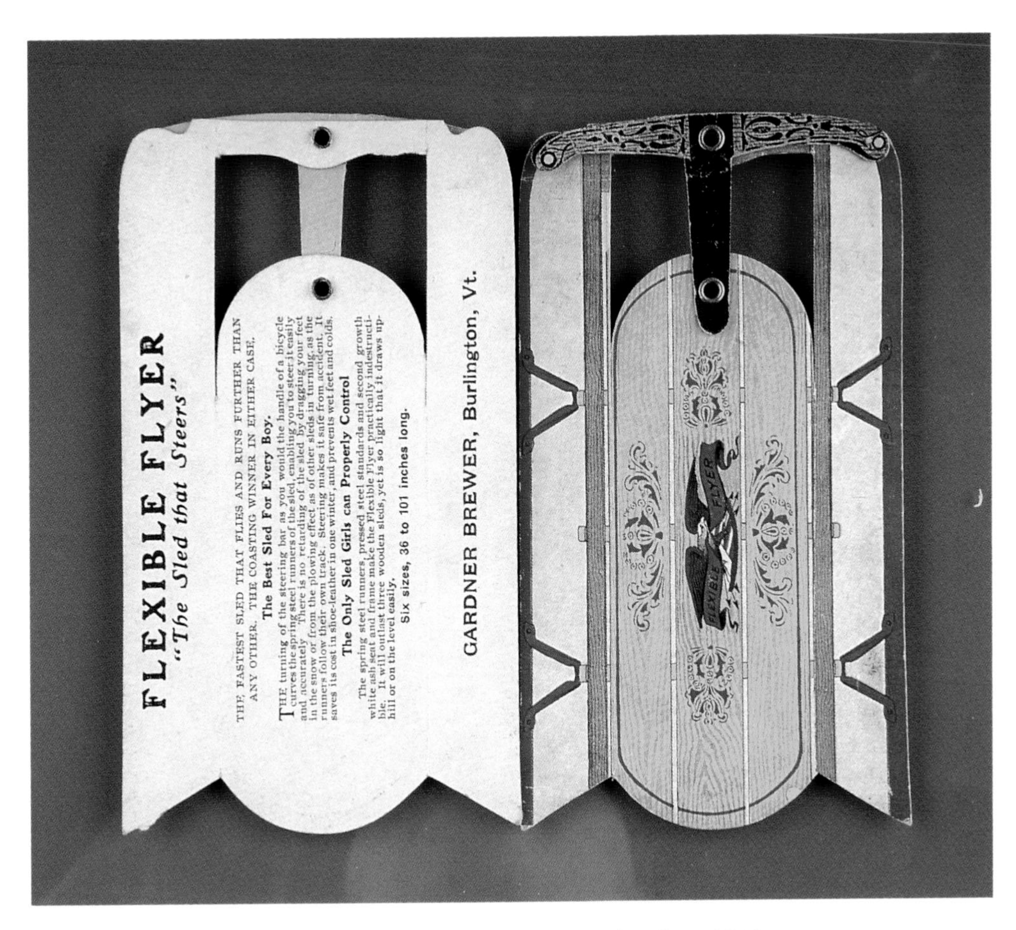

Cardboard models. These earliest models date from 1900-1905.
Note the steering bar placement and design. 6" x 3.5". $25-50.

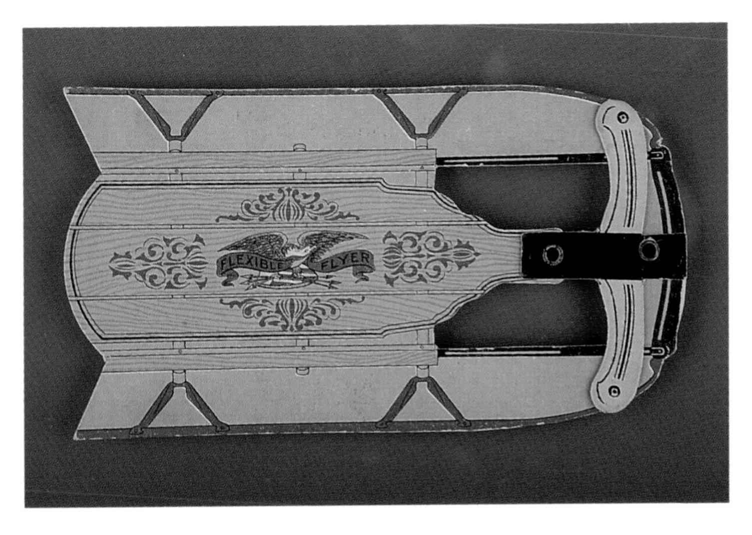

Cardboard model, 1907. Has advertising and patent date on the back, making it rarer. 6" x 3.5". $25-35.

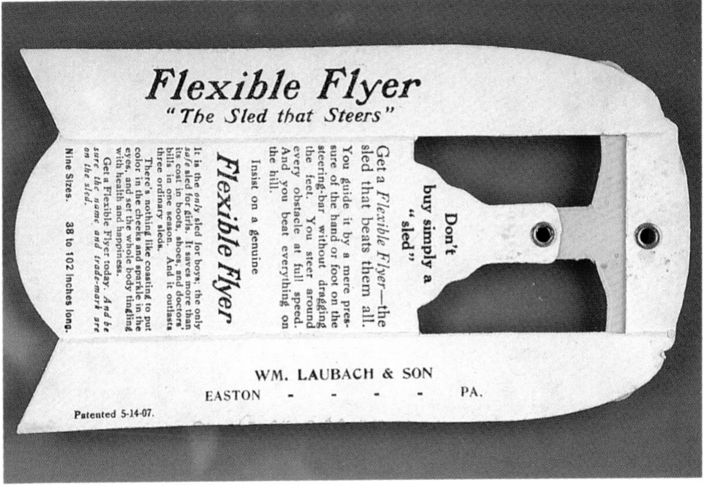

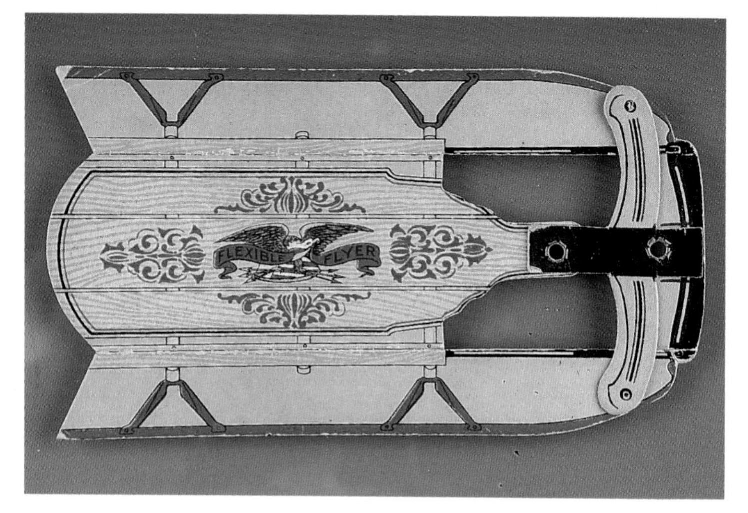

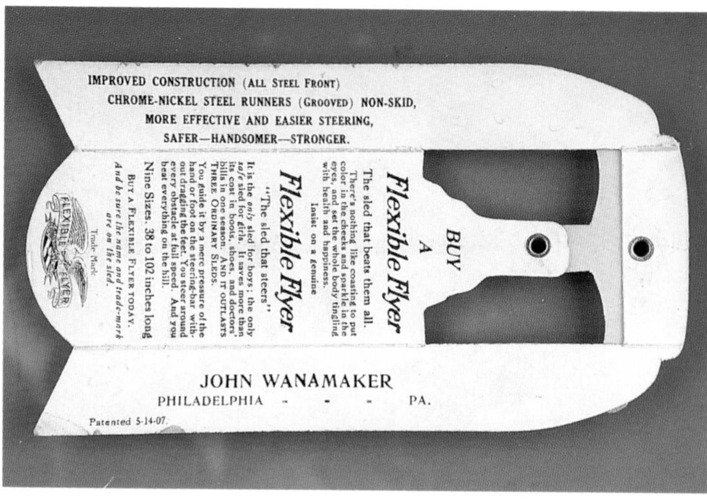

Cardboard model, 1907. Wanamaker advertising on the back. The front is the same as the Laubach model. 6" x 3.5". $50-65.

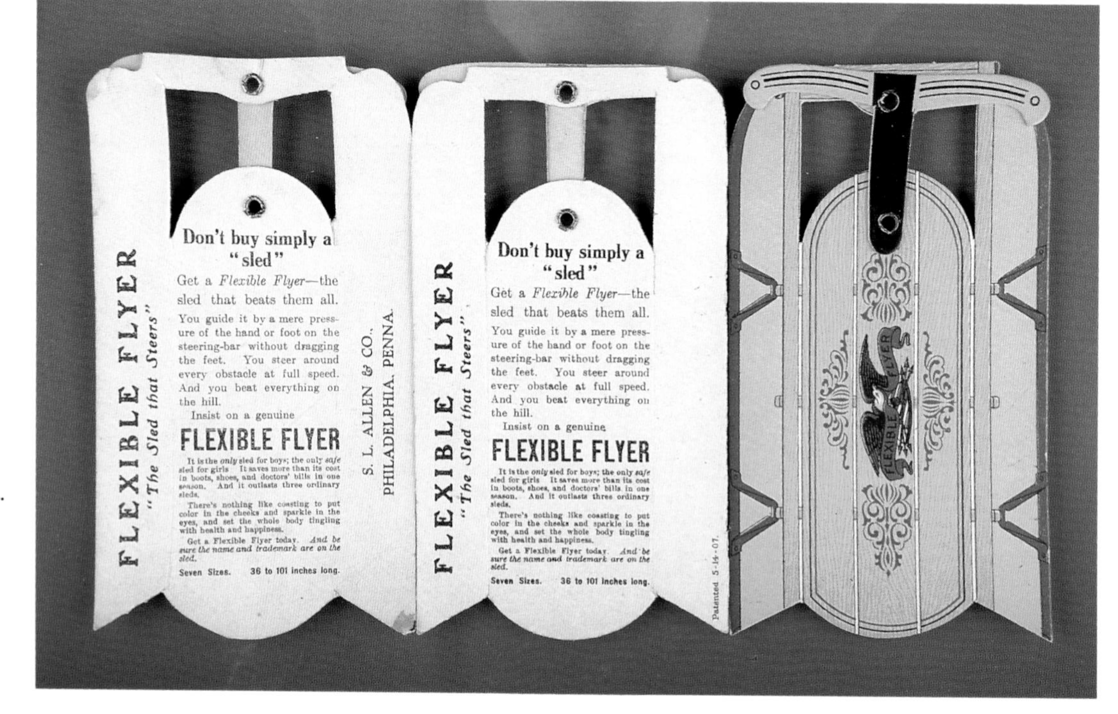

Cardboard models, pre-1920s. The back of the left sled shows advertising for the S.L. Allen Co., the middle one has only a patent number. All have the front shown at the right. 6" x 3.5". S.L. Allen: $25-50; Blank: $25-30.

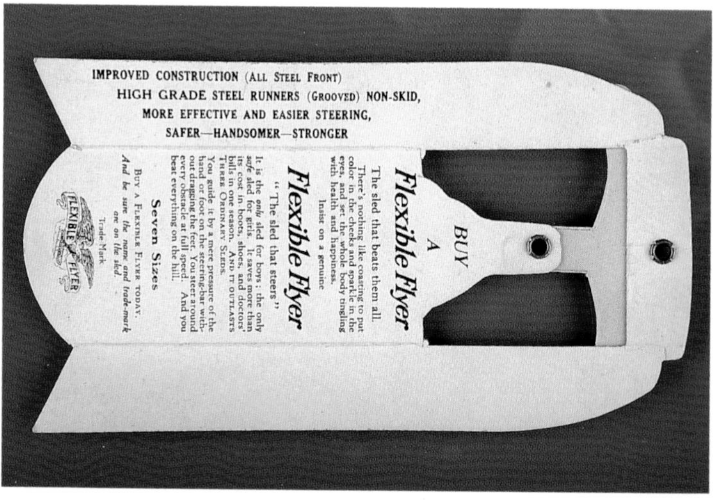

Cardboard model, c. 1920s. The back has no advertising or patent date. The trademark on the front sled dates it to the 1920s. 6" x 3.5". $!5-20.

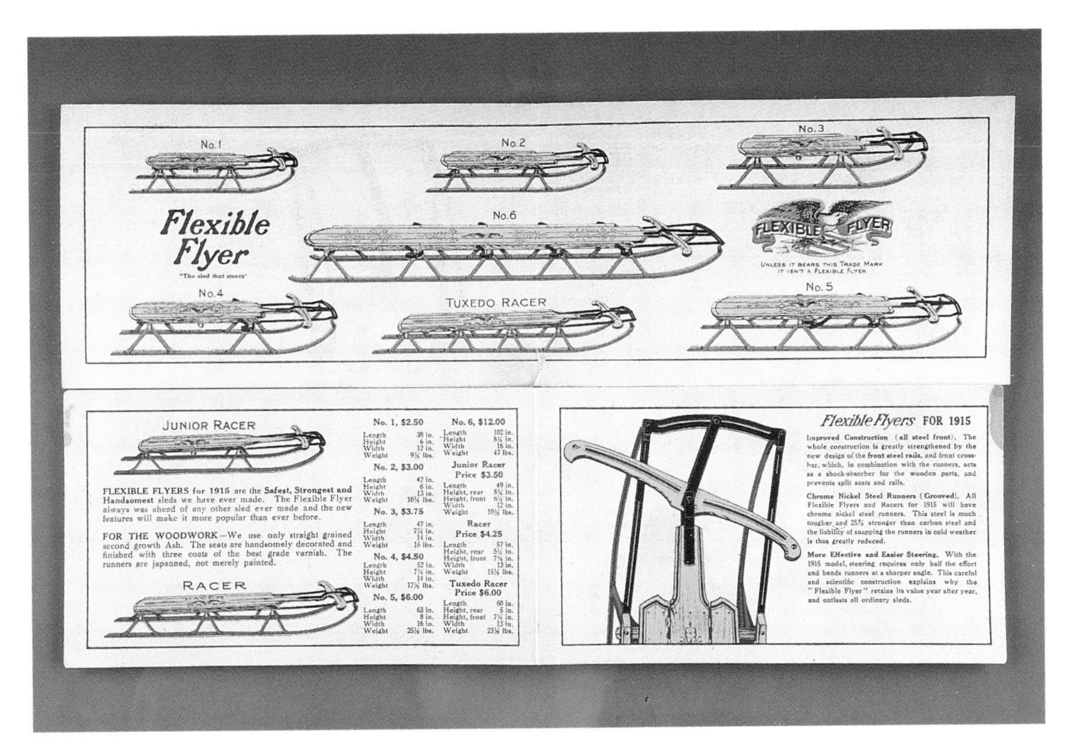

No. 1 No. 2 No. 3

Flexible Flyer

"The sled that steers"

No. 6

FLEXIBLE FLYER

UNLESS IT BEARS THIS TRADE MARK IT ISN'T A FLEXIBLE FLYER

No. 4 TUXEDO RACER No. 5

JUNIOR RACER

FLEXIBLE FLYERS for 1915 are the **Safest, Strongest and Handsomest** sleds we have ever made. The Flexible Flyer always was ahead of any other sled ever made and the new features will make it more popular than ever before.

FOR THE WOODWORK—We use only straight grained second growth Ash. The seats are handsomely decorated and finished with three coats of the best grade varnish. The runners are japanned, not merely painted.

RACER

No. 1, $2.50
Length 38 in.
Height 6 in.
Width 12 in.
Weight 9¾ lbs.

No. 2, $3.00
Length 42 in.
Height 6 in.
Width 13 in.
Weight 10½ lbs.

No. 3, $3.75
Length 47 in.
Height 7¼ in.
Width 14 in.
Weight 16 lbs.

No. 4, $4.50
Length 52 in.
Height 7¼ in.
Width 14 in.
Weight 17½ lbs.

No. 5, $6.00
Length 63 in.
Height 8 in.
Width 16 in.
Weight 25½ lbs.

No. 6, $12.00
Length 102 in.
Height 8¾ in.
Width 16 in.
Weight 47 lbs.

Junior Racer Price $3.50
Length 49 in.
Height, rear 5¾ in.
Height, front 6½ in.
Width 12 in.
Weight 10½ lbs.

Racer Price $4.25
Length 57 in.
Height, rear 5½ in.
Height, front 7⅝ in.
Width 13 in.
Weight 15½ lbs.

Tuxedo Racer Price $6.00
Length 60 in.
Height, rear 5 in.
Height, front 7¼ in.
Width 13 in.
Weight 23½ lbs.

Flexible Flyers FOR 1915

Improved Construction (all steel front). The whole construction is greatly strengthened by the new design of the **front steel rails**, and front cross-bar, which, in combination with the runners, acts as a shock-absorber for the wooden parts, and prevents split seats and rails.

Chrome Nickel Steel Runners (Grooved). All Flexible Flyers and Racers for 1915 will have chrome nickel steel runners. This steel is much tougher and 25% stronger than carbon steel and the liability of snapping the runners in cold weather is thus greatly reduced.

More Effective and Easier Steering. With the 1915 model, steering requires only half the effort and bends runners at a sharper angle. This careful and scientific construction explains why the "Flexible Flyer" retains its value year after year, and outlasts all ordinary sleds.

"Flexible Flyers for 1915," folding paper flyer. Closed: 3.5" x 6", open 3.5" x 12". $15-25.

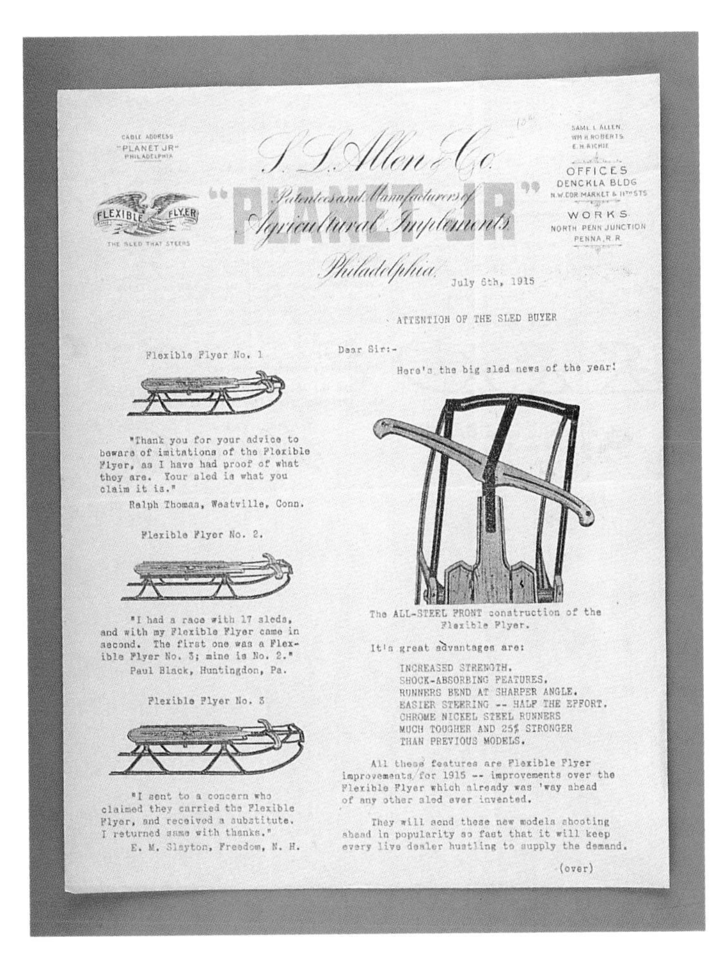

CABLE ADDRESS "PLANET JR" PHILADELPHIA

FLEXIBLE FLYER
THE SLED THAT STEERS

S. L. Allen & Co.
"PLANET JR"
Patentees and Manufacturers of Agricultural Implements
Philadelphia

SAML. L. ALLEN
WM. H. ROBERTS
E. H. RICHIE

OFFICES
DENCKLA BLDG
N.W. COR MARKET & 11TH STS

WORKS
NORTH PENN JUNCTION
PENNA. R.R.

July 6th, 1915

ATTENTION OF THE SLED BUYER

Dear Sir:-

Here's the big sled news of the year!

The ALL-STEEL FRONT construction of the Flexible Flyer.

It's great advantages are:

INCREASED STRENGTH.
SHOCK-ABSORBING FEATURES.
RUNNERS BEND AT SHARPER ANGLE.
EASIER STEERING -- HALF THE EFFORT.
CHROME NICKEL STEEL RUNNERS
MUCH TOUGHER AND 25% STRONGER
THAN PREVIOUS MODELS.

All these features are Flexible Flyer improvements for 1915 -- improvements over the Flexible Flyer which already was 'way ahead of any other sled ever invented.

They will send these new models shooting ahead in popularity so fast that it will keep every live dealer hustling to supply the demand.

(over)

Flexible Flyer No. 1

"Thank you for your advice to beware of imitations of the Flexible Flyer, as I have had proof of what they are. Your sled is what you claim it is."
Ralph Thomas, Westville, Conn.

Flexible Flyer No. 2.

"I had a race with 17 sleds, and with my Flexible Flyer came in second. The first one was a Flexible Flyer No. 3; mine is No. 2."
Paul Black, Huntingdon, Pa.

Flexible Flyer No. 3

"I sent to a concern who claimed they carried the Flexible Flyer, and received a substitute. I returned same with thanks."
E. M. Slayton, Freedom, N. H.

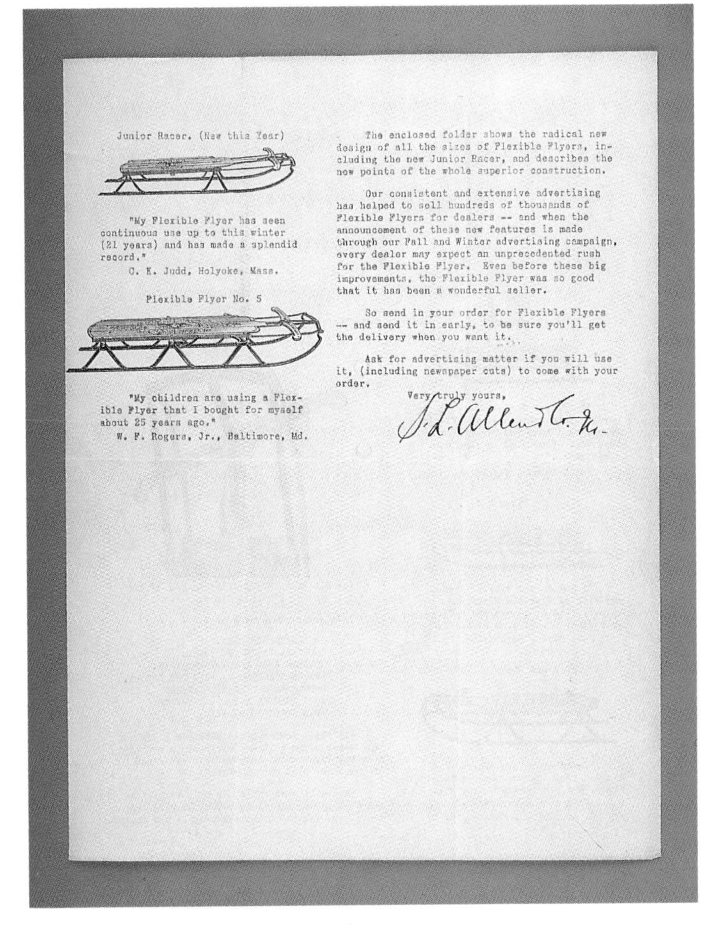

Junior Racer. (New this Year)

"My Flexible Flyer has seen continuous use up to this winter (21 years) and has made a splendid record."
C. K. Judd, Holyoke, Mass.

Flexible Flyer No. 5

"My children are using a Flexible Flyer that I bought for myself about 25 years ago."
W. F. Rogers, Jr., Baltimore, Md.

The enclosed folder shows the radical new design of all the sizes of Flexible Flyers, including the new Junior Racer, and describes the new points of the whole superior construction.

Our consistent and extensive advertising has helped to sell hundreds of thousands of Flexible Flyers for dealers -- and when the announcement of these new features is made through our Fall and Winter advertising campaign, every dealer may expect an unprecedented rush for the Flexible Flyer. Even before these big improvements, the Flexible Flyer was so good that it has been a wonderful seller.

So send in your order for Flexible Flyers -- and send it in early, to be sure you'll get the delivery when you want it.

Ask for advertising matter if you will use it, (including newspaper cuts) to come with your order.

Very truly yours,
S. L. Allen & Co. Inc.

Black and white model information sheet for 1915, introducing the models for the year. 8.5" x 11". $10-20.

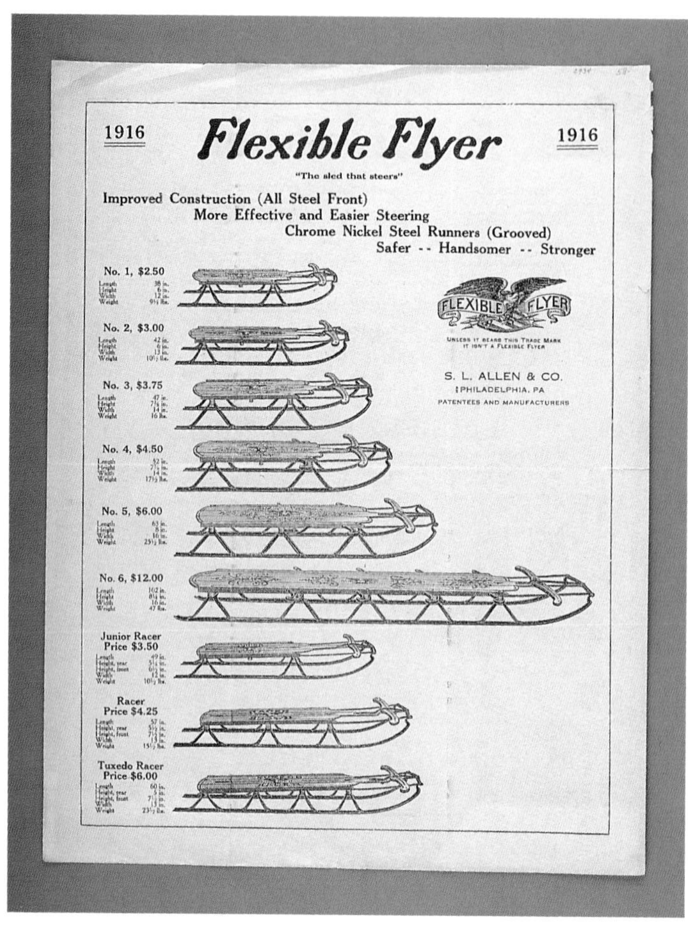

Two-sided black and white advertisement for the Flexible Flyer®, post 1916. The Fire Fly Coaster is featured on the front and the other models are on the back. 8" x 10.5". $10-20.

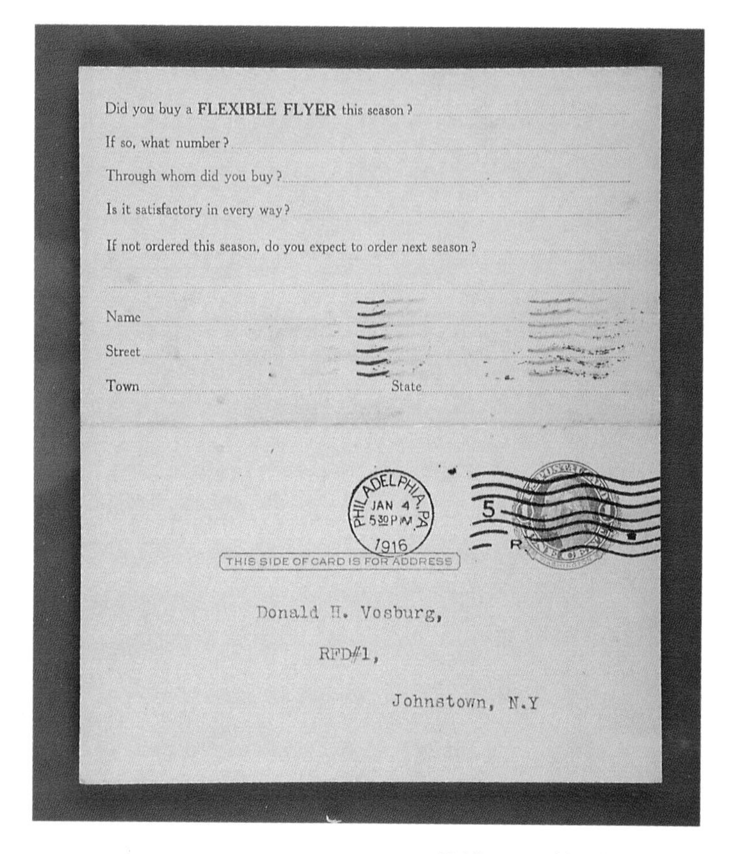

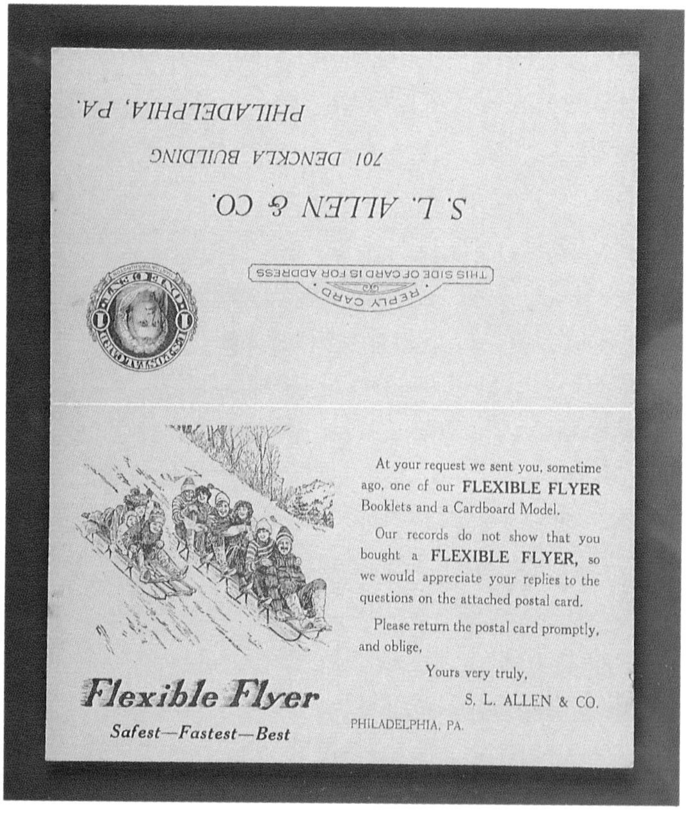

Folding mail-back to assure customer satisfaction, 1916 (the sled is a graphic from pre-1914.) Rare: $50-75.

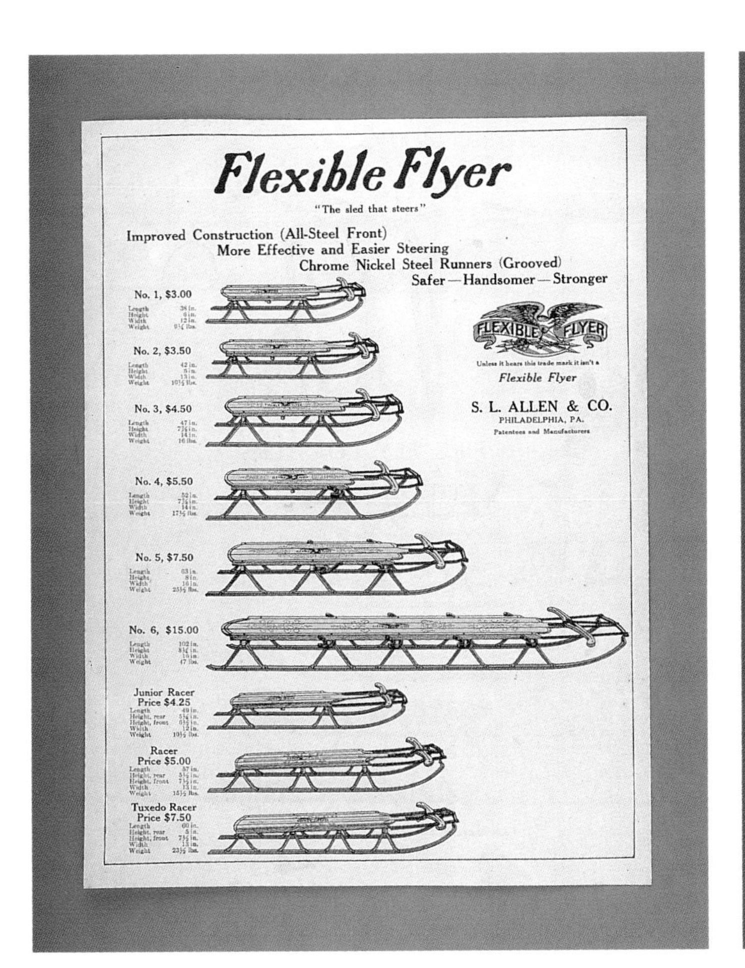

Flexible Flyer

"The sled that steers"

Improved Construction (All-Steel Front)
More Effective and Easier Steering
Chrome Nickel Steel Runners (Grooved)
Safer — Handsomer — Stronger

No. 1, $3.00
Length 38 in.
Height 6 in.
Width 12 in.
Weight 9¼ lbs.

No. 2, $3.50
Length 42 in.
Height 5 in.
Width 13 in.
Weight 10½ lbs.

No. 3, $4.50
Length 47 in.
Height 7¾ in.
Width 14 in.
Weight 16 lbs.

No. 4, $5.50
Length 52 in.
Height 7¾ in.
Width 14 in.
Weight 17½ lbs.

No. 5, $7.50
Length 63 in.
Height 8 in.
Width 16 in.
Weight 25½ lbs.

No. 6, $15.00
Length 102 in.
Height 8¼ in.
Width 16 in.
Weight 47 lbs.

Junior Racer
Price $4.25
Length 49 in.
Height, rear 5¾ in.
Height, front 6½ in.
Width 12 in.
Weight 10½ lbs.

Racer
Price $5.00
Length 57 in.
Height, rear 5½ in.
Height, front 7½ in.
Width 13 in.
Weight 15½ lbs.

Tuxedo Racer
Price $7.50
Length 60 in.
Height, rear 5 in.
Height, front 7½ in.
Width 13 in.
Weight 23½ lbs.

FLEXIBLE FLYER

Unless it bears this trade mark it isn't a
Flexible Flyer

S. L. ALLEN & CO.
PHILADELPHIA, PA.
Patentees and Manufacturers

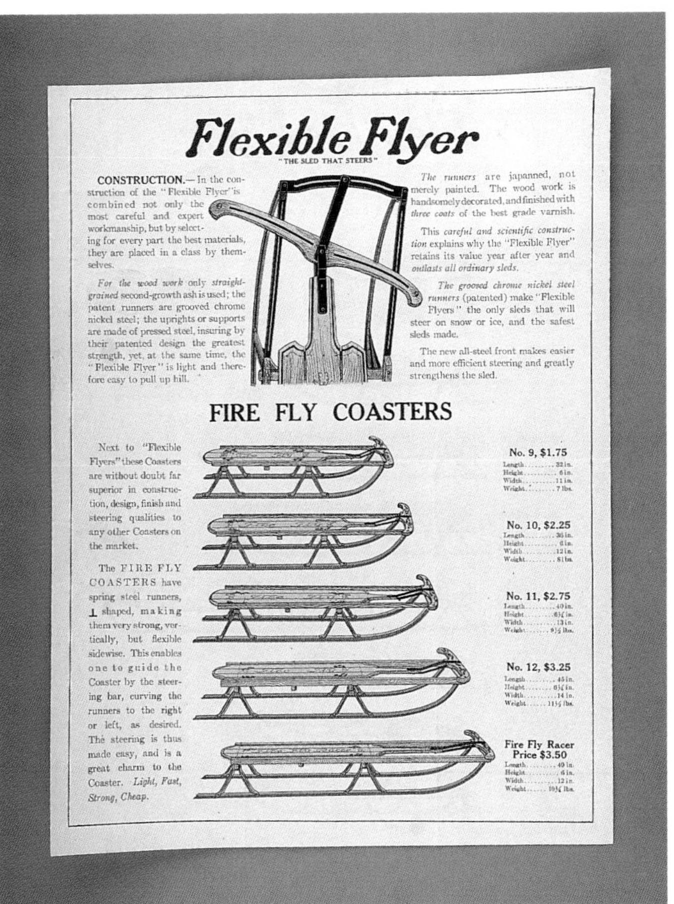

Flexible Flyer

"THE SLED THAT STEERS"

CONSTRUCTION.— In the construction of the "Flexible Flyer" is combined not only the most careful and expert workmanship, but by selecting for every part the best materials, they are placed in a class by themselves.

For the wood work only *straight-grained* second-growth ash is used; the patent runners are grooved chrome nickel steel; the uprights or supports are made of pressed steel, insuring by their patented design the greatest strength, yet, at the same time, the "Flexible Flyer" is light and therefore easy to pull up hill.

The runners are japanned, not merely painted. The wood work is handsomely decorated, and finished with *three coats* of the best grade varnish.

This *careful and scientific construction* explains why the "Flexible Flyer" retains its value year after year and *outlasts all ordinary sleds.*

The grooved chrome nickel steel runners (patented) make "Flexible Flyers" the only sleds that will steer on snow or ice, and the safest sleds made.

The new all-steel front makes easier and more efficient steering and greatly strengthens the sled.

FIRE FLY COASTERS

Next to "Flexible Flyers" these Coasters are without doubt far superior in construction, design, finish and steering qualities to any other Coasters on the market.

The FIRE FLY COASTERS have spring steel runners, ⊥ shaped, making them very strong, vertically, but flexible sidewise. This enables one to guide the Coaster by the steering bar, curving the runners to the right or left, as desired. The steering is thus made easy, and is a great charm to the Coaster. *Light, Fast, Strong, Cheap.*

No. 9, $1.75
Length........32 in.
Height........6 in.
Width........11 in.
Weight........7 lbs.

No. 10, $2.25
Length........36 in.
Height........6 in.
Width........12 in.
Weight........8 lbs.

No. 11, $2.75
Length........40 in.
Height........6¼ in.
Width........13 in.
Weight........9½ lbs.

No. 12, $3.25
Length........45 in.
Height........6¼ in.
Width........14 in.
Weight........11½ lbs.

Fire Fly Racer
Price $3.50
Length........40 in.
Height........6 in.
Width........12 in.
Weight........10¼ lbs.

Color model sheet for Flexible Flyers® Flexible Flyer® Coasters, 1917. 8.5" x 11". $25-50.

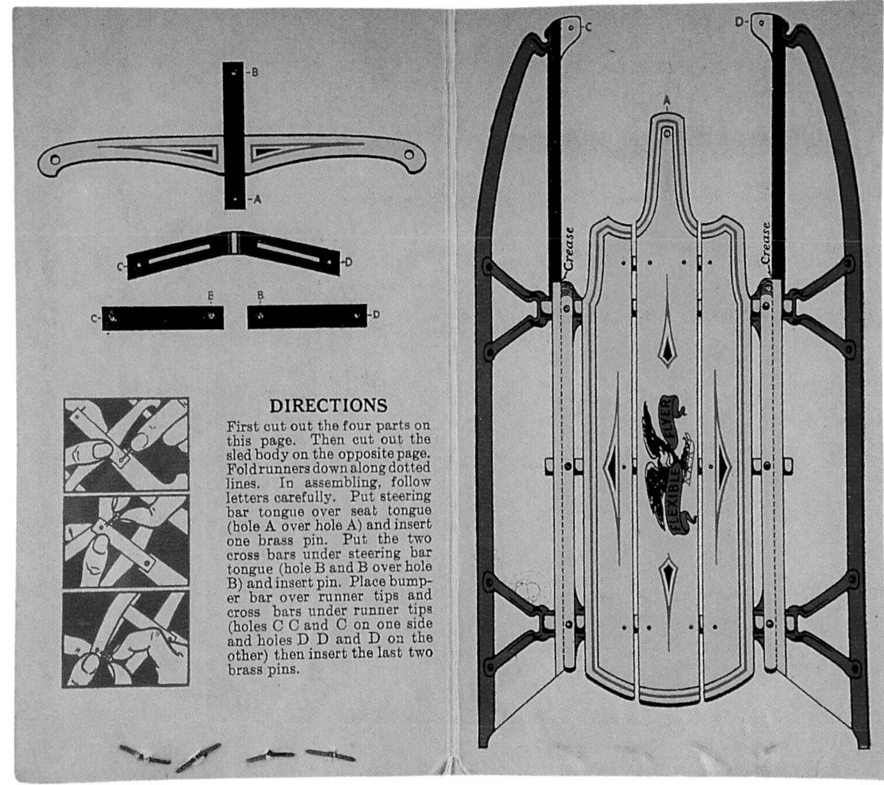

"Build a Model Flexible Flyer," 1920. A cardboard model to be cut out and assembled. 6" x 7" $50-65.

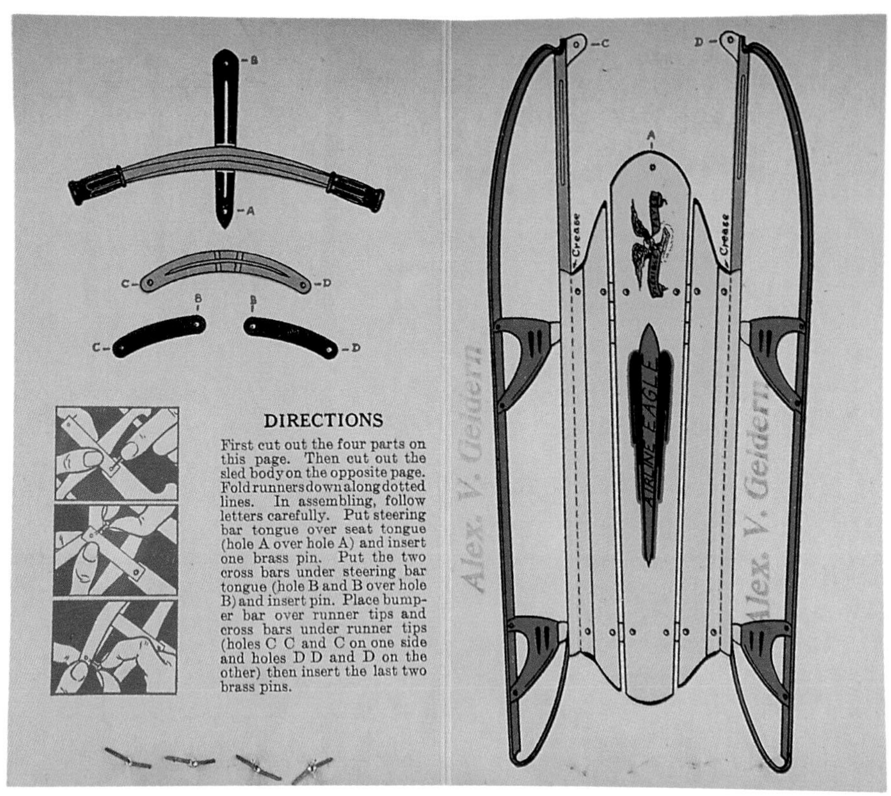

"Build a Model Flexible Flyer," 1935.. A later cardboard model to be cut out and assembled. 6" x 7". $15-25.

Salesman's pencil, 1920-40. $15-25.

Trademark decal, 1900. 3.25" x 7". $50-75.

Decal of the trademark used from 1921-1965 when they reintroduced the old trademark. $25-50.

1924 Expert Coaster Button. $25-50.

1925 Expert Sledder Button, obtained by writing the company. $20-35.

"Plucky Lindy" pin, 1927. $20-45.

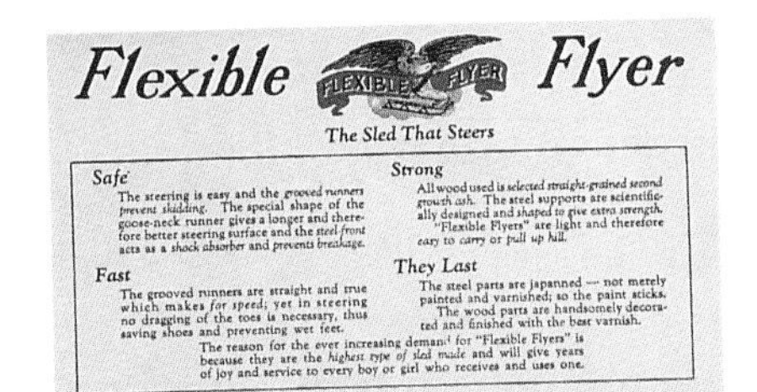

Tri-fold flyer with advertisement for Fire Fly Coasters, 1925. The Fire Fly Coaster was made before 1900. While the other sleds were given metal frames in 1914, the Fire Fly was not given a metal frame until 1935. Open size: 3" x 18". $15-25.

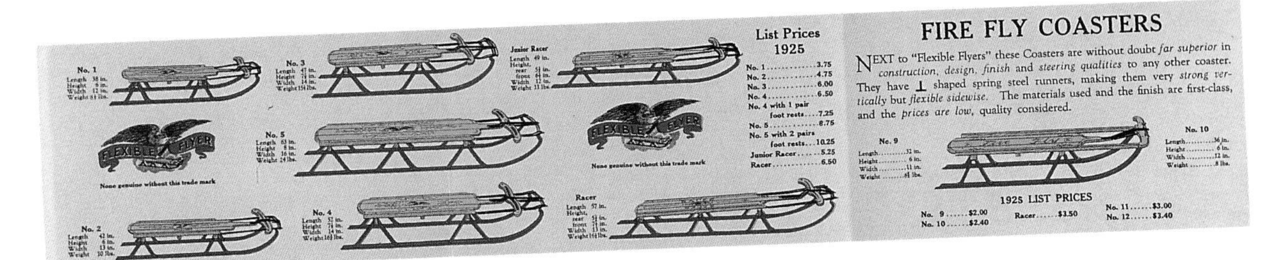

"The Sled of the Nation" tri-fold advertisement, 1928. Open size 6" x 10". $20-45.

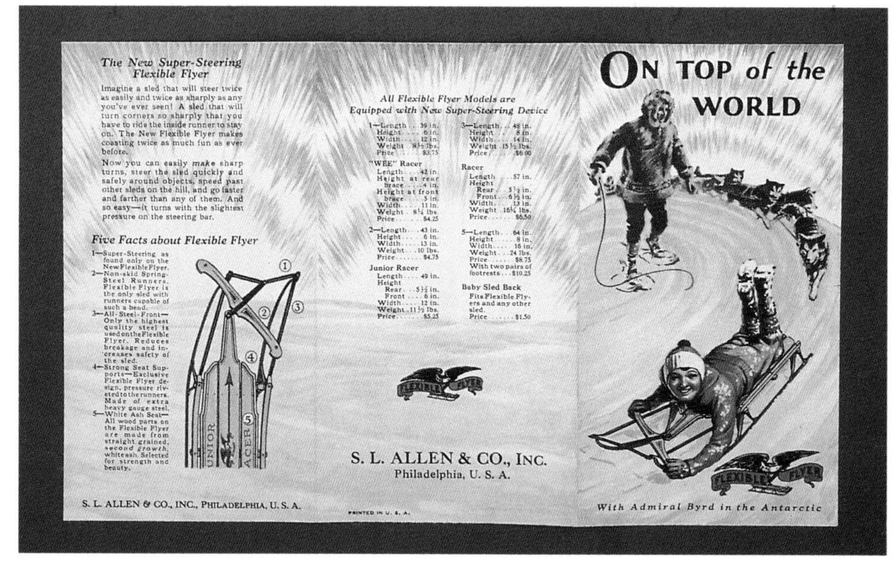

"On top of the world" with Admiral Byrd appears on the back of the 1928 tri-fold advertisement

Left & Below Right: "The Sled of the Nation" tri-fold advertisement, 1929. Open size: 6" x 10". $20-45.

Flexible Flyer® blotter, c. 1930. 3.5" x 5.5". $15-25.

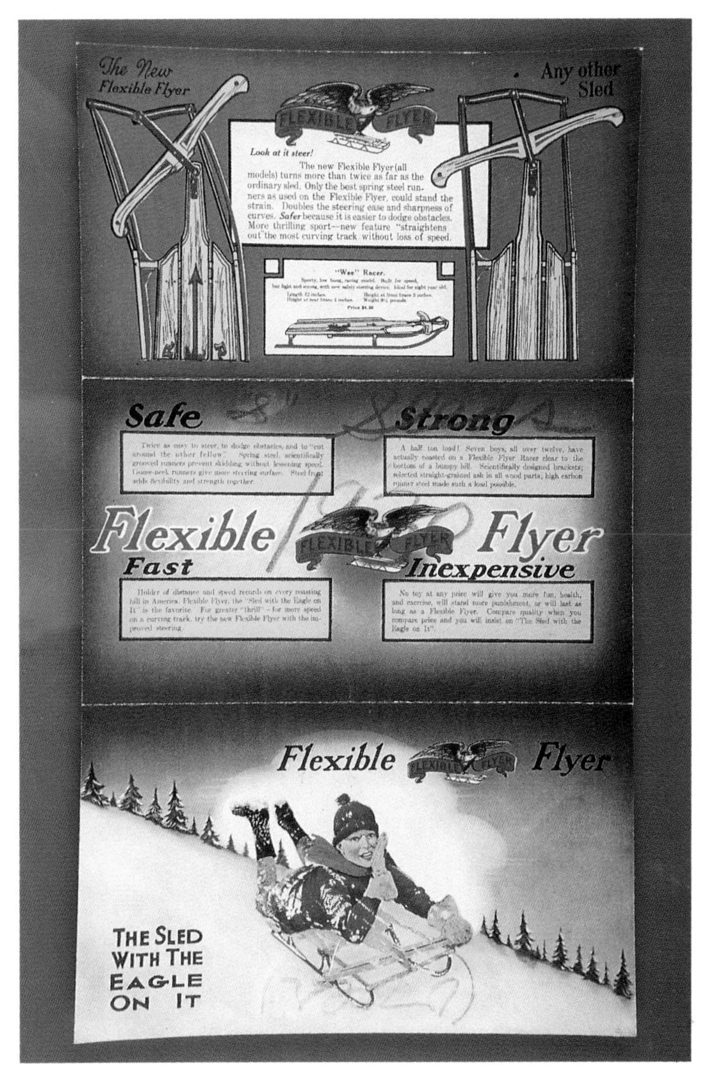

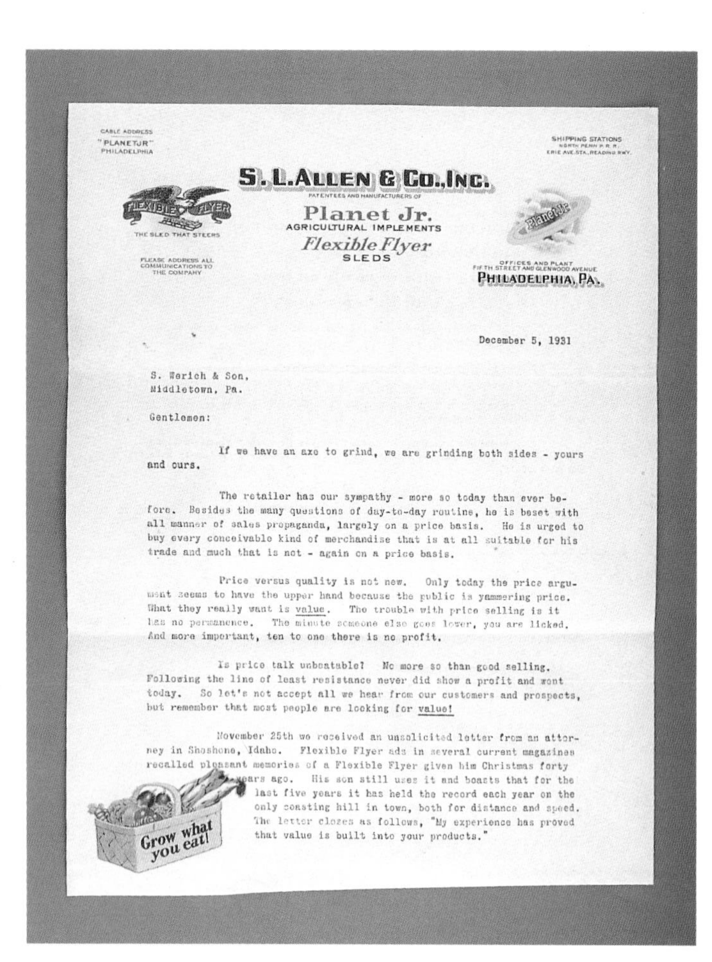

CABLE ADDRESS
"PLANETJR"
PHILADELPHIA

SHIPPING STATIONS
NORTH PENN P.R.R.
ERIE AVE. STA., READING RWY.

S. L. ALLEN & CO., INC.
PATENTEES AND MANUFACTURERS OF
Planet Jr.
AGRICULTURAL IMPLEMENTS
Flexible Flyer
SLEDS

THE SLED THAT STEERS

PLEASE ADDRESS ALL COMMUNICATIONS TO THE COMPANY

OFFICES AND PLANT
FIFTH STREET AND GLENWOOD AVENUE
PHILADELPHIA, PA.

December 5, 1931

S. Werich & Son,
Middletown, Pa.

Gentlemen:

If we have an axe to grind, we are grinding both sides - yours and ours.

The retailer has our sympathy - more so today than ever before. Besides the many questions of day-to-day routine, he is beset with all manner of sales propaganda, largely on a price basis. He is urged to buy every conceivable kind of merchandise that is at all suitable for his trade and much that is not - again on a price basis.

Price versus quality is not new. Only today the price argument seems to have the upper hand because the public is yammering price. What they really want is value. The trouble with price selling is it has no permanence. The minute someone else goes lower, you are licked. And more important, ten to one there is no profit.

Is price talk unbeatable? No more so than good selling. Following the line of least resistance never did show a profit and wont today. So let's not accept all we hear from our customers and prospects, but remember that most people are looking for value!

November 25th we received an unsolicited letter from an attorney in Shoshone, Idaho. Flexible Flyer ads in several current magazines recalled pleasant memories of a Flexible Flyer given him Christmas forty years ago. His son still uses it and boasts that for the last five years it has held the record each year on the only coasting hill in town, both for distance and speed. The letter closes as follows, "My experience has proved that value is built into your products."

Grow what you eat!

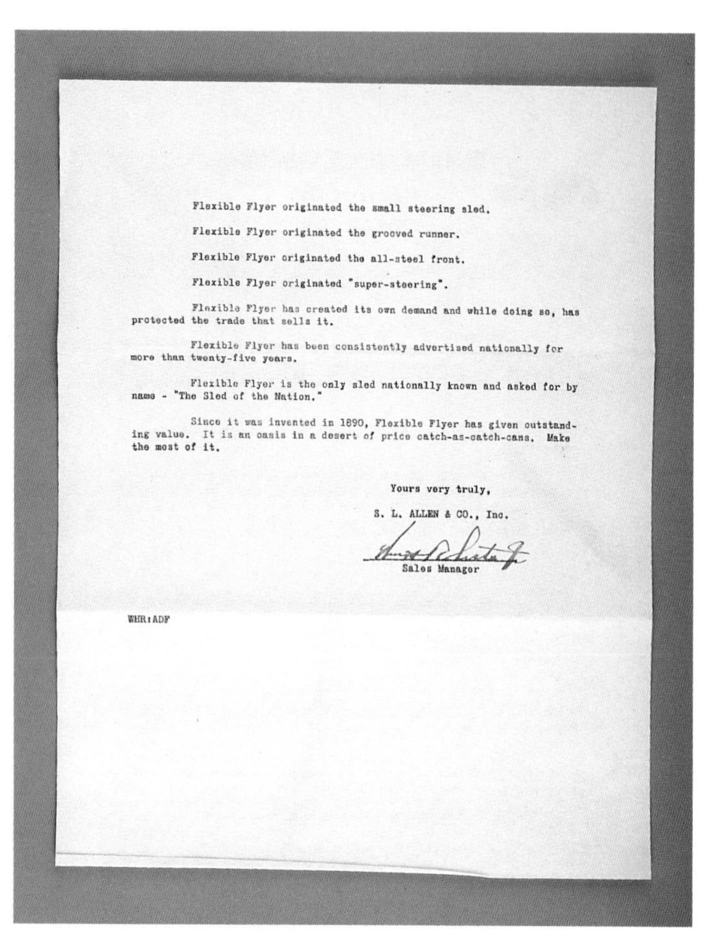

Flexible Flyer originated the small steering sled.

Flexible Flyer originated the grooved runner.

Flexible Flyer originated the all-steel front.

Flexible Flyer originated "super-steering".

Flexible Flyer has created its own demand and while doing so, has protected the trade that sells it.

Flexible Flyer has been consistently advertised nationally for more than twenty-five years.

Flexible Flyer is the only sled nationally known and asked for by name - "The Sled of the Nation."

Since it was invented in 1890, Flexible Flyer has given outstanding value. It is an oasis in a desert of price catch-as-catch-cans. Make the most of it.

Yours very truly,

S. L. ALLEN & CO., Inc.

Sales Manager

WHR:ADF

"If we have an axe to grind it is both sides, your and ours..." begins this letter in response to a customer complaint. On the back is listed some of the firsts of the Flexible Flyer® and a strong statement of the products quality. 1931. $10-15.

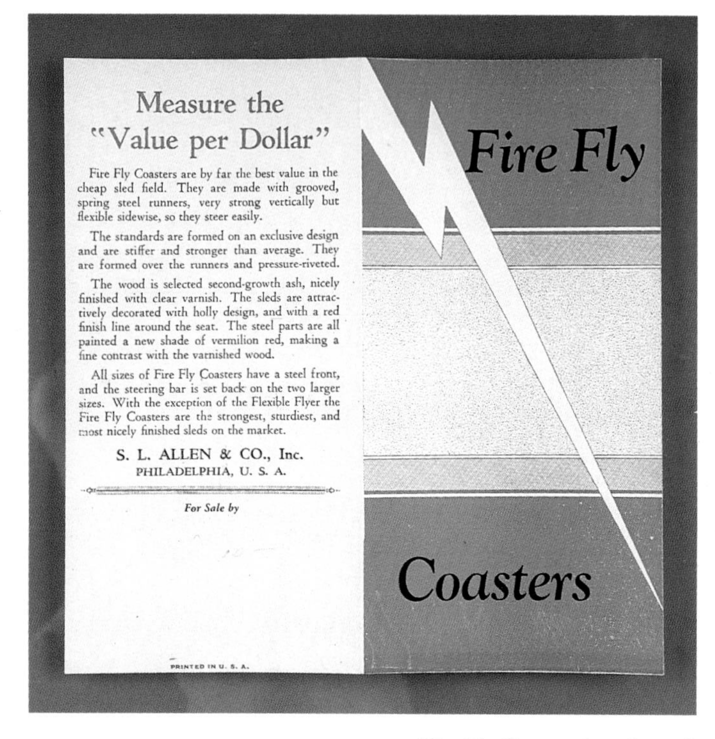

Fire Fly Coaster flyer from the 1930s. The Fire Fly was reintroduced in 1935 with a metal frame. 6" x 6". $10-20.

Advertising brochure for the Airline series, 1935. Open size: 12.5" x 12.5". $25-50.

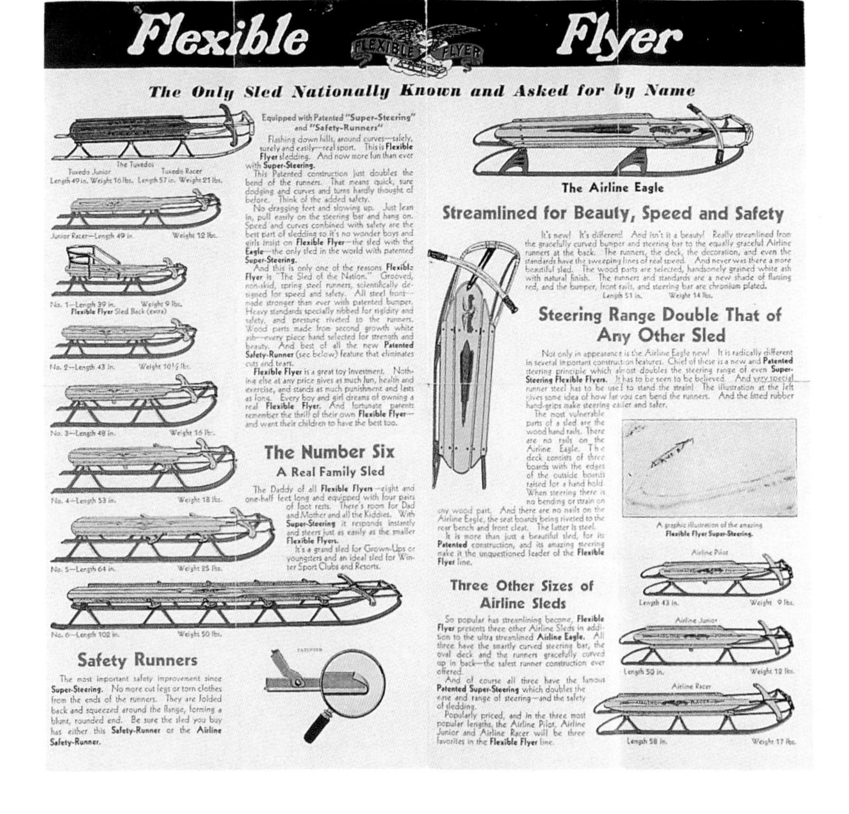

Flexy Line brochure with the Airline Eagle and Tuxedo Racer, 1935. 12.5" x 12.75". $20-45.

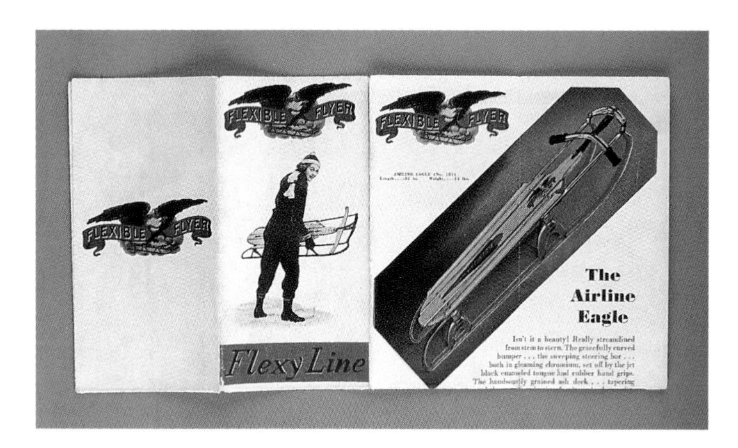

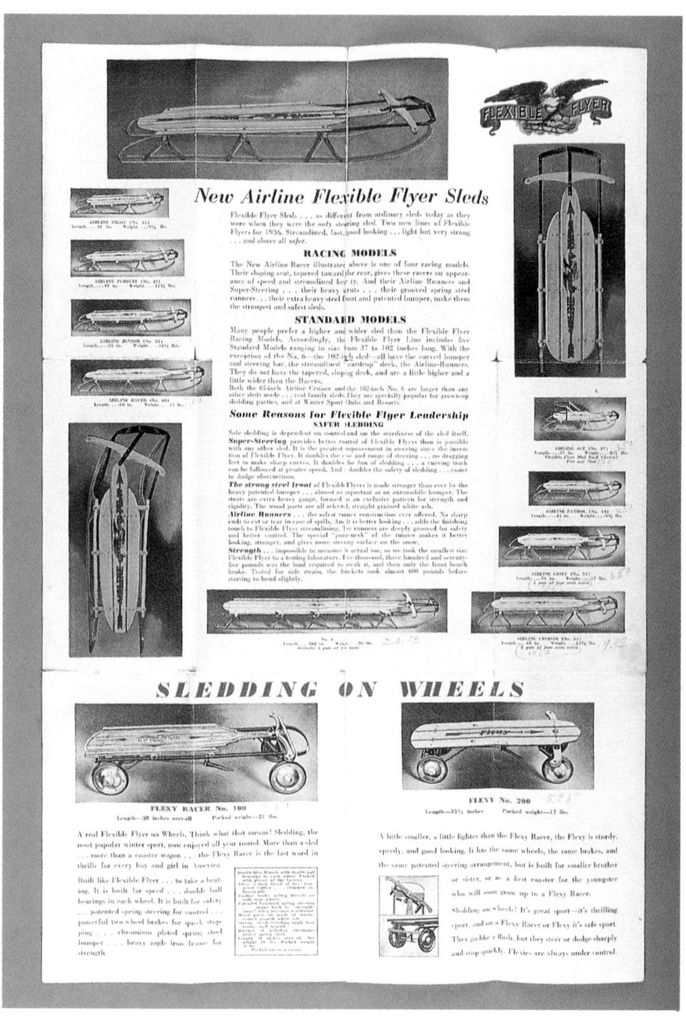

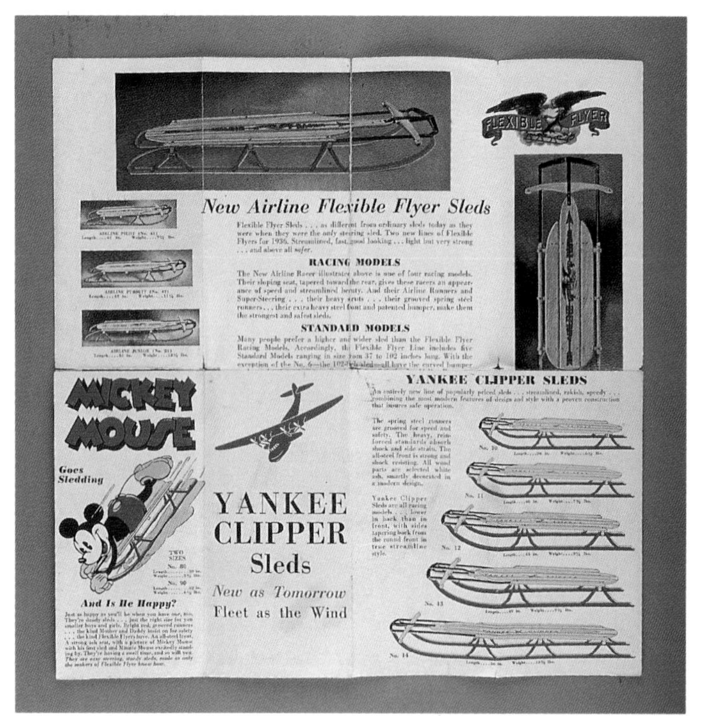

Advertising brochure for the Flexy Line with Mickey Mouse advertising on inside flap. 1935. Open size: 19" x 12.5". $50-65.

Ashtray from the S.L. Allen Company from the 1930s. $100-150.

Mickey Mouse flyer, 1935. Open size: 6.25" x 6.5". $50-75.

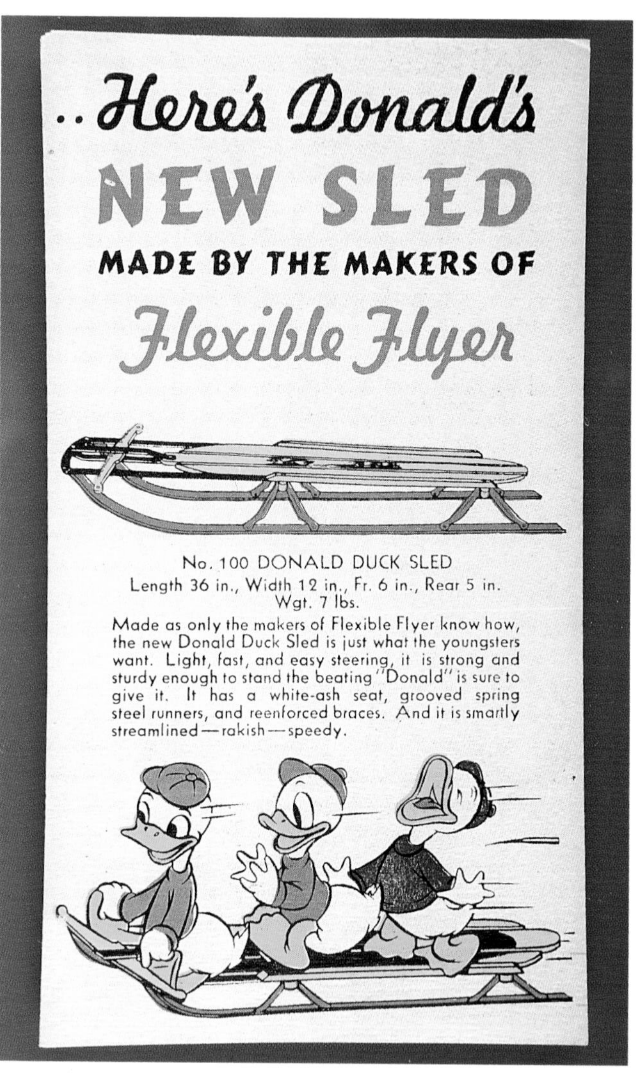

Donald Duck flyer, 1938. 6" x 3.5". $50-75.

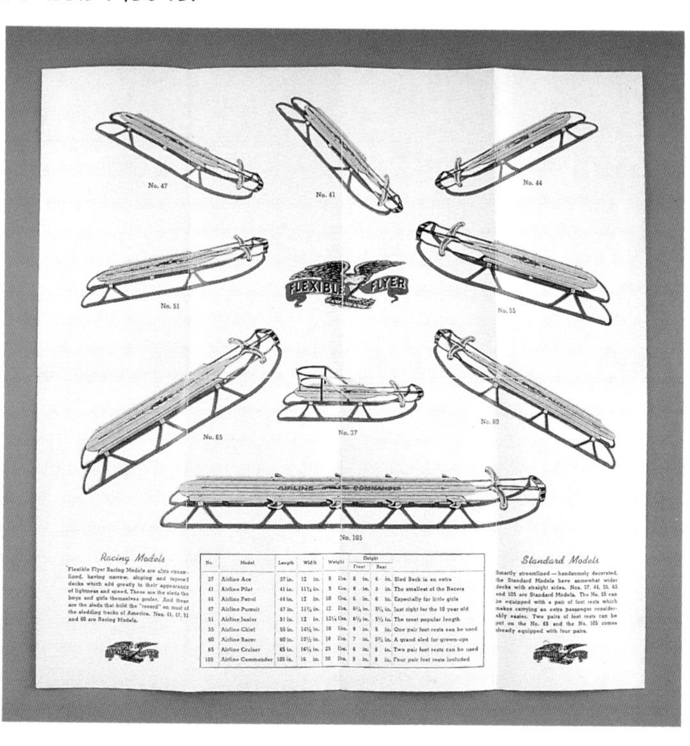

Folded advertising brochure, 1930s. 12" x 12.5". $25-50.

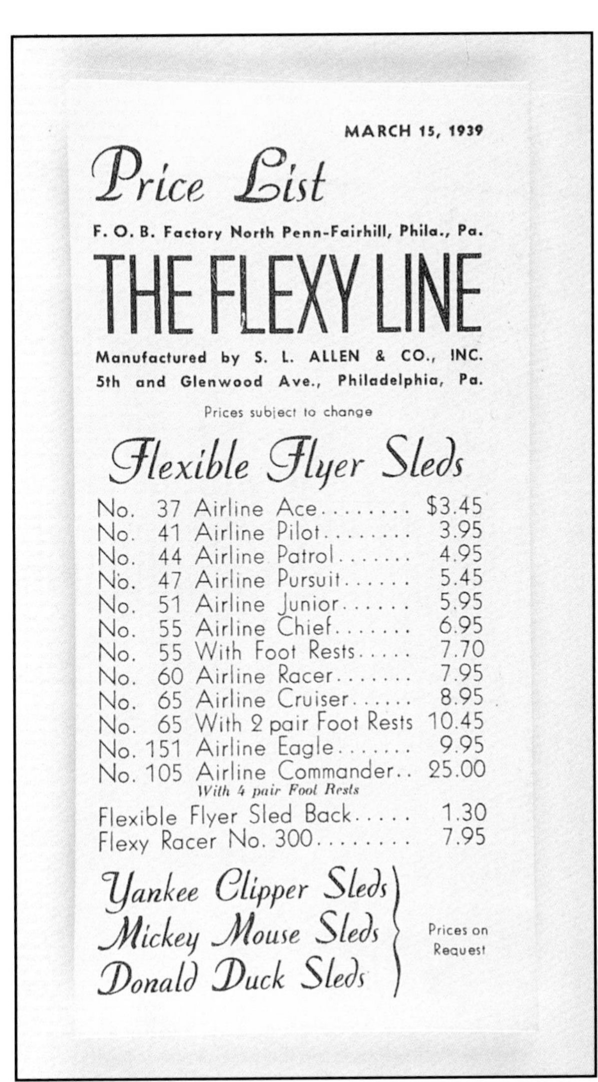
MARCH 15, 1939

Price List

F. O. B. Factory North Penn-Fairhill, Phila., Pa.

THE FLEXY LINE

Manufactured by S. L. ALLEN & CO., INC.
5th and Glenwood Ave., Philadelphia, Pa.

Prices subject to change

Flexible Flyer Sleds

No. 37 Airline Ace........	$3.45
No. 41 Airline Pilot........	3.95
No. 44 Airline Patrol.......	4.95
No. 47 Airline Pursuit......	5.45
No. 51 Airline Junior......	5.95
No. 55 Airline Chief.......	6.95
No. 55 With Foot Rests.....	7.70
No. 60 Airline Racer.......	7.95
No. 65 Airline Cruiser......	8.95
No. 65 With 2 pair Foot Rests	10.45
No. 151 Airline Eagle.......	9.95
No. 105 Airline Commander.. *With 4 pair Foot Rests*	25.00
Flexible Flyer Sled Back.....	1.30
Flexy Racer No. 300........	7.95

Yankee Clipper Sleds
Mickey Mouse Sleds
Donald Duck Sleds
Prices on Request

Left:
Price list, 1939. 5.5" x 3". $10-15.
Below:
Advertising brochure for the 50th anniversary of the Flexible Flyer®, 1939. Included is an offering of running lights for the sled. 12" x 12". $25-50.

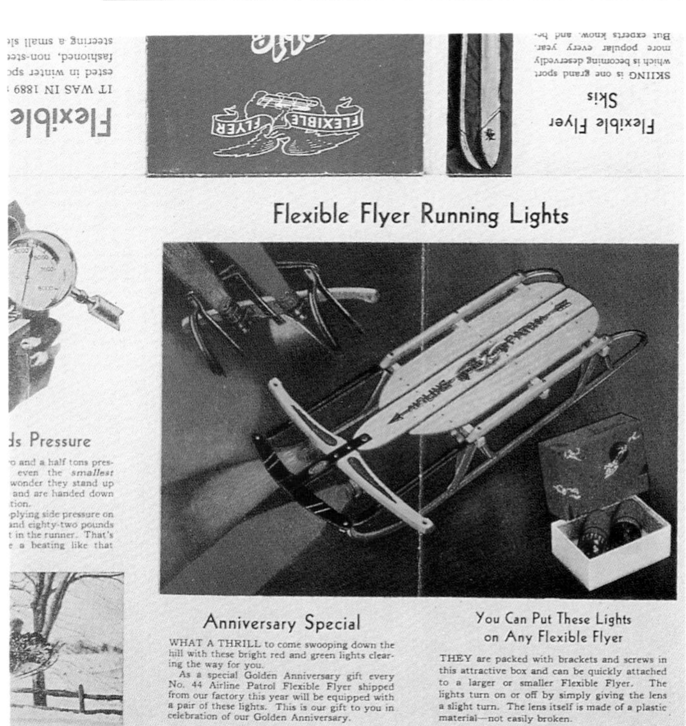

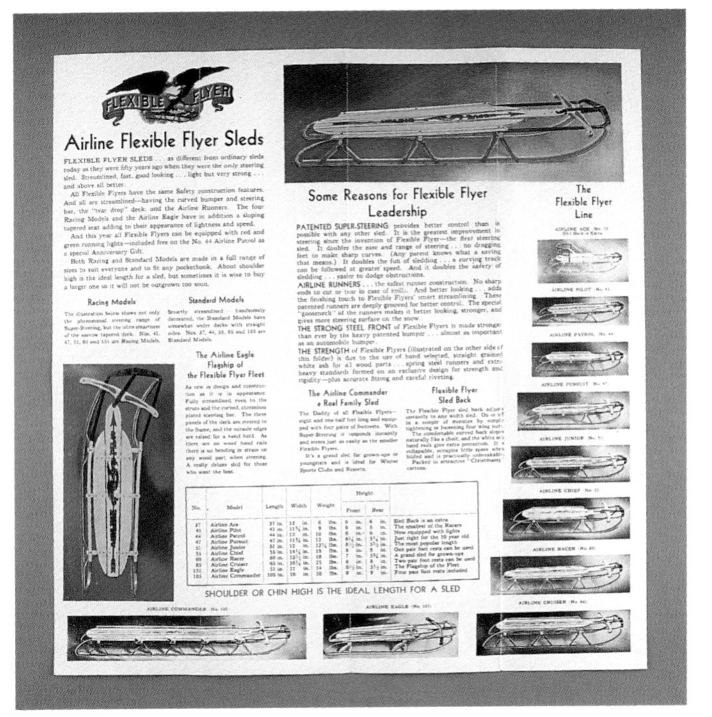

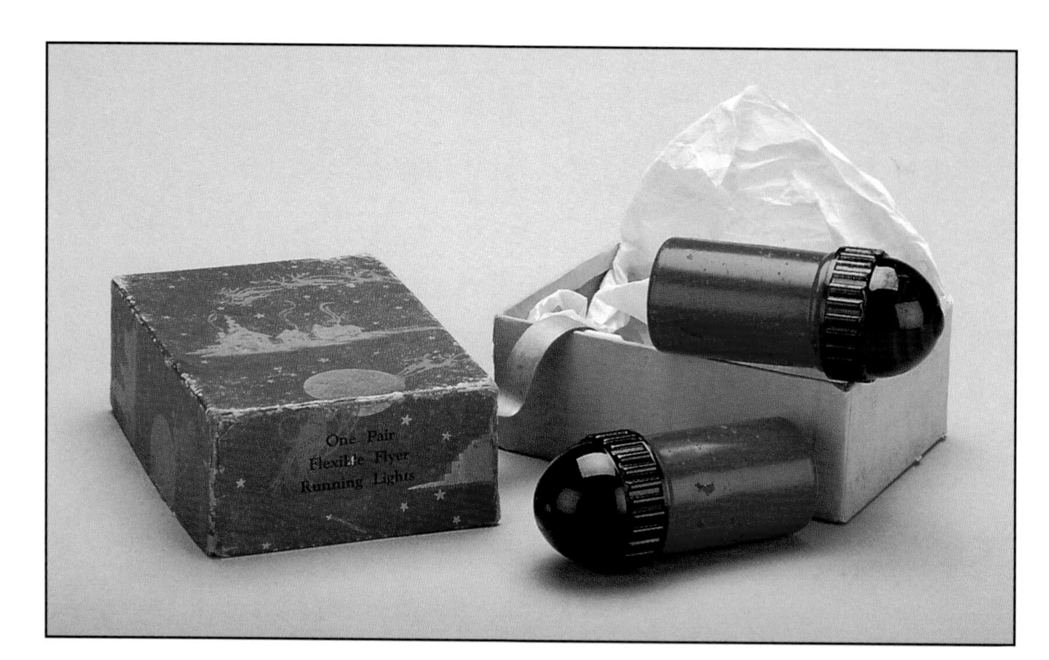

Flexible Flyer®
50th Anniversary
Special Gift.
Running lights for
the Airline Patrol
sleds that left the
factory. 1939.
$100-150.

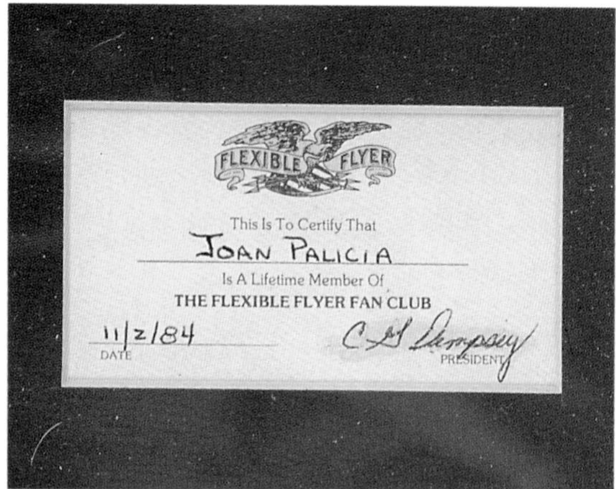

1939 membership card for the Flexible Flyers® Fan Club. $15-25.

"Sure Sledders Membership Pin," 1939. Went with the fan club card. $25-50.

Flexible Flyer® Baby Sled Back, 1940s. $50-75, mint in box.

YANKEE CLIPPER

Sleds

● New as tomorrow . . . fleet as the wind . . . Yankee Clipper sleds are streamlined, rakish, speedy . . . combining the most modern features of design and style with a proven construction for safety and dependability.

The spring steel runners are grooved for speed and safety. The heavy, reinforced standards absorb shock and side strain. The all-steel front is strong and shock-resisting.

All wood parts are selected white ash, smartly decorated in a modern design, finished with clear varnish. The runners and standards are a new shade of chinese red, making a handsome contrast with the black enameled steel front and light wood.

With the exception of the two smaller sizes, Yankee Clipper sleds are all racing models . . . lower in back than in front . . . and tapering back from the rounded bumper in true streamline style. Handsome and modern in appearance . . . light in weight but strong and sturdy in construction . . . Yankee Clipper sleds set a new standard for popular priced sleds.

Flexible Flyer

Skis

● Skiing is one grand sport which is becoming deservedly more popular every year. But experts know, and beginners are rapidly discovering that proper skis and bindings are necessary for real enjoyment.

Flexible Flyer Skis are quality skis and all that the name implies — made of the best straight grained white hickory — by skilled craftsmen, who appreciate the importance of matched balance, flexibility, weight and "camber."

Whether expert or beginner, there is a Flexible Flyer ski to suit. Send for special Flexible Flyer Ski folder — gladly mailed on request.

S. L. ALLEN & CO., INC.
5th and Glenwood Ave.
Philadelphia, Pa.

●

Reilly Bros. & Raub, Inc.
Lancaster, Pa.

Printed in U. S. A.

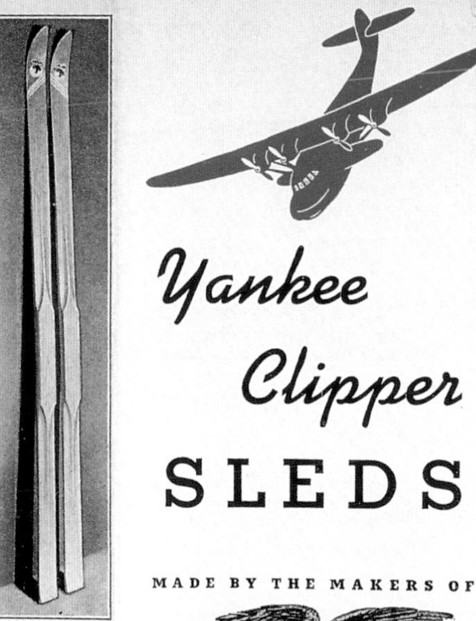

Yankee Clipper

SLEDS

MADE BY THE MAKERS OF

SLEDS AND SKIS

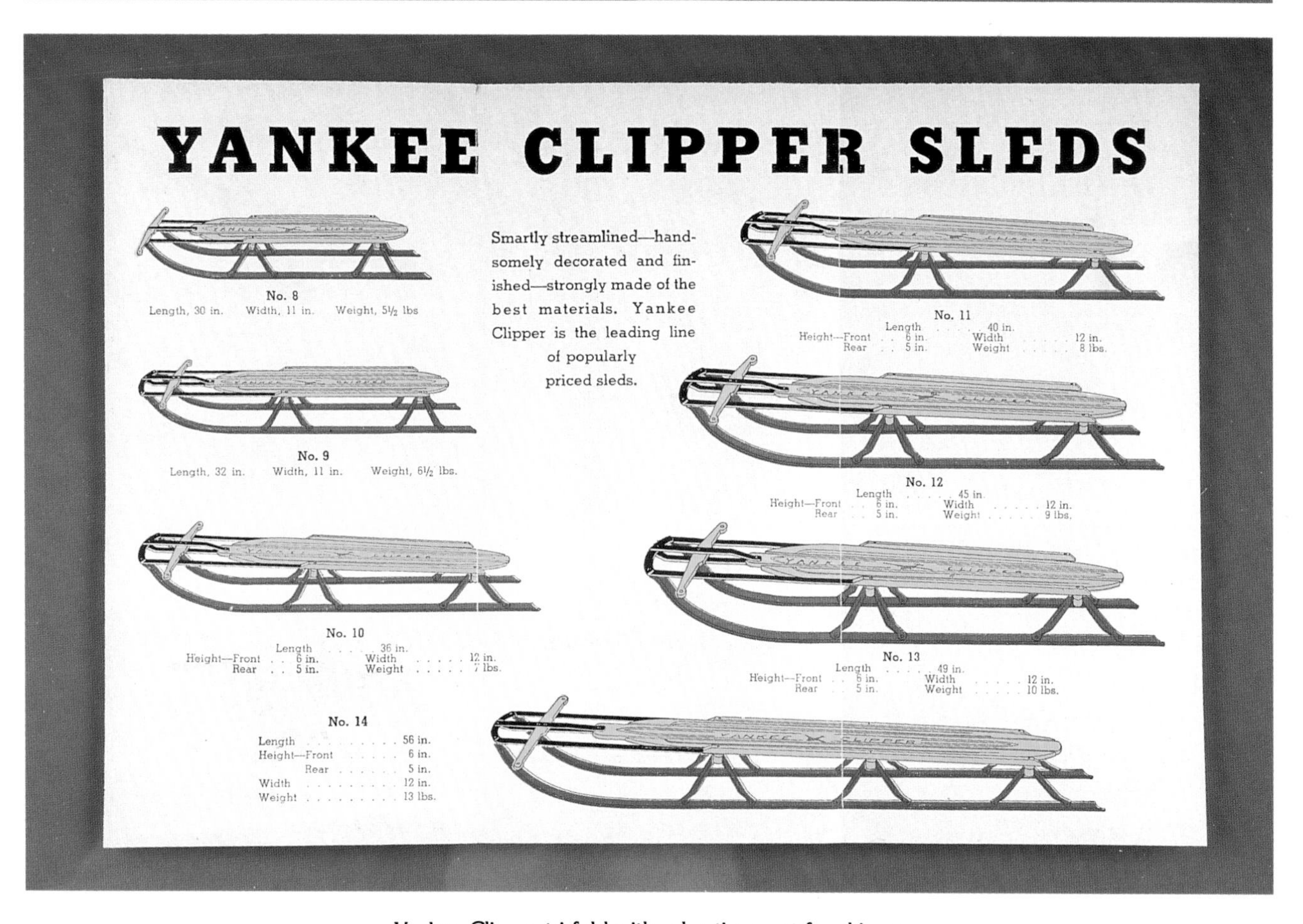

YANKEE CLIPPER SLEDS

Smartly streamlined—handsomely decorated and finished—strongly made of the best materials. Yankee Clipper is the leading line of popularly priced sleds.

No. 8
Length, 30 in. Width, 11 in. Weight, 5½ lbs

No. 9
Length, 32 in. Width, 11 in. Weight, 6½ lbs.

No. 10
Length . . . . 36 in.
Height—Front . . 6 in. Width . . . . . 12 in.
Rear . . 5 in. Weight . . . . . 7 lbs.

No. 11
Length . . . . 40 in.
Height—Front . . 6 in. Width . . . . . 12 in.
Rear . . 5 in. Weight . . . . . 8 lbs.

No. 12
Length . . . . 45 in.
Height—Front . . 6 in. Width . . . . . 12 in.
Rear . . 5 in. Weight . . . . . 9 lbs.

No. 13
Length . . . . 49 in.
Height—Front . . 6 in. Width . . . . . 12 in.
Rear . . 5 in. Weight . . . . . 10 lbs.

No. 14
Length . . . . 56 in.
Height—Front . . . . . 6 in.
Rear . . . . . 5 in.
Width . . . . 12 in.
Weight . . . . 13 lbs.

Yankee Clipper tri-fold with advertisement for skis on the back, 1940s. Open size: 6.5" x 9.5". $20-35.

Brochure for the Ski Racer, c. 1940. 8.5" x 11". $50-65.

FLEXIBLE FLYER AND FLEXIBLE FLYER-SPLITKEIN SKIS

Skiing is one grand sport which is becoming deservedly more popular every year. But experts know, and beginners are rapidly discovering that proper skis and bindings are necessary for real enjoyment.

Flexible Flyer Skis are quality skis and all that the name implies—made of the best straight grained white hickory—by skilled craftsmen who appreciate the importance of matched balance, flexibility, weight and "camber."

In addition Flexible Flyer has acquired the U. S. manufacturing rights for the world famous Splitkein Laminated Ski. This is the ski which since its introduction in 1935 has become the most popular ski in Europe among both experts and ordinary skiers. In fact most of the world's records in jumping, cross-country, down hill and slalom have been won on Splitkein Skis.

Whether expert or beginner, child or adult, there is a Flexible Flyer Ski to suit. Send for special Flexible Flyer Ski folders—gladly mailed on request.

FLEXIBLE FLYER SLEDS *Streamlined for Beauty, Safety and Speed*

All Flexible Flyer Sleds are streamlined and all have the exclusive features of Super-Steering and Airline-Safety-Runners. There is a size for every boy and girl, priced to fit any pocketbook—from the Airline Ace—37 inches long—to the Airline Commander—8½ feet long, a real family sled. There are five Racing Models with sloping, tapered decks—very speedy. The five Standard Models are a little roomier—higher and wider. Send for illustrated folder and price list.

FLEXY RACER

Sledding the Year 'Round

Flashing speed... thrilling turns... Real Sledding! All yours with a Flexy Racer... and all under safety control. Double disc—double ball bearing wheels with tempered rubber tires. Patented balanced spring steering. Powerful brakes operated by merely twisting the steering bar. Rubber hand grips and steel bumper.

INDEX

48

1948 catalog for S.L. Allen. while the catalog deals principally with farm implements, the last page features advertisements for skis, sleds and the Flexy Racer. $10-15.

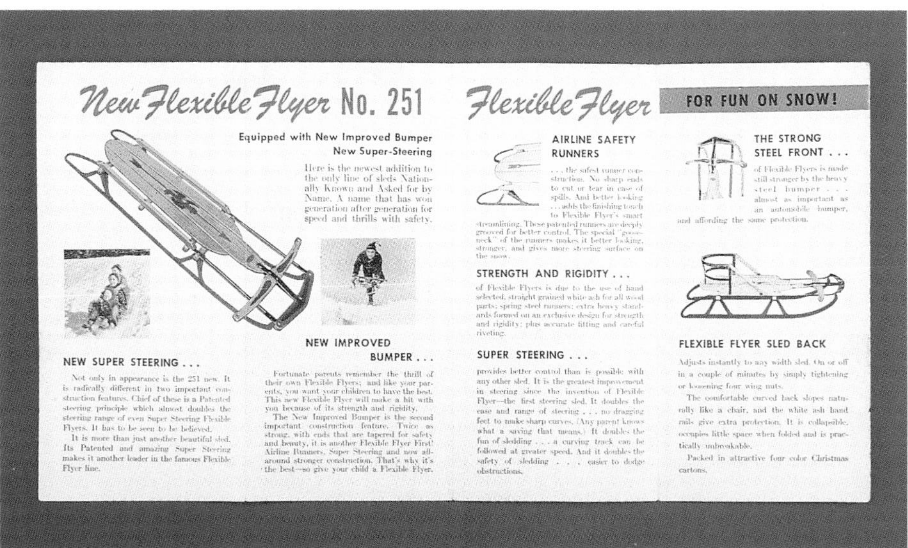

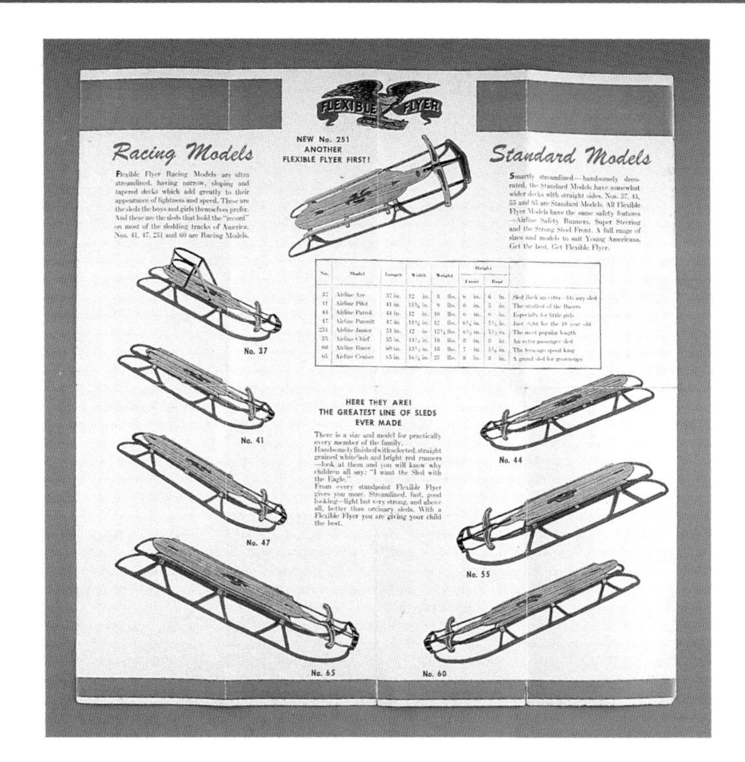

Brochure from 1950, when they introduced the new three digit identification number and stopped putting the model name on the sled's deck. Open size: 12" x 12". $10-20.

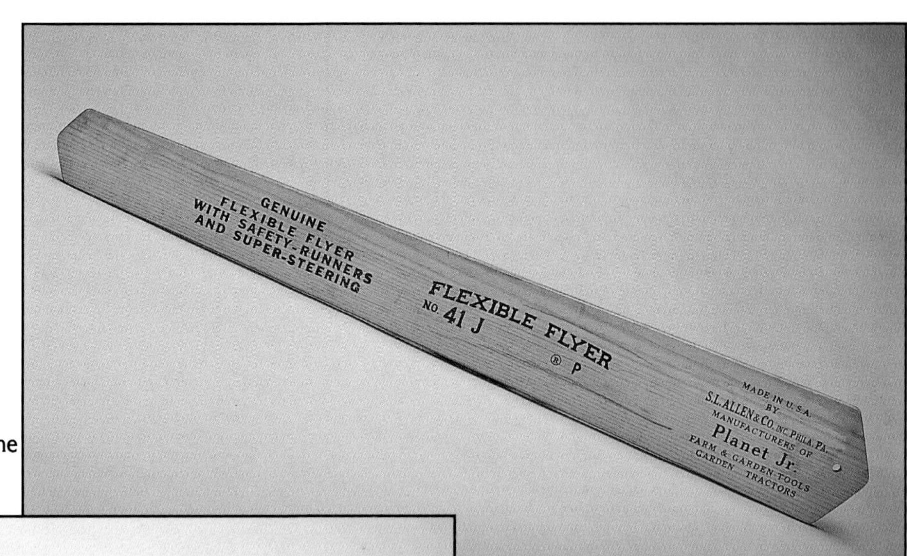

Board made to be hung on the wall as a sign. $25-50.

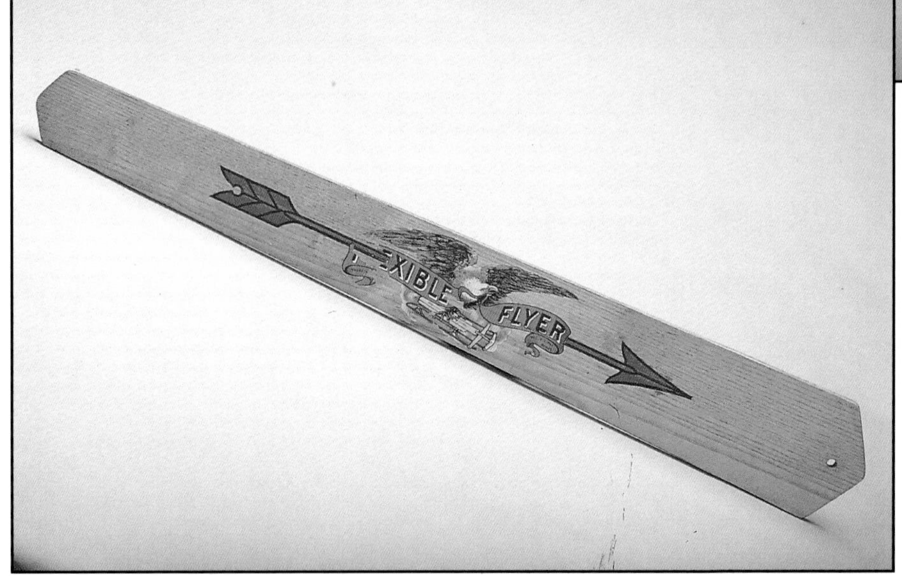

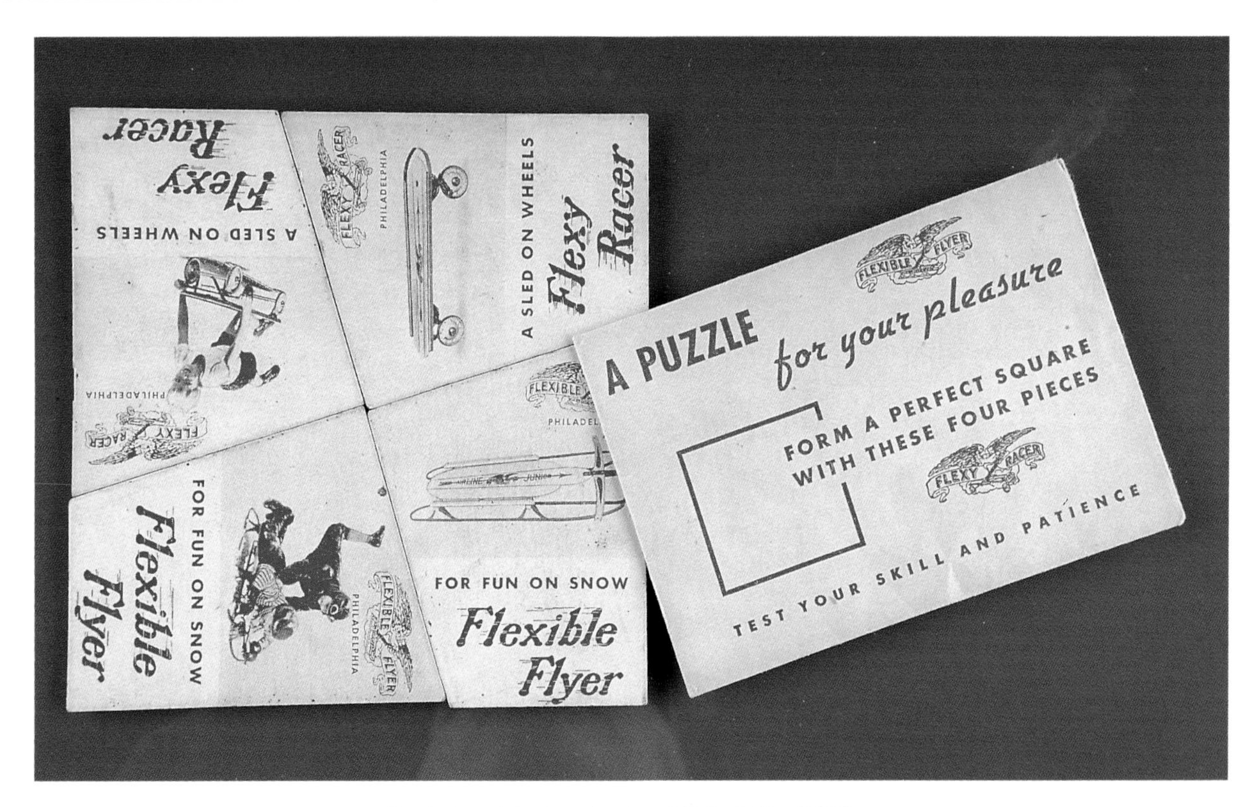

Puzzle in original packaging, 1951. 4.5" x 3.25". Complete puzzle size: 5" x 5". $15-25.

1962 advertisements for the sled back and the bob sled conversion kit. $10-15.

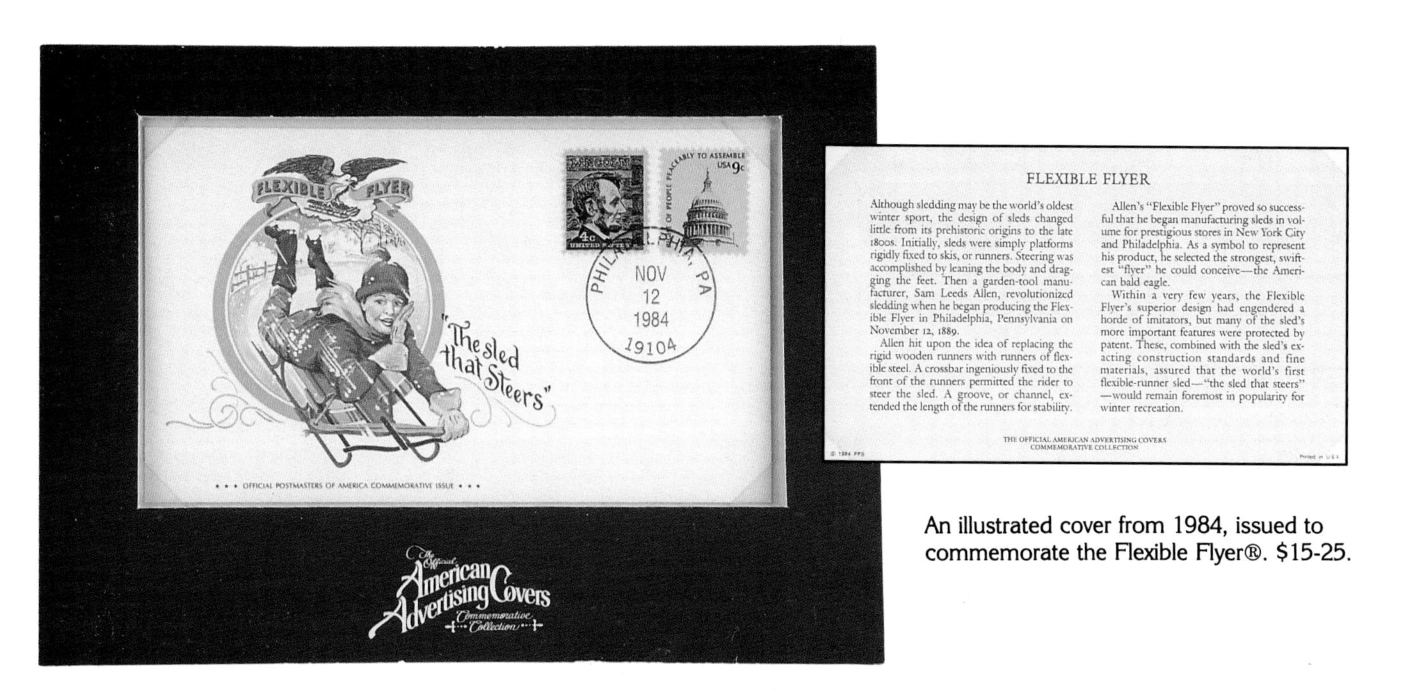

FLEXIBLE FLYER

Although sledding may be the world's oldest winter sport, the design of sleds changed little from its prehistoric origins to the late 1800s. Initially, sleds were simply platforms rigidly fixed to skis, or runners. Steering was accomplished by leaning the body and dragging the feet. Then a garden-tool manufacturer, Sam Leeds Allen, revolutionized sledding when he began producing the Flexible Flyer in Philadelphia, Pennsylvania on November 12, 1889.

Allen hit upon the idea of replacing the rigid wooden runners with runners of flexible steel. A crossbar ingeniously fixed to the front of the runners permitted the rider to steer the sled. A groove, or channel, extended the length of the runners for stability.

Allen's "Flexible Flyer" proved so successful that he began manufacturing sleds in volume for prestigious stores in New York City and Philadelphia. As a symbol to represent his product, he selected the strongest, swiftest "flyer" he could conceive—the American bald eagle.

Within a very few years, the Flexible Flyer's superior design had engendered a horde of imitators, but many of the sled's more important features were protected by patent. These, combined with the sled's exacting construction standards and fine materials, assured that the world's first flexible-runner sled—"the sled that steers"—would remain foremost in popularity for winter recreation.

THE OFFICIAL AMERICAN ADVERTISING COVERS COMMEMORATIVE COLLECTION

© 1984 PPS

Printed in U.S.A.

An illustrated cover from 1984, issued to commemorate the Flexible Flyer®. $15-25.

December, 1920 23

Flexible Flyer

The original steering sled that made coasting popular and safe

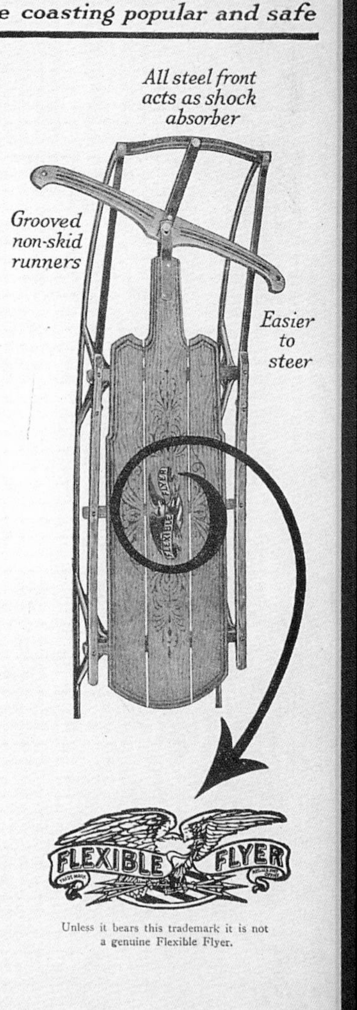

REMEMBER the first sled you got for Christmas? Didn't it thrill you and make you dance with joy, even though it was one of those old-fashioned rigid types?

Imagine how happy you can make your girl or boy this Christmas with a Flexible Flyer—"the sled that steers."

It is the safest, speediest sled. Patented non-skid runners make steering easy and coasting safe, comfortable and swift. Built like an airplane—sturdy, strong yet light in weight.

Outlasts 3 ordinary sleds

The new steel front takes up shock, adds strength and prevents splitting of seat and rails. Seven sizes 38 to 63 inches. Look for the name and eagle trademark on seat. None genuine without it.

S. L. Allen & Co., Inc.

Box 1100D **Philadelphia**

FREE cardboard model showing how the Flexible Flyer steers. Write for it.

Flexible Flyer® advertisement, December 1920. $10-20.

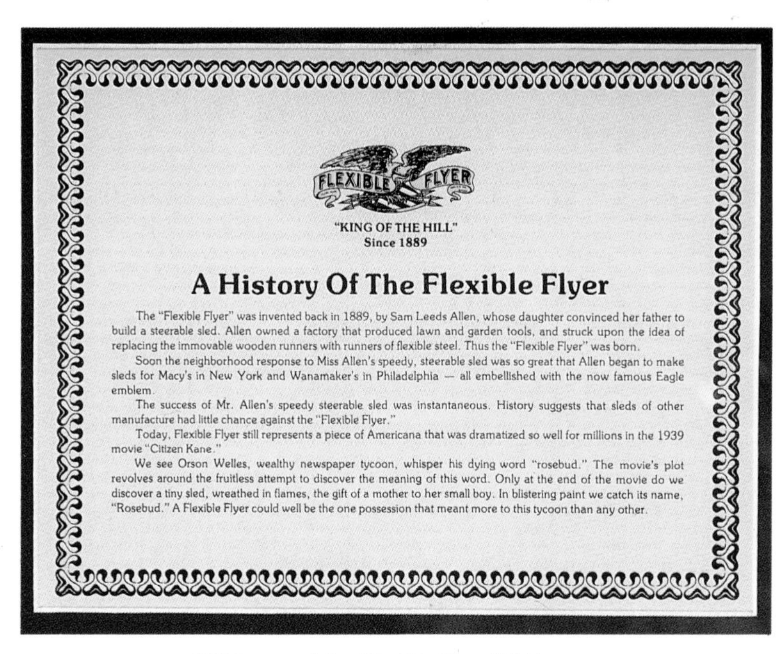

FLEXIBLE FLYER

"KING OF THE HILL"
Since 1889

A History Of The Flexible Flyer

The "Flexible Flyer" was invented back in 1889, by Sam Leeds Allen, whose daughter convinced her father to build a steerable sled. Allen owned a factory that produced lawn and garden tools, and struck upon the idea of replacing the immovable wooden runners with runners of flexible steel. Thus the "Flexible Flyer" was born.

Soon the neighborhood response to Miss Allen's speedy, steerable sled was so great that Allen began to make sleds for Macy's in New York and Wanamaker's in Philadelphia — all embellished with the now famous Eagle emblem.

The success of Mr. Allen's speedy steerable sled was instantaneous. History suggests that sleds of other manufacture had little chance against the "Flexible Flyer."

Today, Flexible Flyer still represents a piece of Americana that was dramatized so well for millions in the 1939 movie "Citizen Kane."

We see Orson Welles, wealthy newspaper tycoon, whisper his dying word "rosebud." The movie's plot revolves around the fruitless attempt to discover the meaning of this word. Only at the end of the movie do we discover a tiny sled, wreathed in flames, the gift of a mother to her small boy. In blistering paint we catch its name, "Rosebud." A Flexible Flyer could well be the one possession that meant more to this tycoon than any other.

"A History of the Flexible Flyer." Written for the Smithsonian Museum, 1986. $15-25.

Flexible Flyer® advertisement from *Youth's Companion.* Top: 1924; bottom: 1927. $3-10, unframed.

Flexible Flyer® advertisement.
Top: 1924; bottom: 1930. $3-10.

Flexible Flyer® advertisement, 1927.

Flexible Flyer® advertisement, 1929.

Flexible Flyer® advertisement from *American Boy*, 1930s.

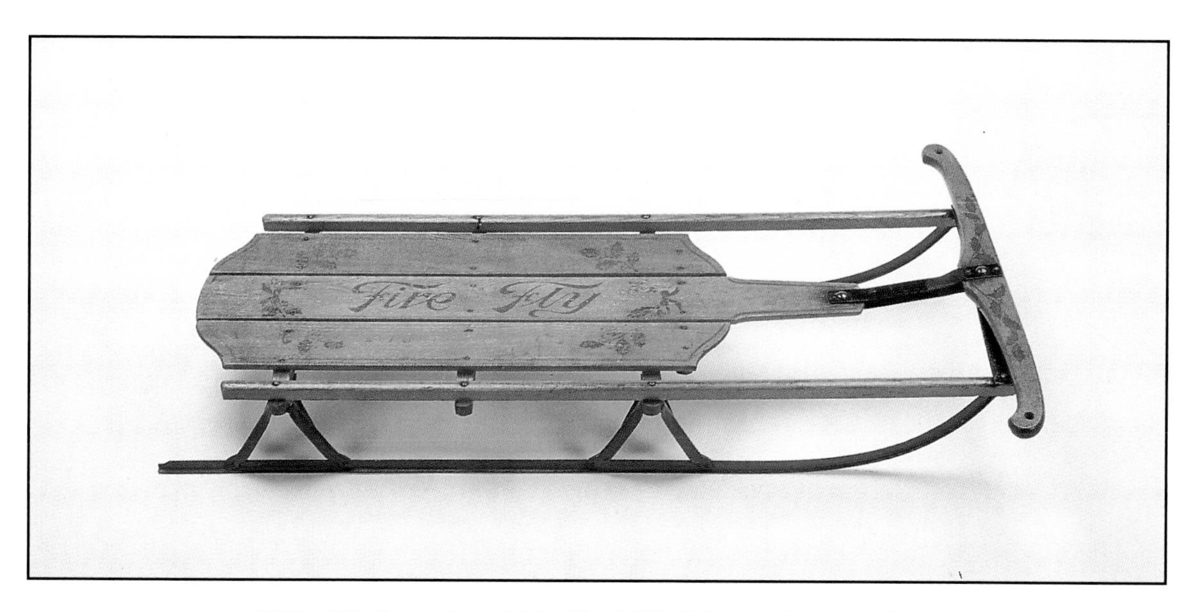

1890-1900. Pre-trademark Fire Fly #12B. Holly motif, named after Planet Jr. Seeder. All wood frame. Note placement of steering bar, no bumper. L: 46", W: 14" H: 6.25". Original price, $2.50. $150-300

1900, #5B. All wood frame with two pair of boot rests. The bumper is also wood. Original trademark (eagle, shield, and ribbon), intricate scrollwork. Note steering bar placement, 8" from the bumper.; L: 62", W: 16", H: 8.5". Original price, $6.00. $500-1500.

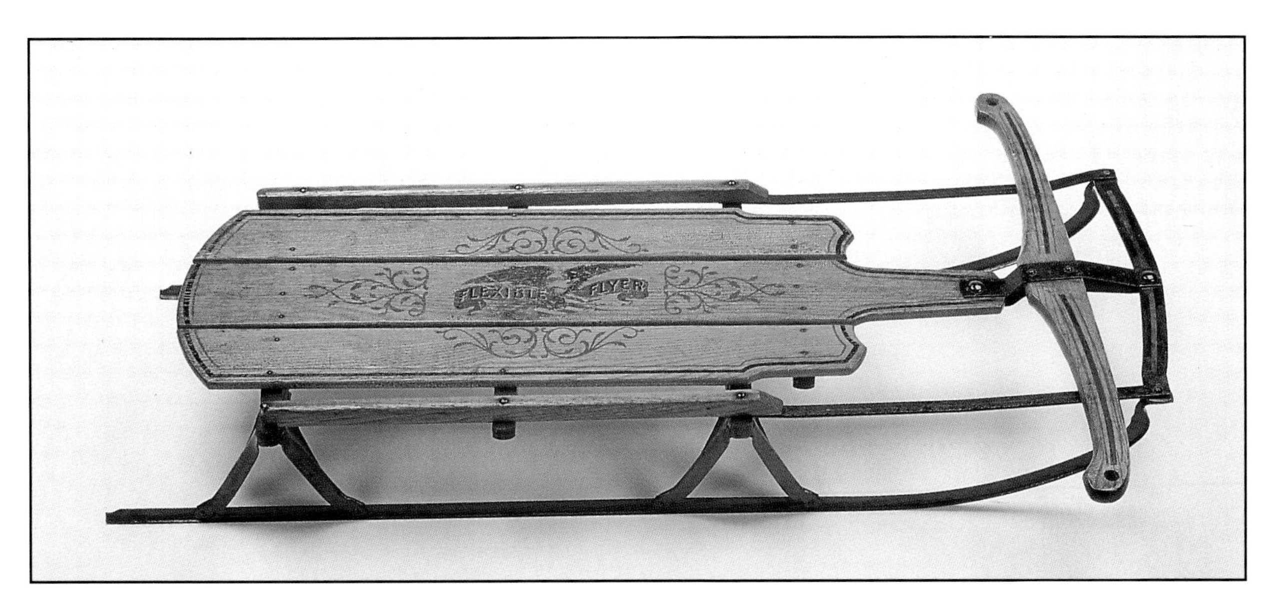

1914. #2C. all steel frame, original trademark (eagle, shield, & ribbon), scroll work on deck, steering bar pinstriped. Note placement of steering bar 3" down from bumper. L: 42", w: 13" H: 6". Original price $3.50. $100-150

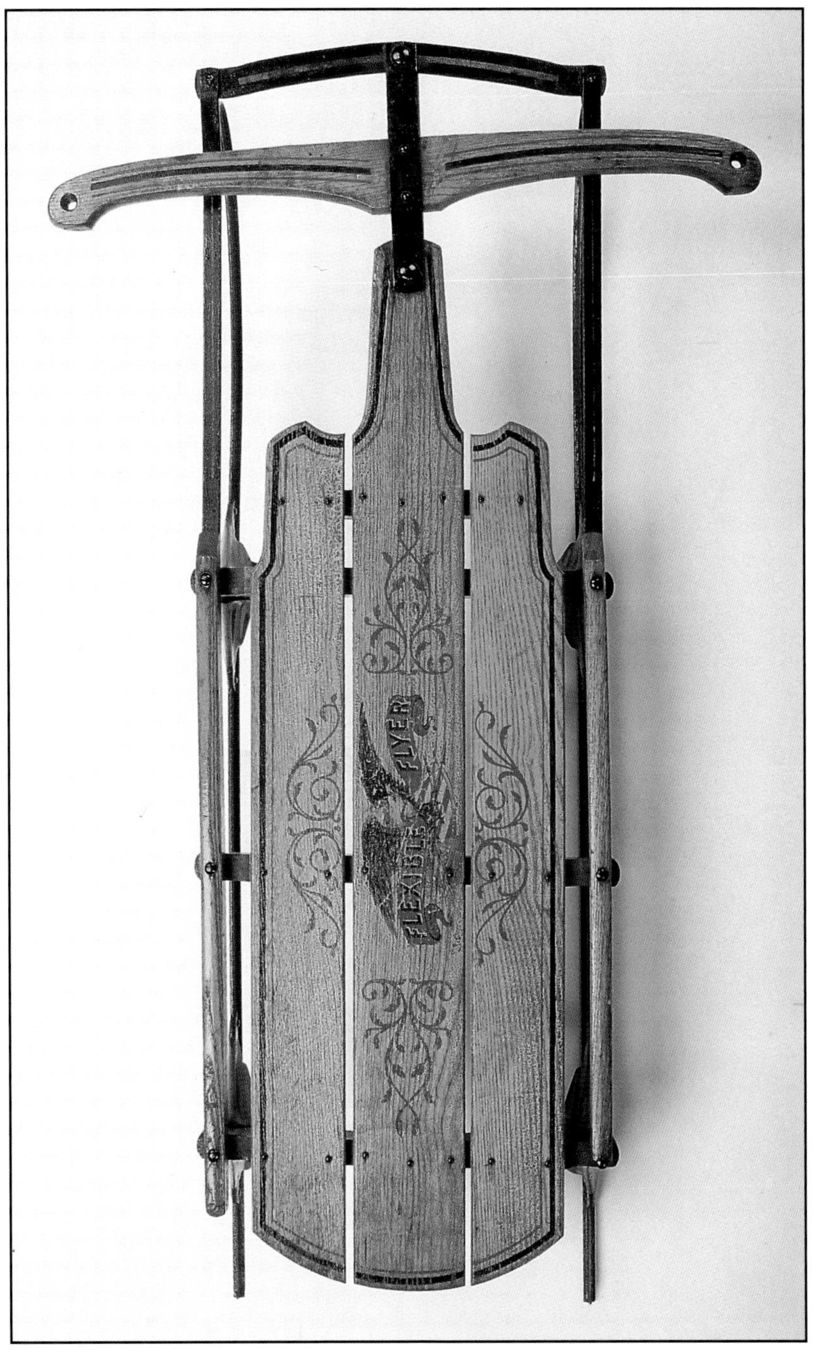

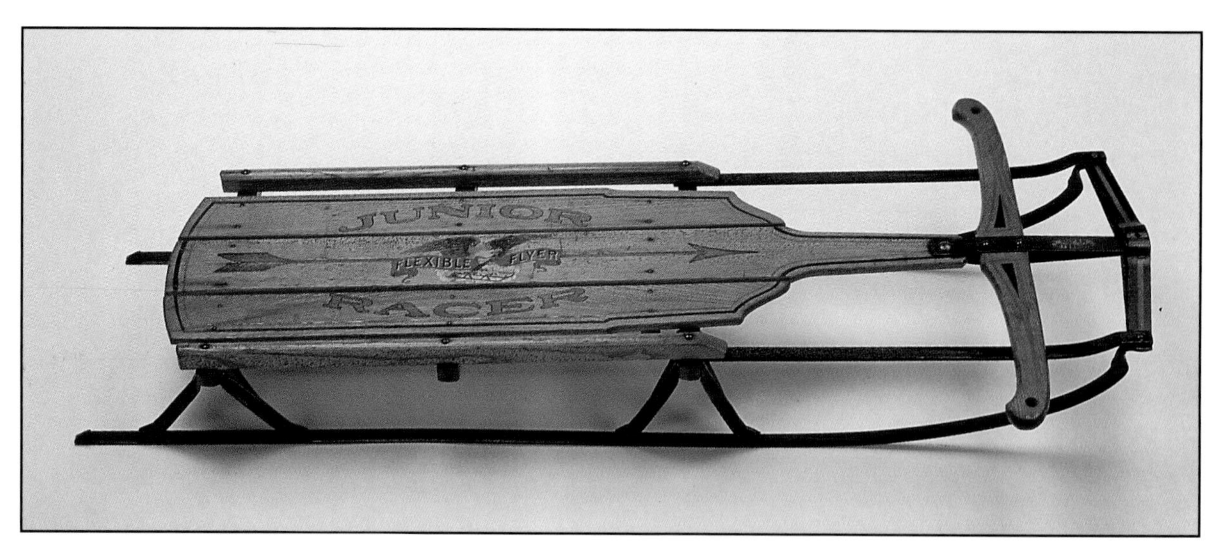

1915. Junior Racer. New in 1915. L: 49", W: 12", H: 6.5" front, 5.25" back. Original price, $4.25. $100-200.

1921 Model # 3E. Trademark change with new diamond pattern.
L: 47", W: 14", H: 7.875". Original price $4.50. $75-150.

1921. New Diamond Pattern No. 6. No letter "C". First year of release. Six man sled with four pairs of boot rests. L:" 102", W: 16", H: 8.5". Original price. $15.00. $750-2500.

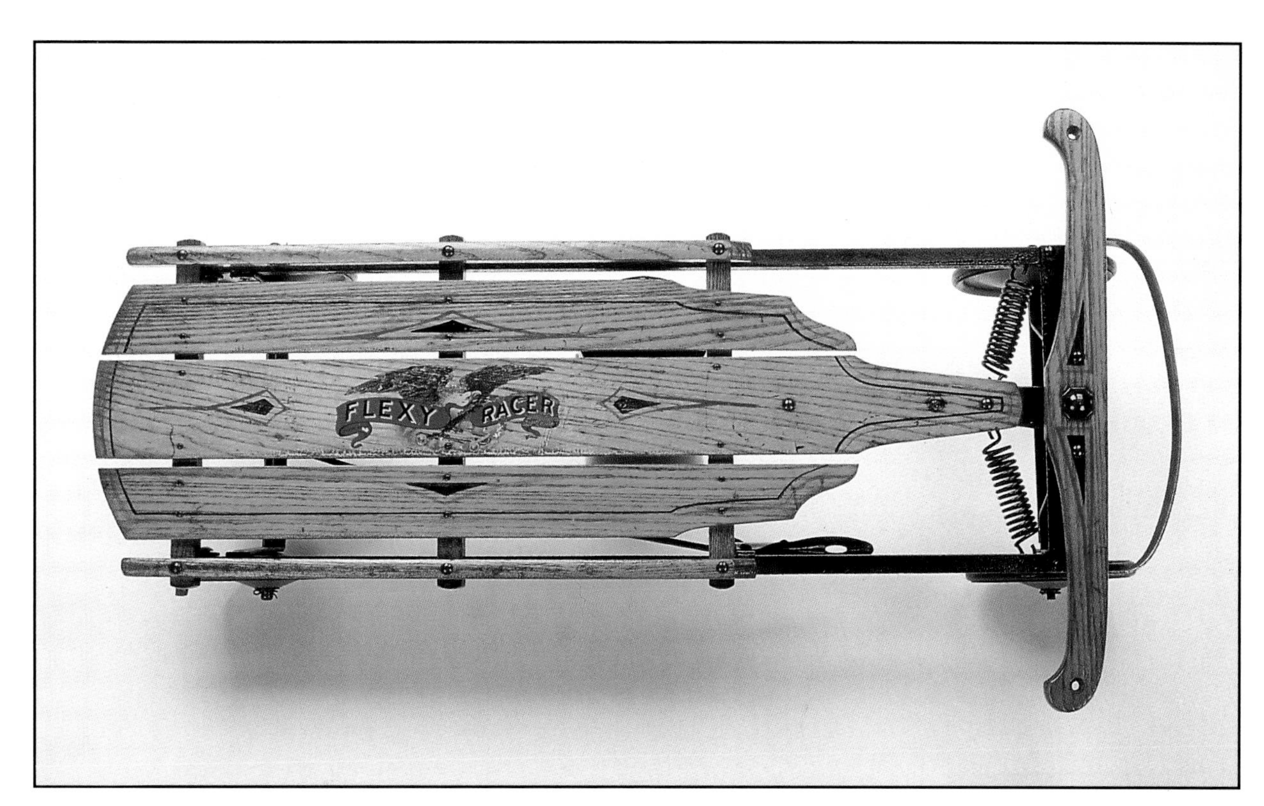

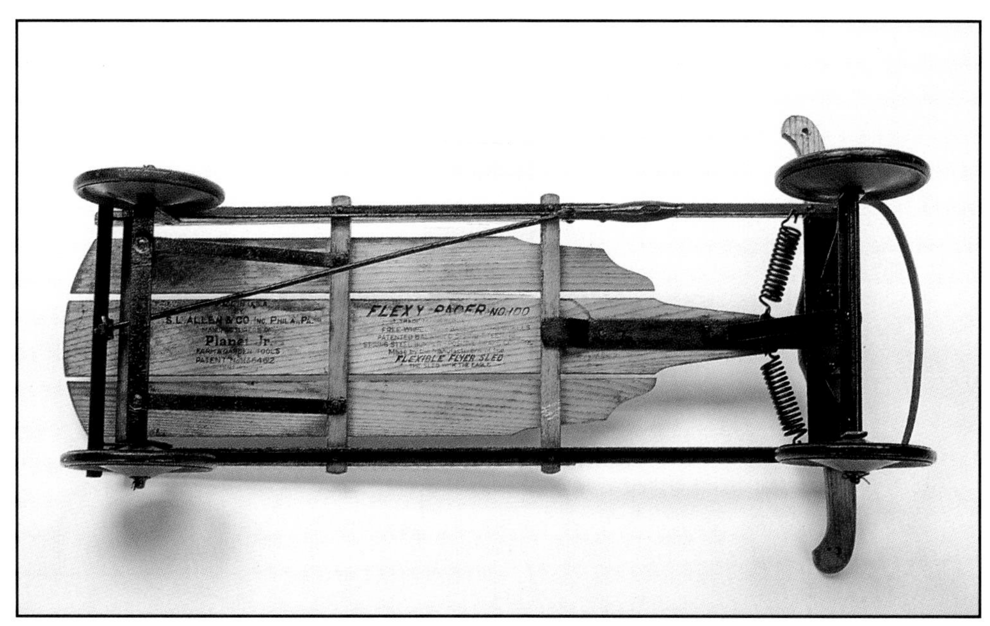

1921. Flexy Racer #100 with a hand brake on the side. $500-750.

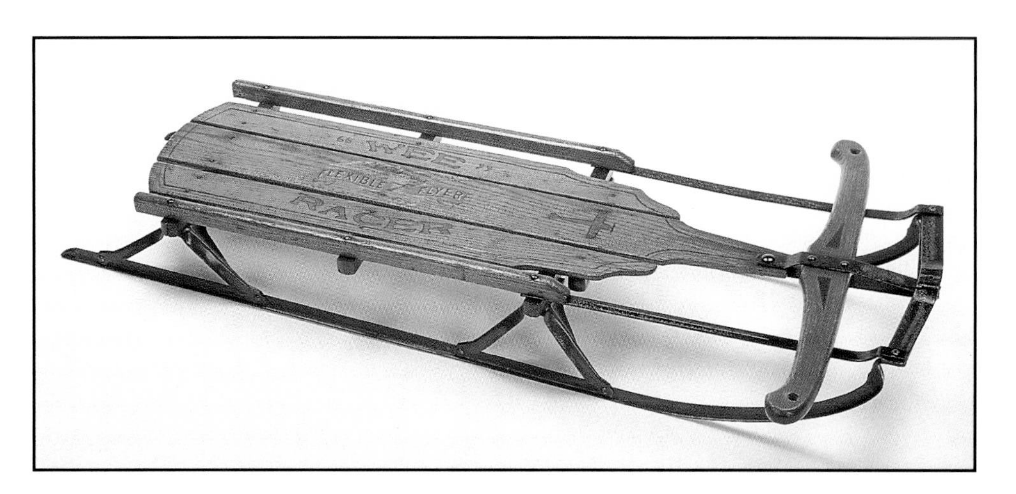

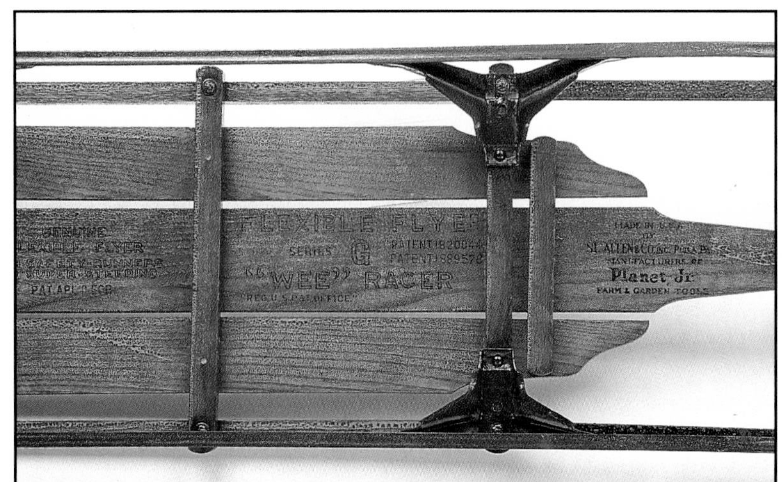

1927 "Wee" Racer. L: 42", W: 11", H: 4" rear, 5" front. Rare with plane impressed and painted on top. This is the only model released with this graphic in this position on the deck. Yankee Clipper sleds have the plane on the center of the deck. $100-200.

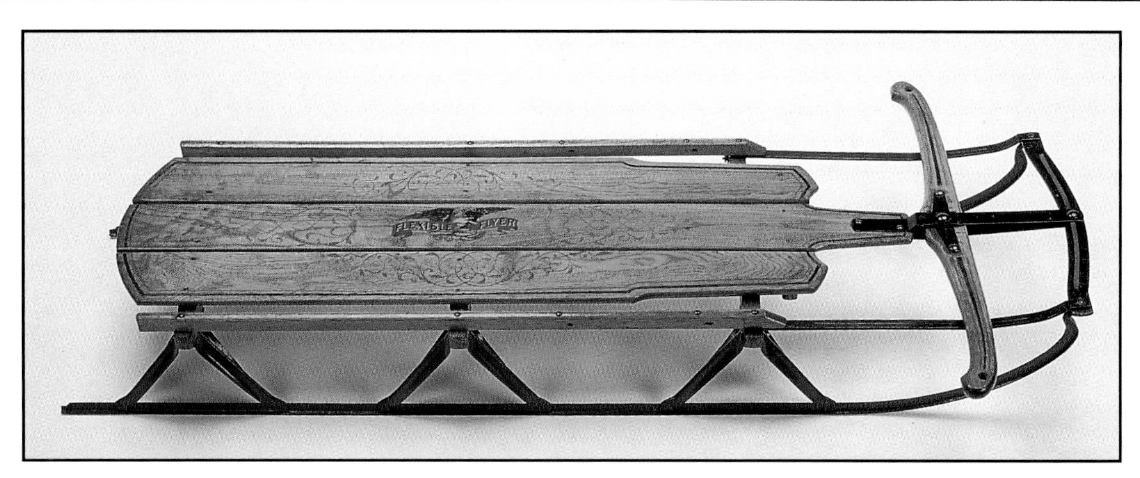

1928-5G Admiral Byrd Model. L: 63", W: 16", H: 8" Original price $8.75. Ad reads "You to can buy from your dealer a Flexible Flyer exactly the same as Commander Byrd is using. He took six #5's to the South Pole. $100-200

Flexible Flyer® advertisement from *Youth's Companion*, 1928.

Flexible Flyer® advertisement from *Saturday Evening Post*, December, 1928.

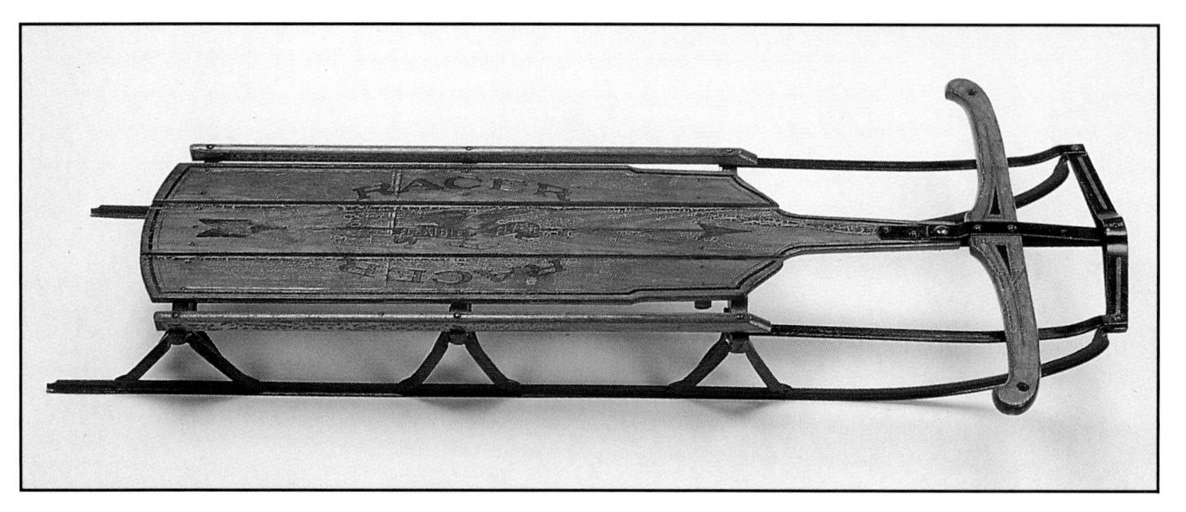

1930s Racer-Racer-Series G. L: 57", w: 13", H: 6.75" front, 6" rear.
Original price: $5.00. $75-125.

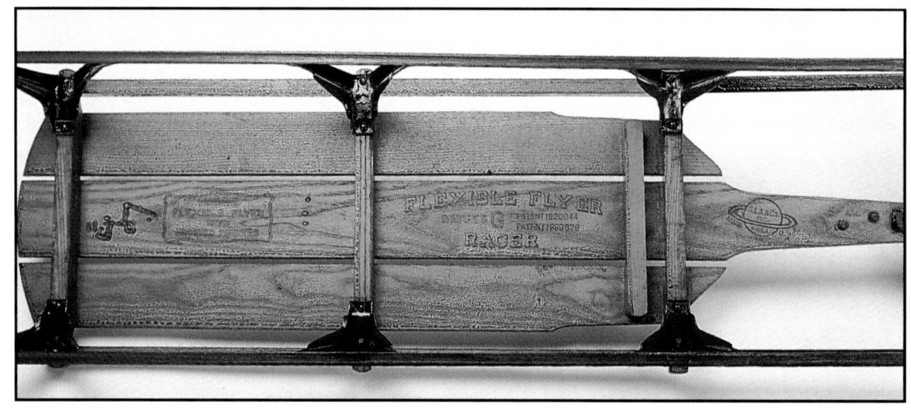

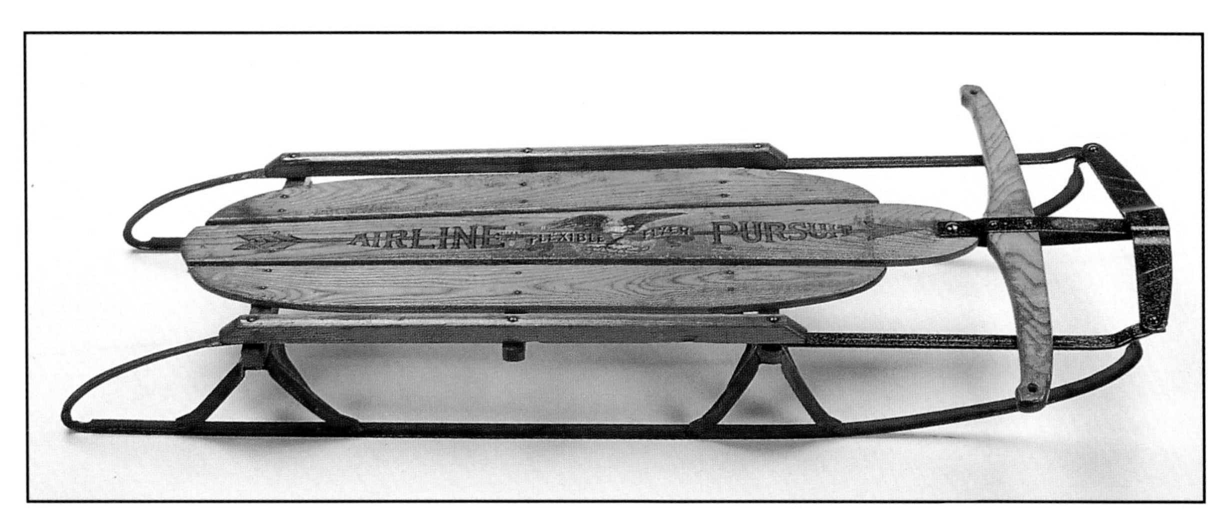

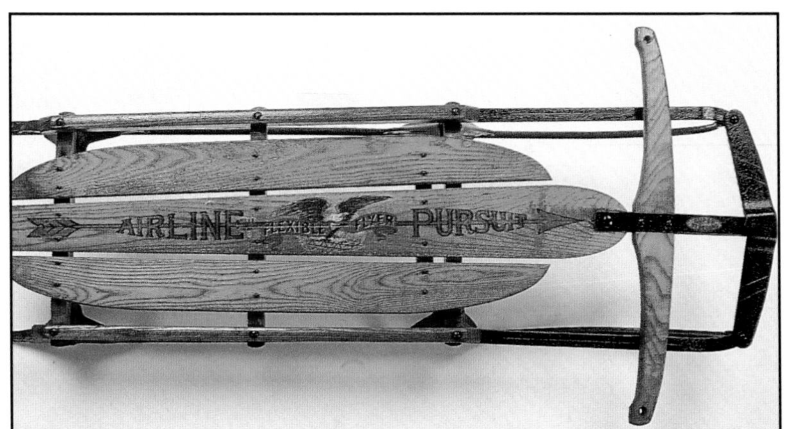

Above: 1935. Airline Pursuit (first year). Arrow on decks, no decoration (wings) on steering bar. Safety runners. L: 47", W: 11.75", H: 6.25" front, 5.25" back. Original price. $5.00. $100.

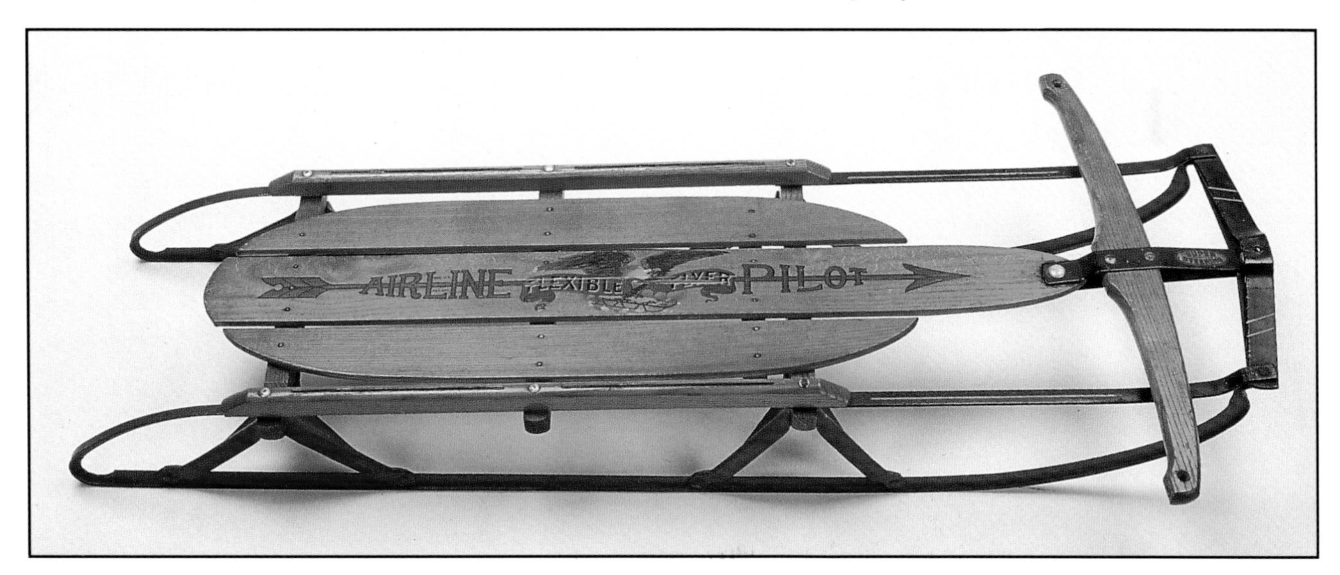

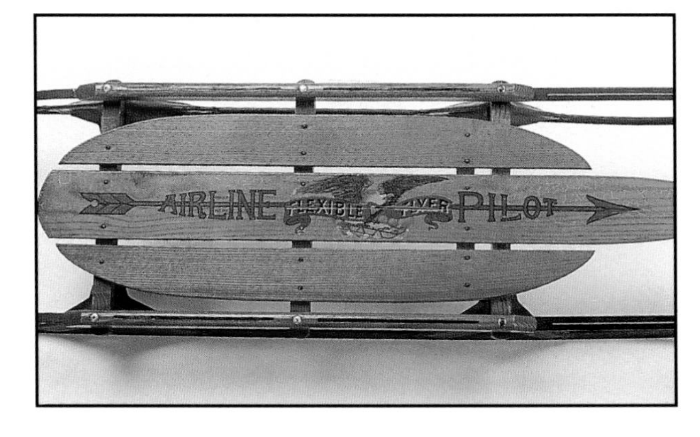

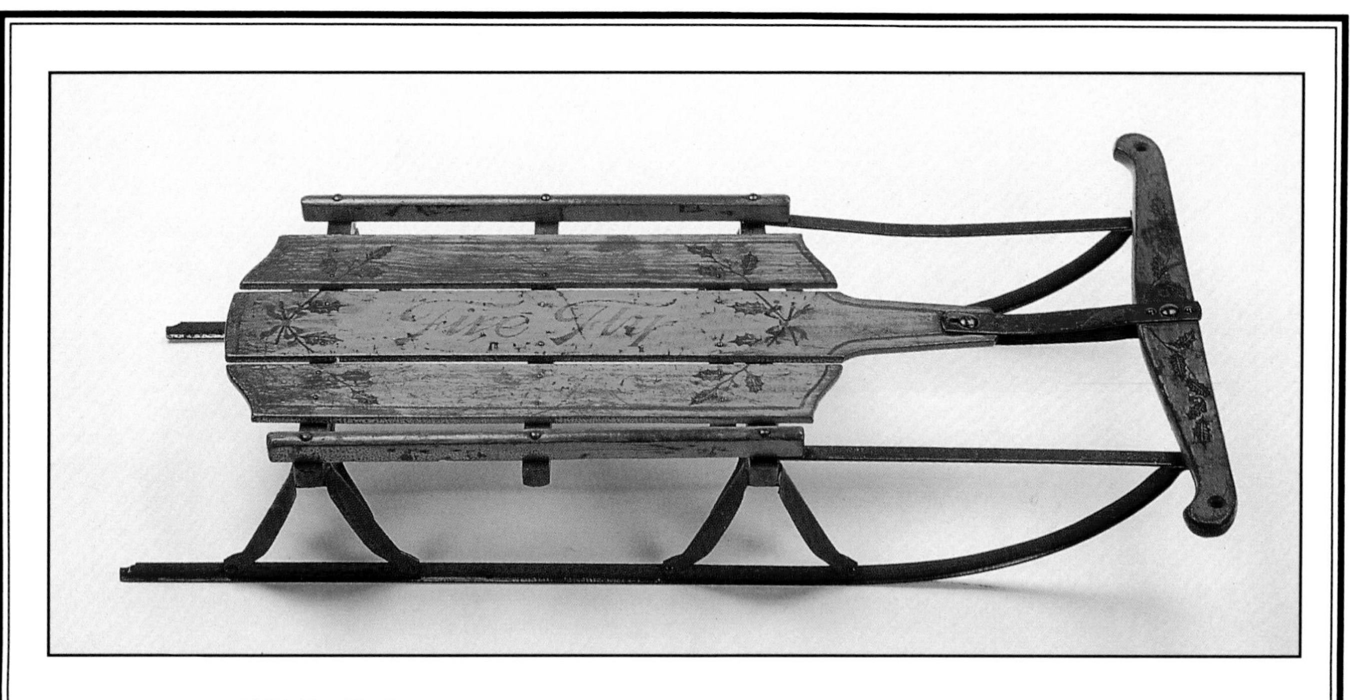

1935. Fire Fly Coaster #10 reintroduced. All steel frame. L: 36", W: 12", H: 6". $50-75.

Previous Page, Bottom Grouping: 1935. Airline Pilot, first year. Series A. No decoration (wings) on steering bar. No number on back. L: 41", W: 11.75", H: 6" front, 5" back. Original price, $4.00. $75.

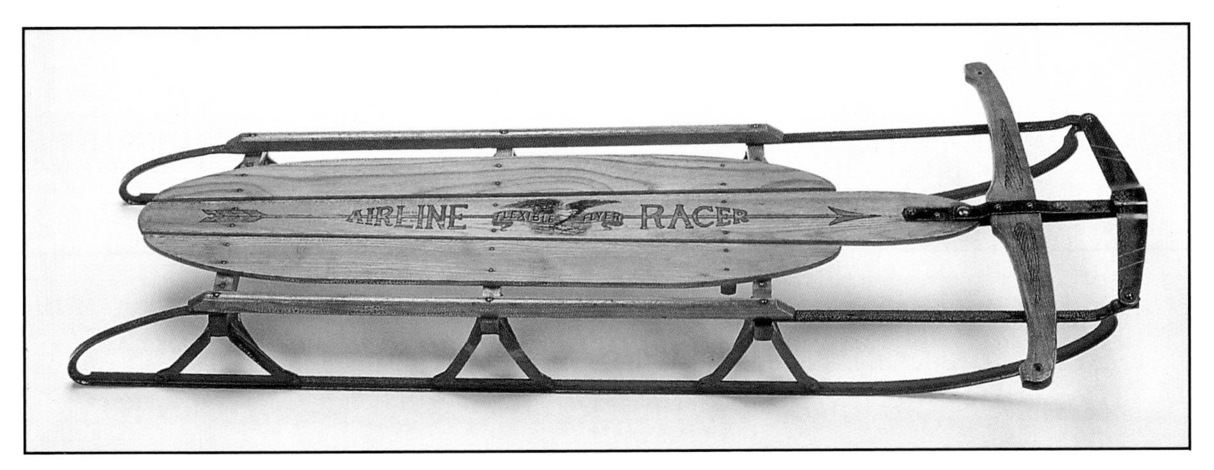

1936 Airline Racer #60. Steering bar decorated with wings. L: 60", W: 13.5", H: 7" front, 5.75" rear. $50-100.

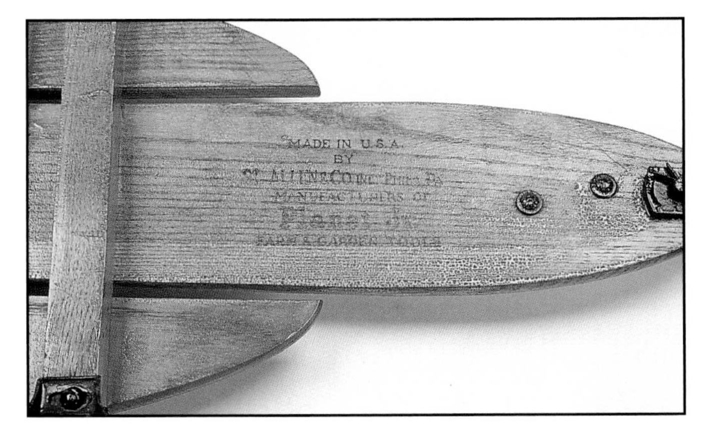

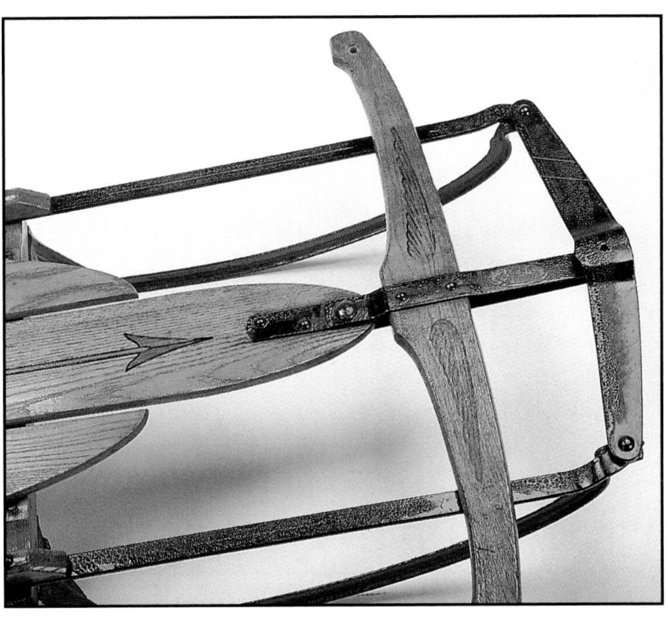

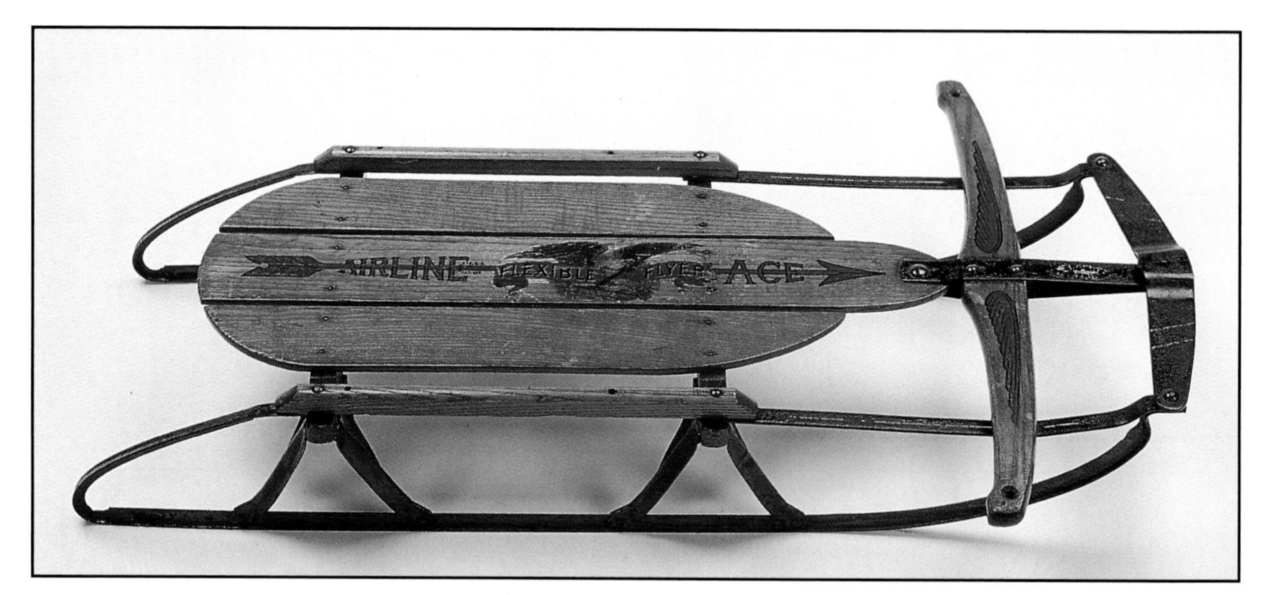

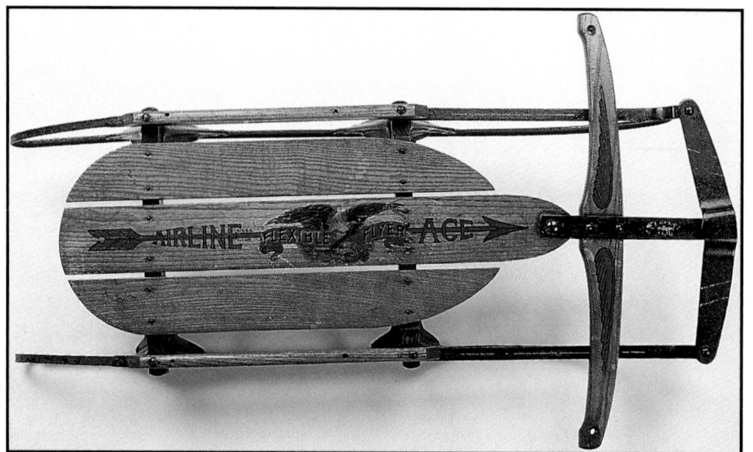

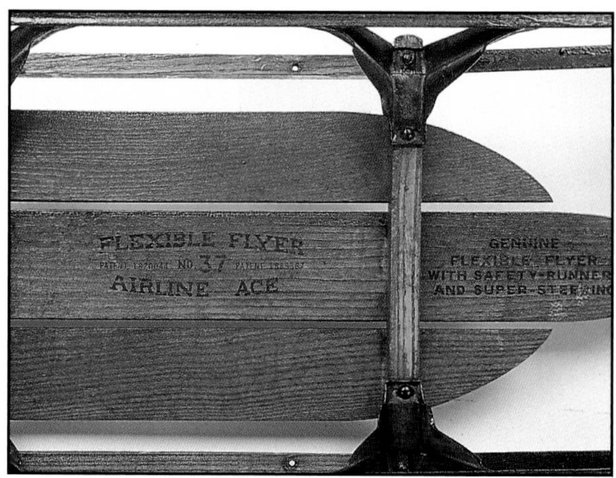

1936. Airline Ace, No. 37. The smallest sled in the Airline series. L: 37", W: 12", H: 6". $50-75.

Baby Sleigh #28W-2. Not a sled back attached to sled. No steering bar, straight runners & mechanism to allow wheels down when there is no snow. $100-150.

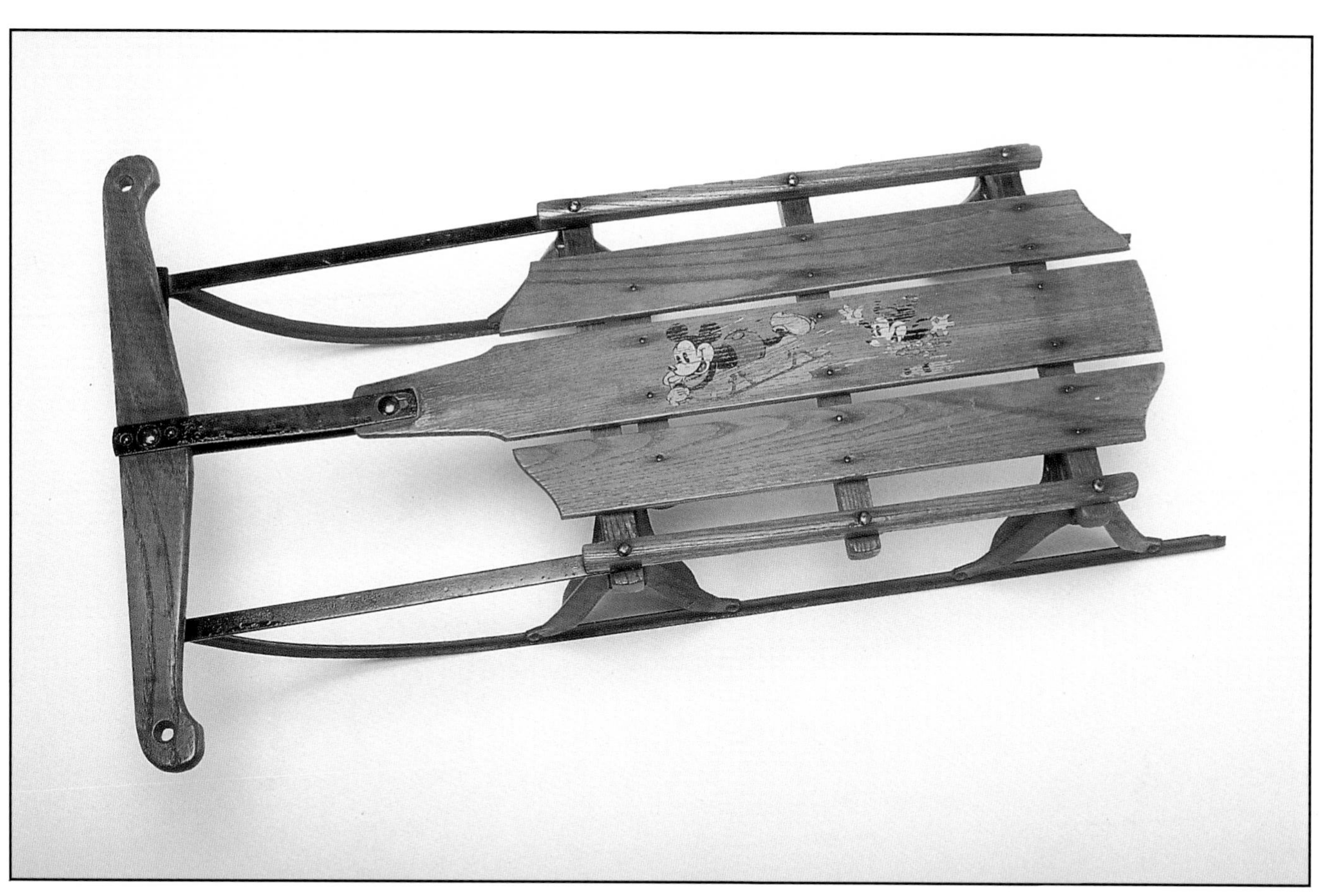

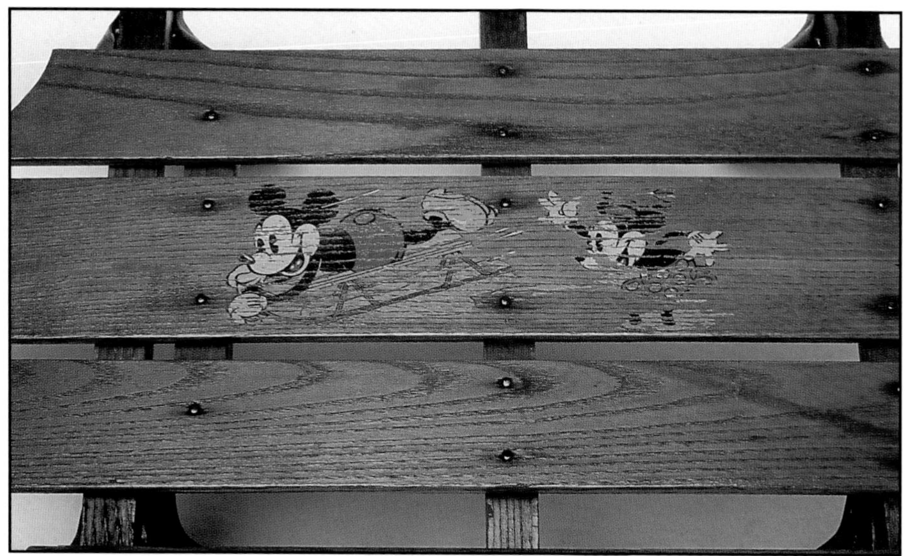

Mickey Mouse #90J. $300-500.

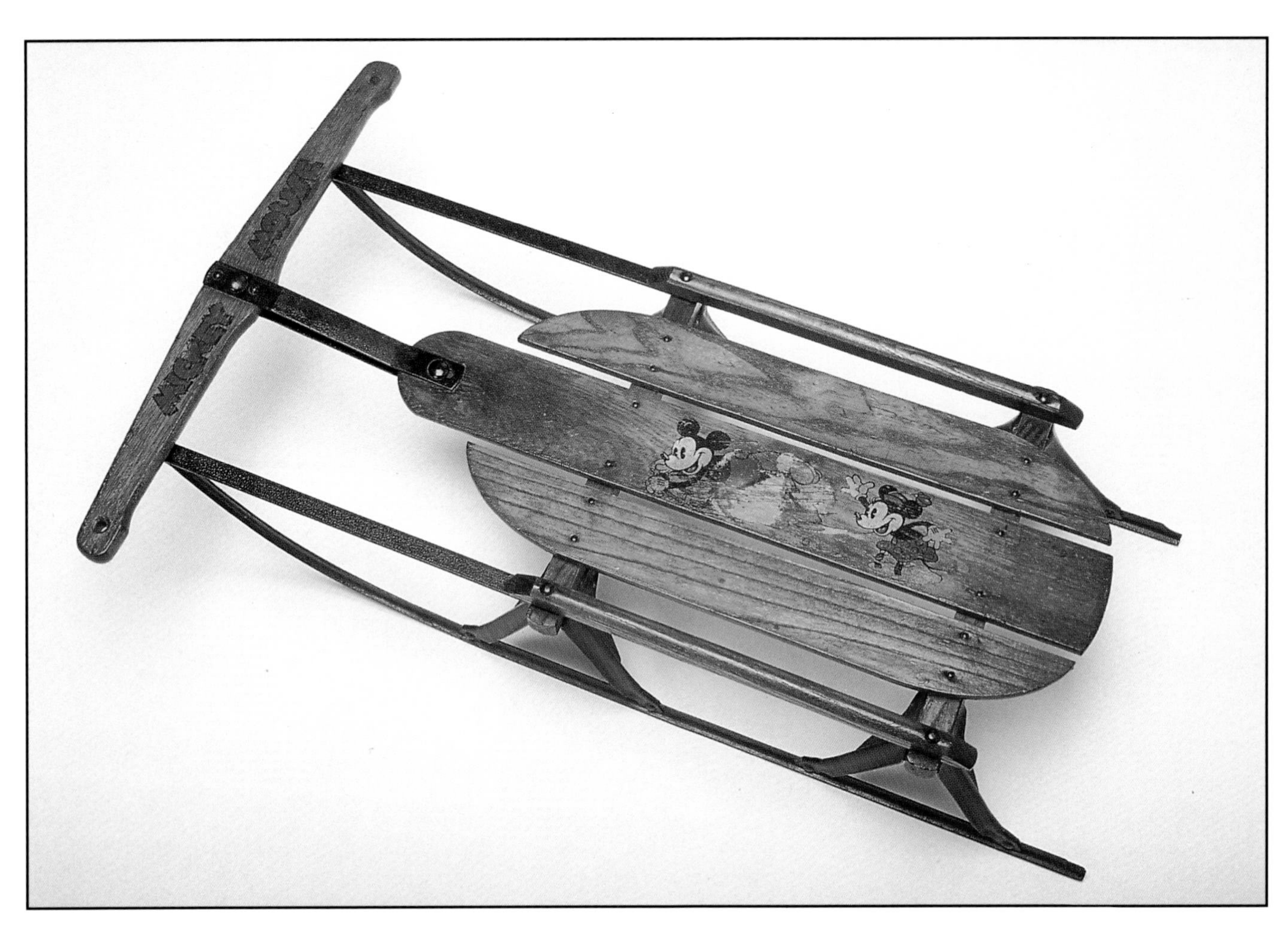

Mickey Mouse #80.
$250-450.

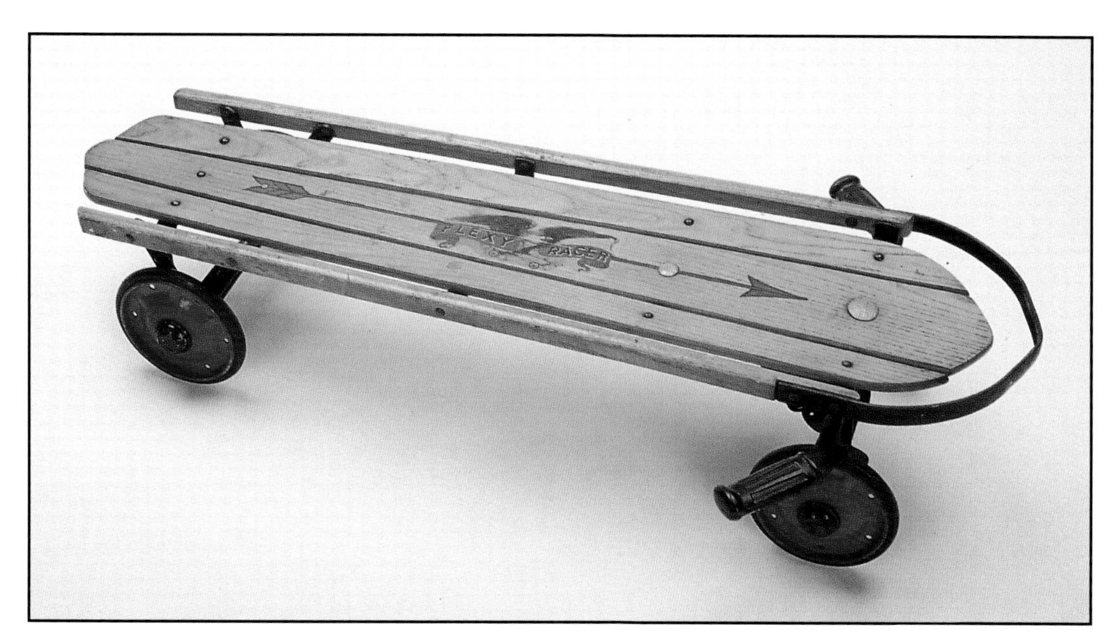

1935. Flexy Racer #300. Introduced in 1935. New braking system in hand grippers. Natural wood deck and siderails with red wheels and hubcaps. Red siderails and silver wheels introduced in 1937. Flexy Racer #300 manufactured until 1968. Blue wheels introduced in the late 1950s. $150-350.

Tri-fold flyer for the Flexy Racer No. 300, 1937. Open size: 6.25" x 9.25". $25-35

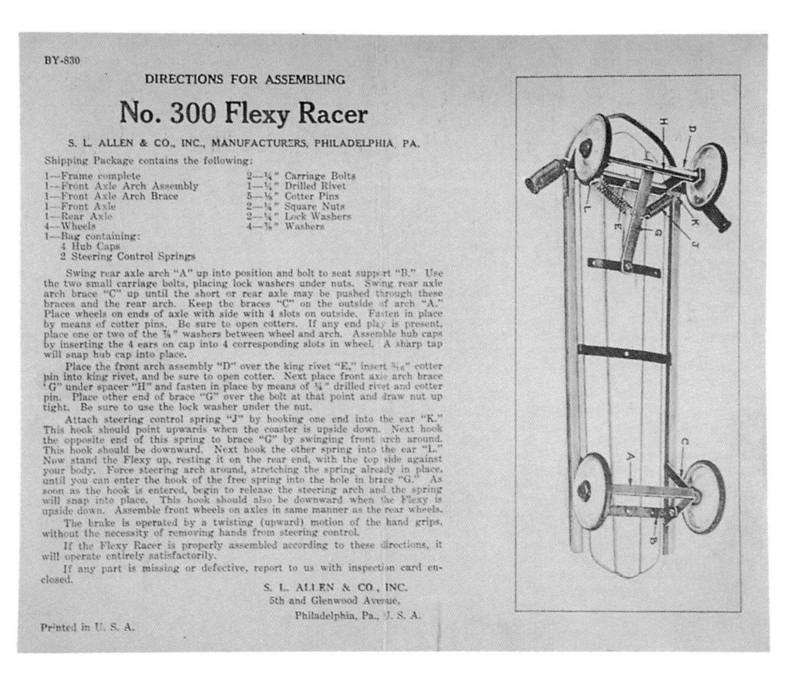

BY-830

DIRECTIONS FOR ASSEMBLING

No. 300 Flexy Racer

S. L. ALLEN & CO., INC., MANUFACTURERS, PHILADELPHIA, PA.

Shipping Package contains the following:

1—Frame complete	2—1/4" Carriage Bolts
1—Front Axle Arch Assembly	1—1/4" Drilled Rivet
1—Front Axle Arch Brace	5—1/8" Cotter Pins
1—Front Axle	2—1/4" Square Nuts
1—Rear Axle	2—1/4" Lock Washers
4—Wheels	4—7/8" Washers
1—Bag containing: 4 Hub Caps, 2 Steering Control Springs	

Swing rear axle arch "A" up into position and bolt to seat support "B." Use the two small carriage bolts, placing lock washers under nuts. Swing rear axle arch brace "C" up until the short or rear axle may be pushed through these braces and the rear arch. Keep the braces "C" on the outside of arch "A." Place wheels on ends of axle with side with 4 slots on outside. Fasten in place by means of cotter pins. Be sure to open cotters. If any end play is present, place one or two of the 7/8" washers between wheel and arch. Assemble hub caps by inserting the 4 ears on cap into 4 corresponding slots in wheel. A sharp tap will snap hub cap into place.

Place the front arch assembly "D" over the king rivet "E," insert 3/16" cotter pin into king rivet, and be sure to open cotter. Next place front axle arch brace "G" under spacer "H" and fasten in place by means of 1/4" drilled rivet and cotter pin. Place other end of brace "G" over the bolt at that point and draw nut up tight. Be sure to use the lock washer under the nut.

Attach steering control spring "J" by hooking one end into the ear "K." This hook should point upwards when the coaster is upside down. Next hook the opposite end of this spring to brace "G" by swinging front arch around. This hook should be downward. Next hook the other spring into the ear "L." Now stand the Flexy up, resting it on the rear end, with the top side against your body. Force steering arch around, stretching the spring already in place, until you can enter the hook of the free spring into the hole in brace "G." As soon as the hook is entered, begin to release the steering arch and the spring will snap into place. This hook should also be downward when the Flexy is upside down. Assemble front wheels on axles in same manner as the rear wheels.

The brake is operated by a twisting (upward) motion of the hand grips, without the necessity of removing hands from steering control.

If the Flexy Racer is properly assembled according to these directions, it will operate entirely satisfactorily.

If any part is missing or defective, report to us with inspection card enclosed.

S. L. ALLEN & CO., INC.
5th and Glenwood Avenue,
Philadelphia, Pa., U. S. A.

Printed in U. S. A.

The instructions for putting a Flexy Racer together, c. 1939. 7" x 8.75". $10-15.

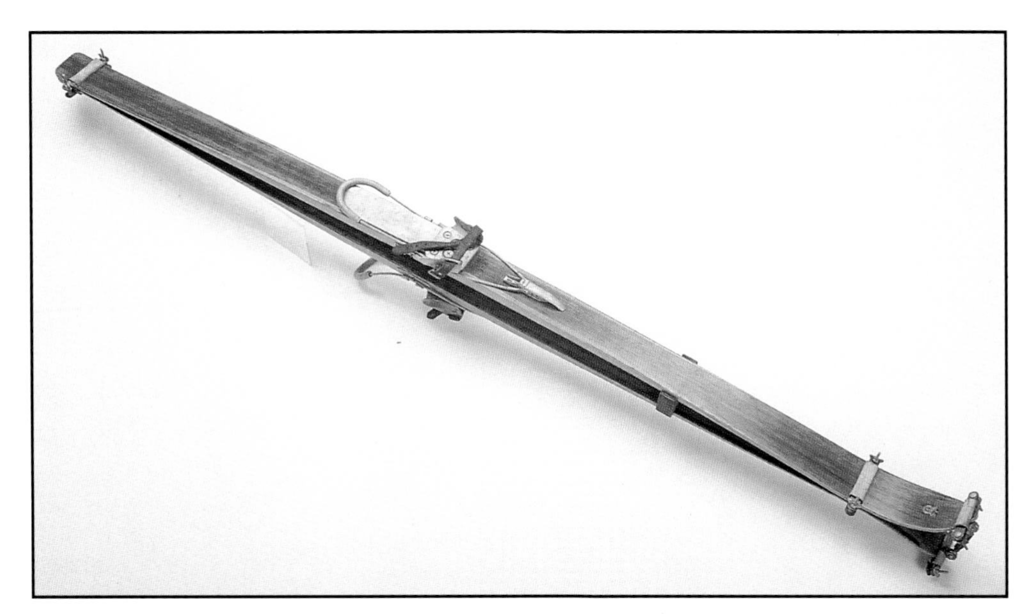

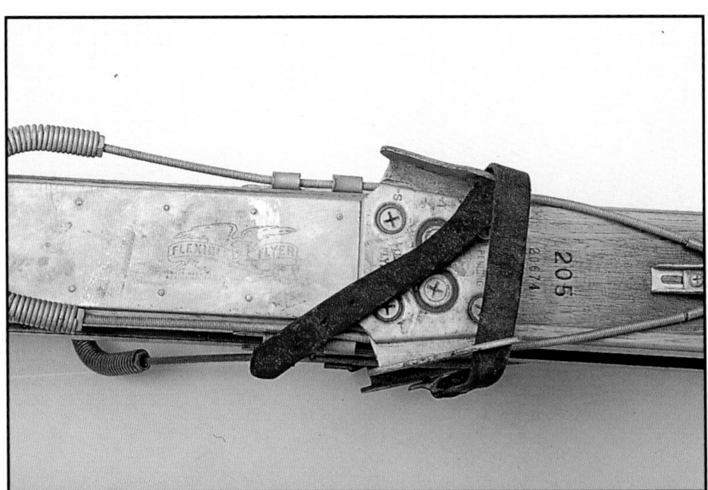

Above:
Flexible Flyer® Splitkein Skis. All original. Foot plate shows FF trademark. Tip of skis also **has trademark. $50-100.**

Flexible Flyer® "Splitkein" skis poster. c. 1935. $500-700 unframed.

Poster for Flexible Flyer® skis, c.1948. $300-500 unframed.

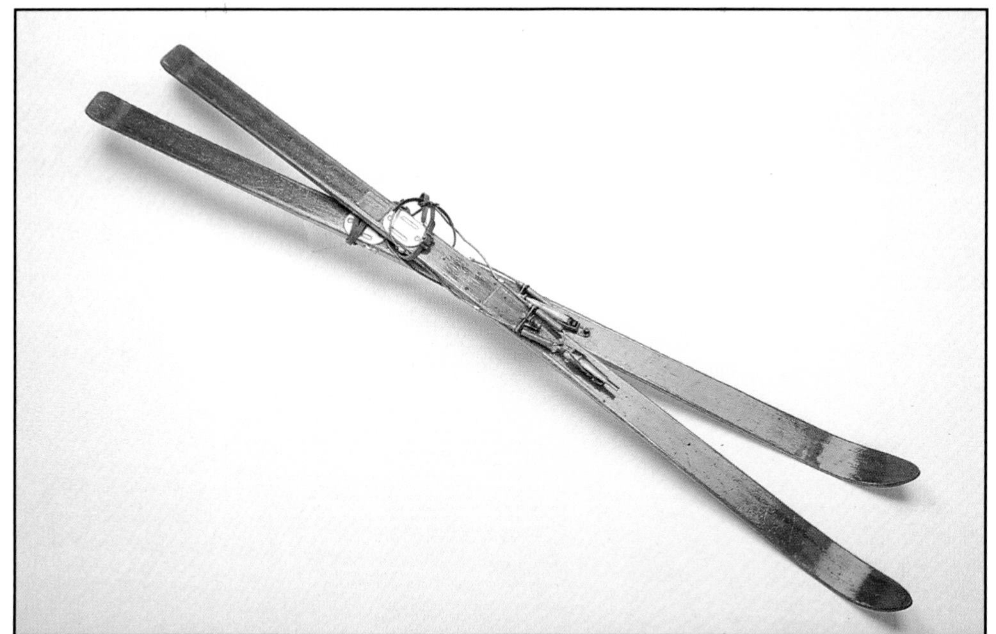

Flexible Flyer® Splitkein Skis. Trademark on ski tips. Brass plate identifies this with the National Ski Patrol System, 1948. Bindings not original. $50-75.

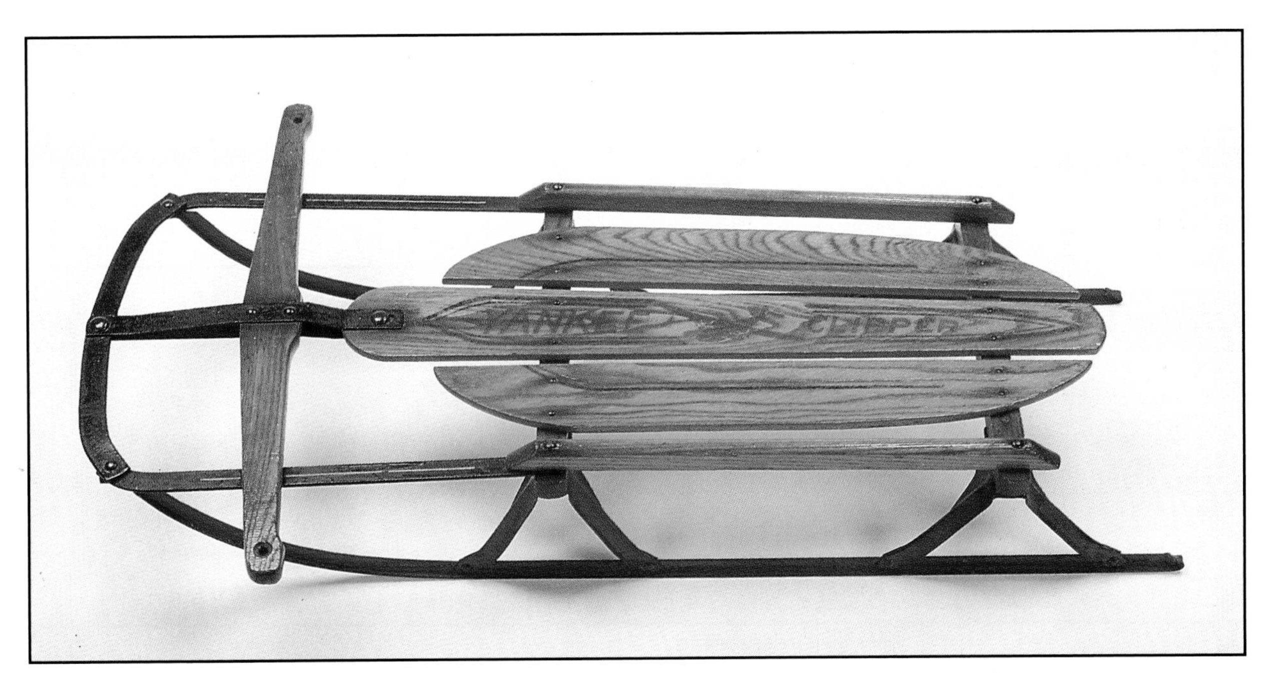

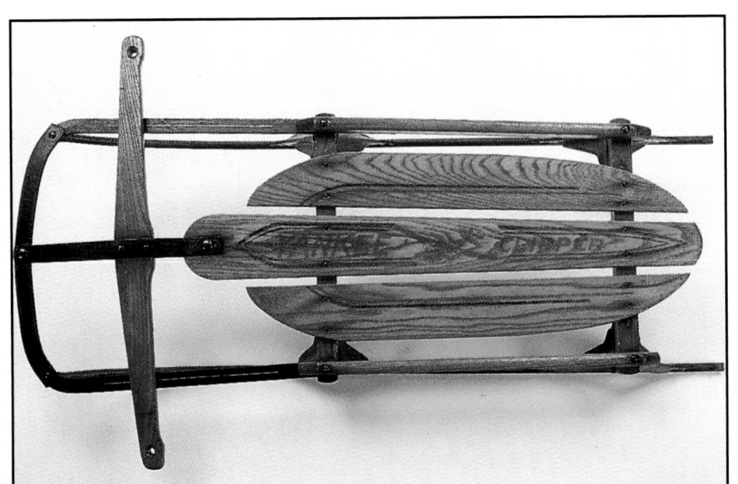

1936. Yankee Clipper No. 10. New, popularly priced. All racing models are lower in the back than in the front. L: 36", W: 11.25", H: 6" front, 5.5" back. $20-35.

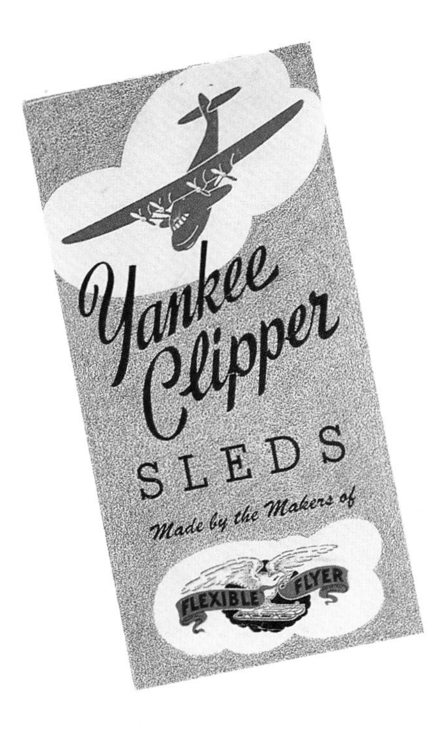

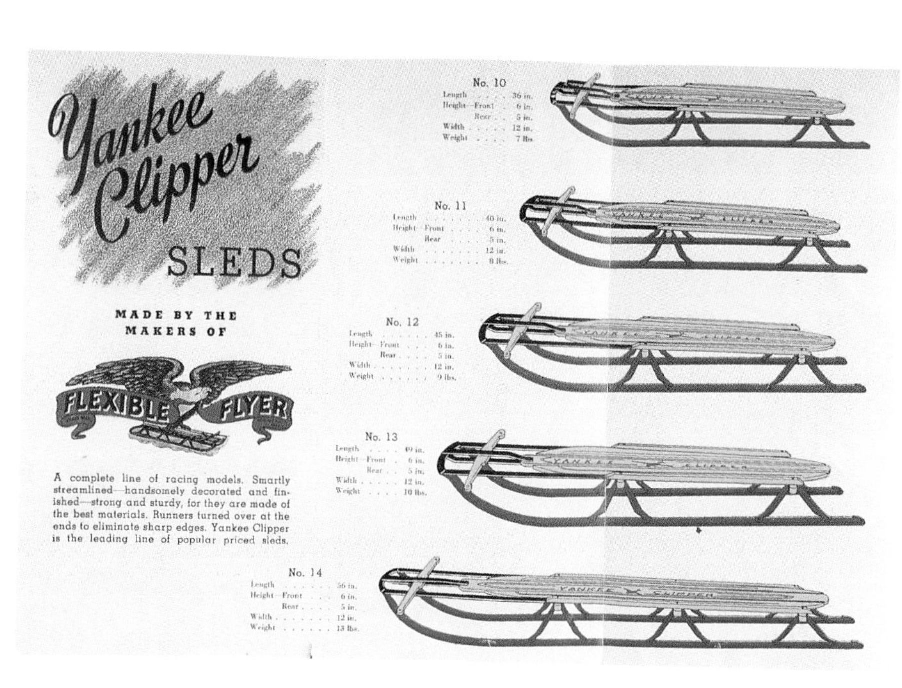

Tri-fold flyer for Yankee Clipper, the economy line, c. 1935. Open size: 6.5" x 9.5". $15-25.

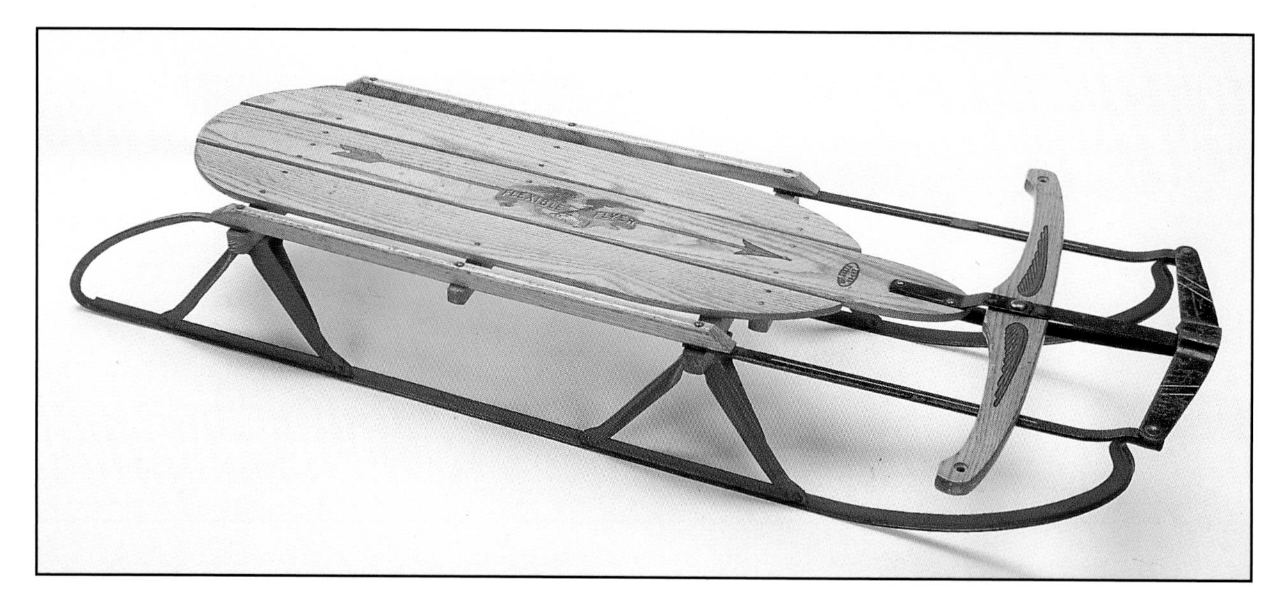

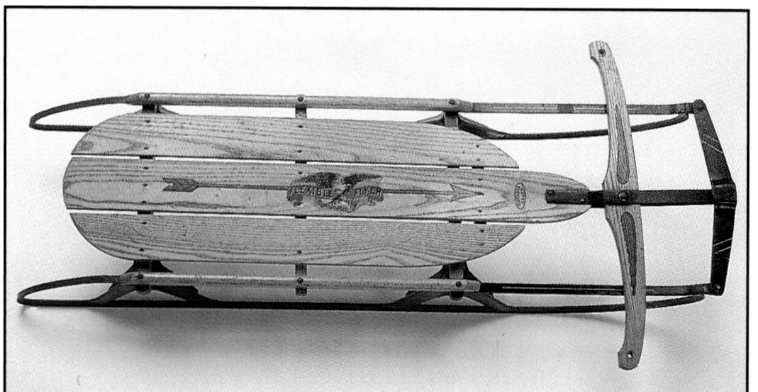

1951 *Boys' Life* advertisement. The deck is plain. At the bottom is an advertising for puzzle.

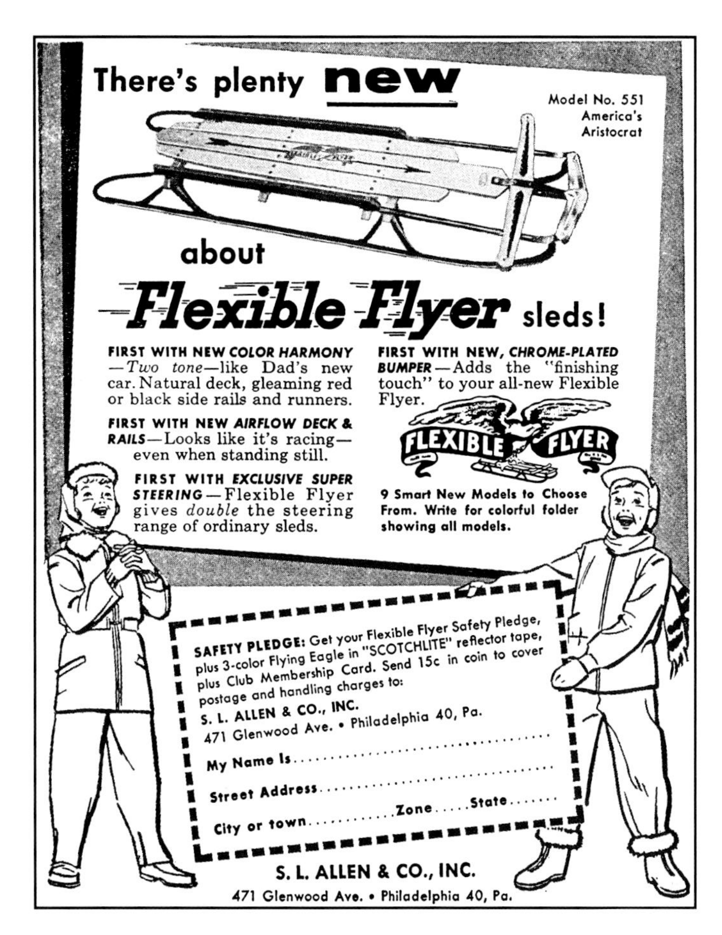

1955 *Boys' Life* advertisement featuring Model No. 551, "America's Aristocrat." This was the first year for the chrome bumper, Airflow deck, and two-toned color.

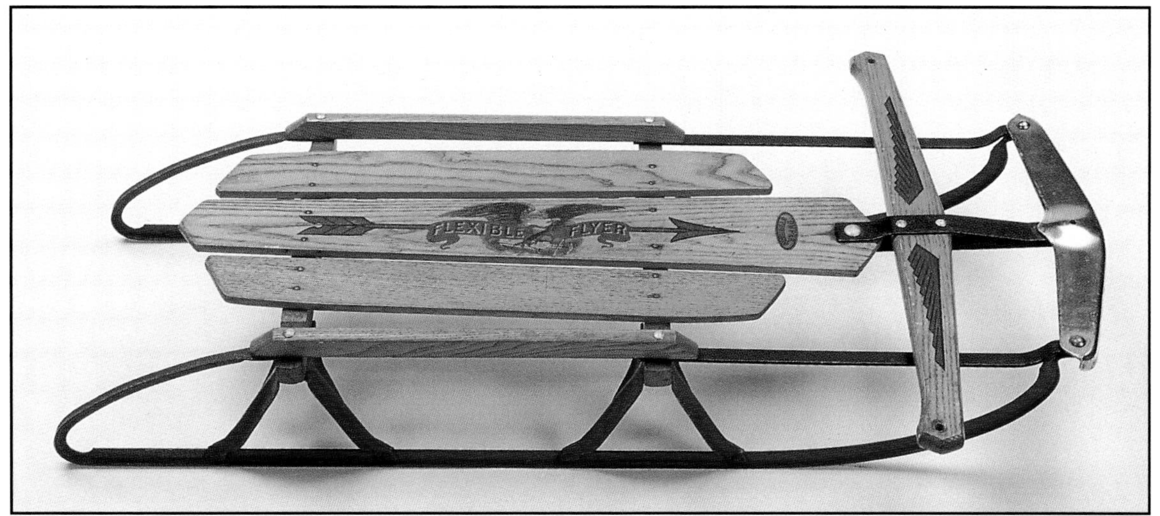

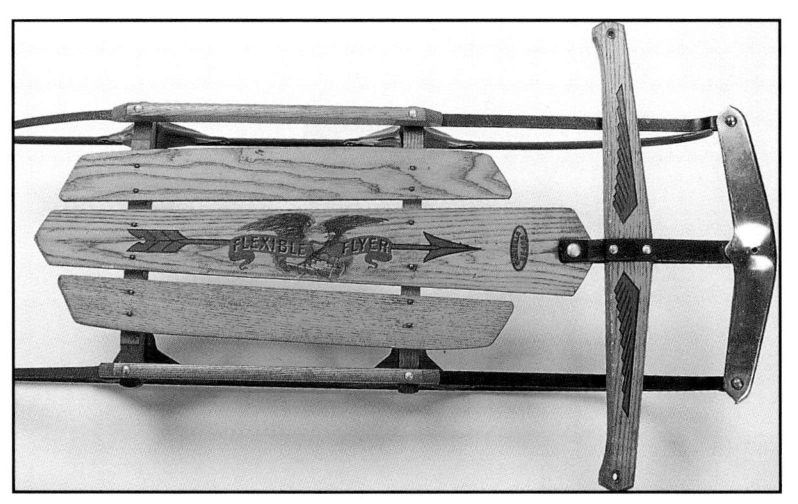

1955. Airline Ace #37H. Airflow deck, chrome bumper, new. L: 37", W: 12", H: 6". $25-50.

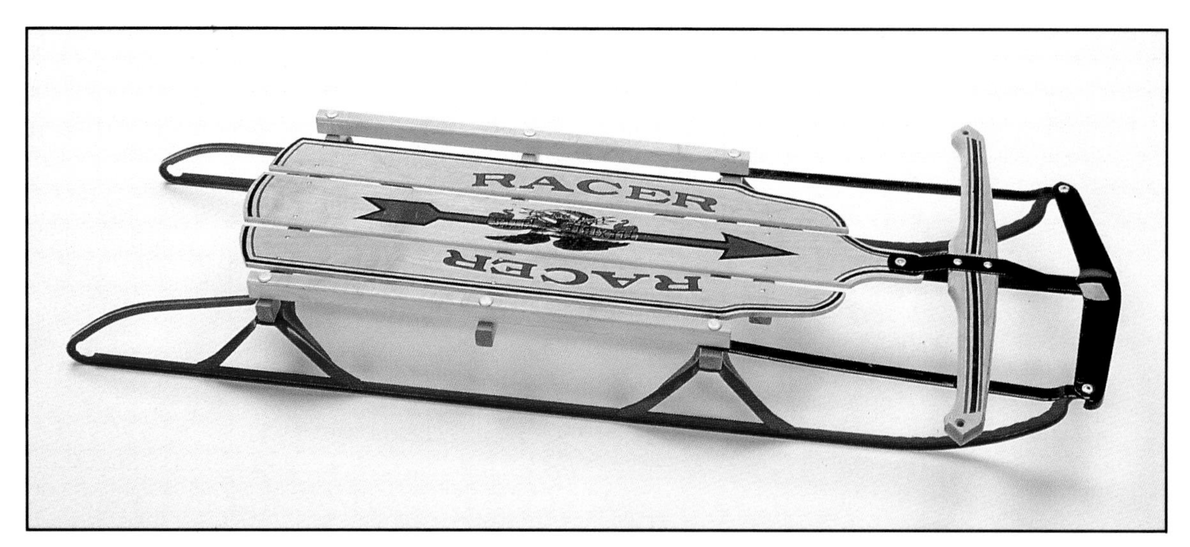

1986. Smithsonian Edition, original prototype. Gift of Blazon Flexible Flyer®, Glen Montgomery. $500-750.

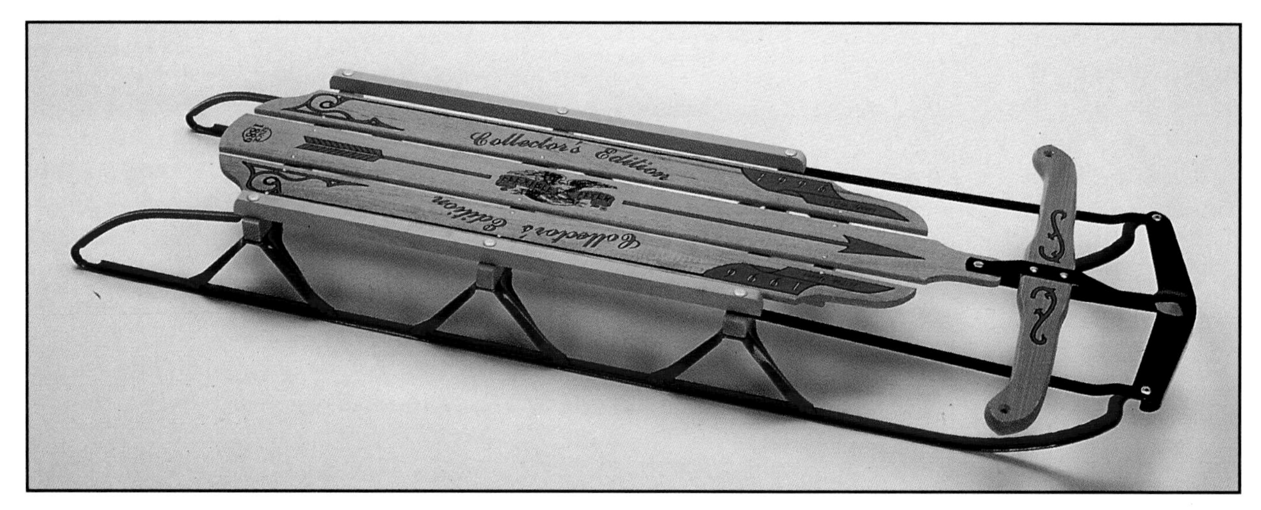

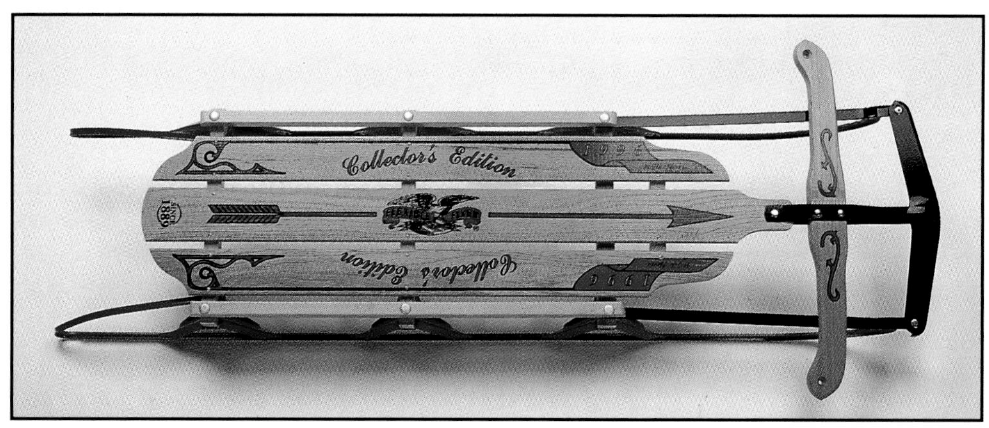

1996. Collector's Series. Prototype to be released in fall, 1996. L: 54", W: 12.5", H: 6".

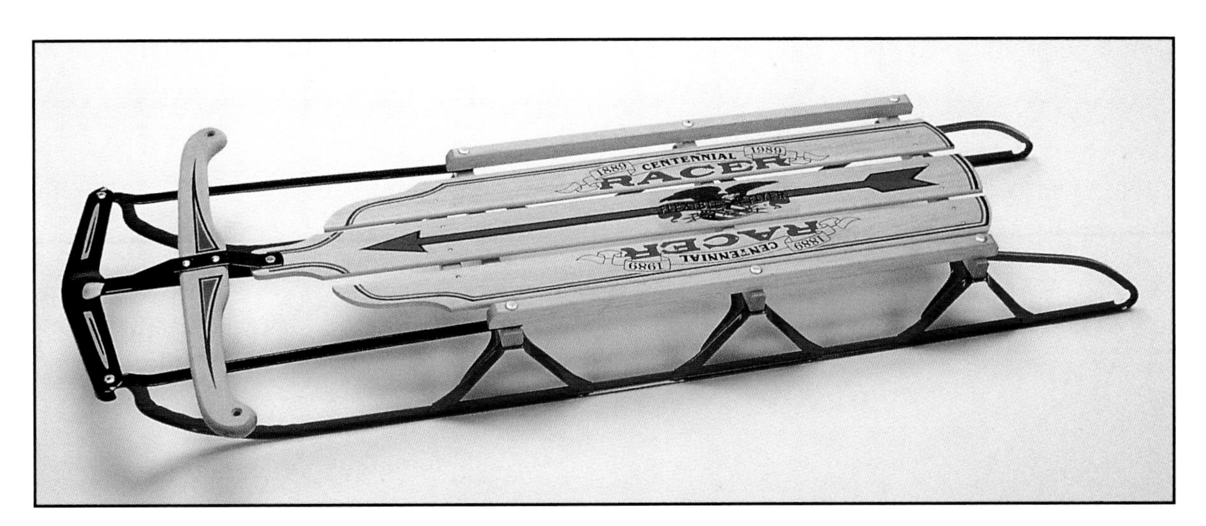

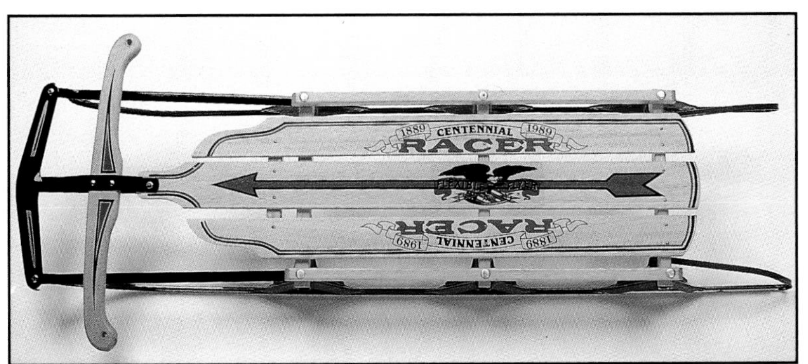

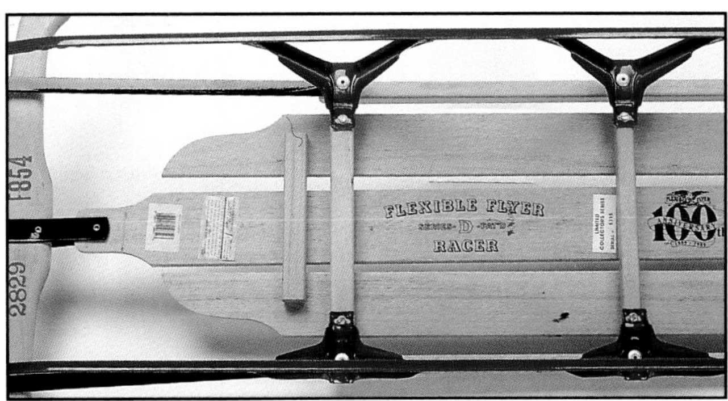

1989. Centennial Racer with box. $150-200, $250-350 mint in box.

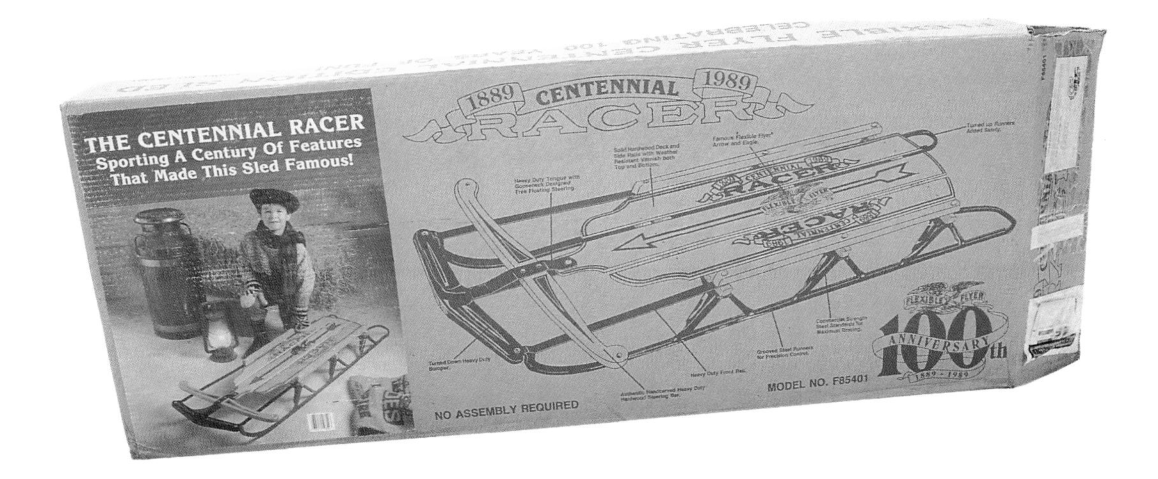

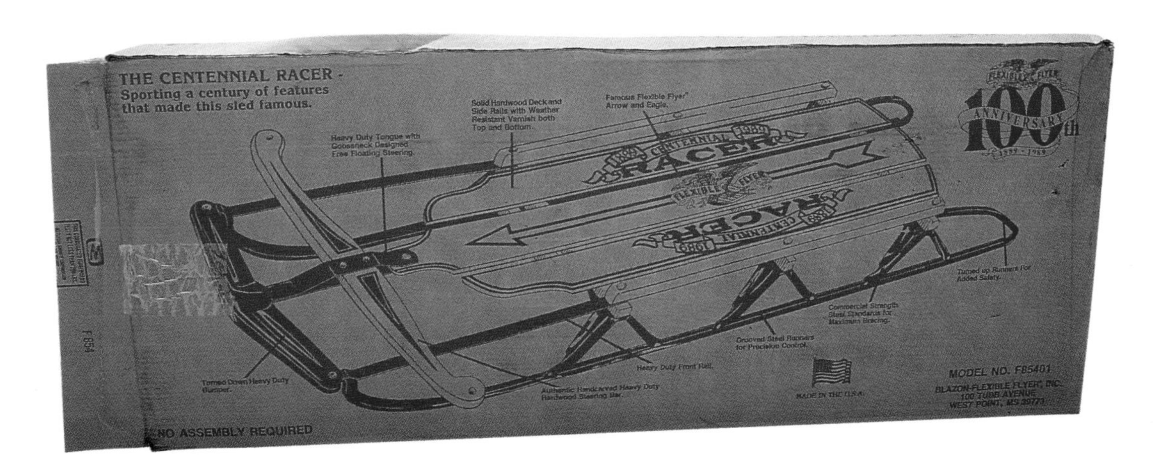

II. Other American Manufacturers

American Toy and Novelty Works - American Acme

The American Toy and Novelty Works was founded by Casper H. Oermann in 1917. Mr. Oermann was a prominent builder and contractor operating a lumber yard on West Poplar St. in York, Pennsylvania. The lumber business was seasonal and he needed a business to keep going all year.

Mr. Oermann established a toy business in 1917, calling it "The American Toy and Novelty Works."

In 1927, American Toy and Novelty Works merged with the Acme Wagon Company in Emigsville, York County, Pennsylvania. The new company was called American Acme.

According to the Gazette and Daily Newspaper, the American Acme Company was gutted by fire on April 2, 1950. Approximately a year and a half later on November 30, 1951, American Acme was ready to ship hundreds of new sleds for the Christmas Season.

Business operations were suspended some time later.

Chronology of American Toy and Novelty Works/ American Acme

1917: Company established by Casper H. Oermann. All sleds manufactured with hardwoods and nickel plated steel bumpers and grooved runners. Models included: Royal Plane, Royal Racer with spring bumpers, Monoplane, Speedplane, Snow King, and Eskimo

1927: Company merges with Acme Wagon Company and becomes American Acme Company, Emigsville, Pennsylvania. Models: Flexoplane #445, Polar Plane, Speed Plane, Rocket Plane, Ice Plane, Royal Plane, and Sky Plane.

April 2, 1950: Fire guts American Acme.

November 30, 1951: Hundreds of Sky Planes, Snow Kings, and Monoplanes ready for Christmas delivery.

19??: Operations Suspended.

Advertising

The advertising in this section is courtesy of the Agricultural and Industrial Museum of York County, Pennsylvania.

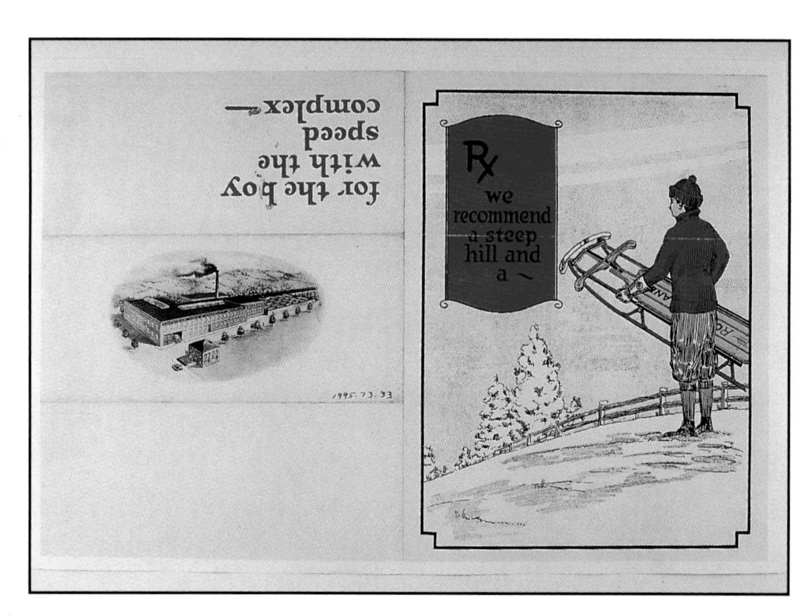

American Toy and Novelty Works brochure showing original factory, 1917.

American Toy and Novelty Works brochure, c. 1917, with the Speedplane.

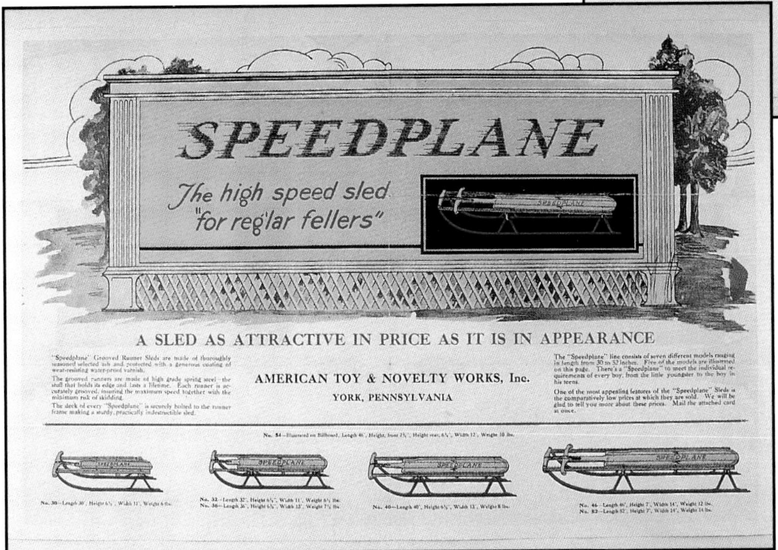

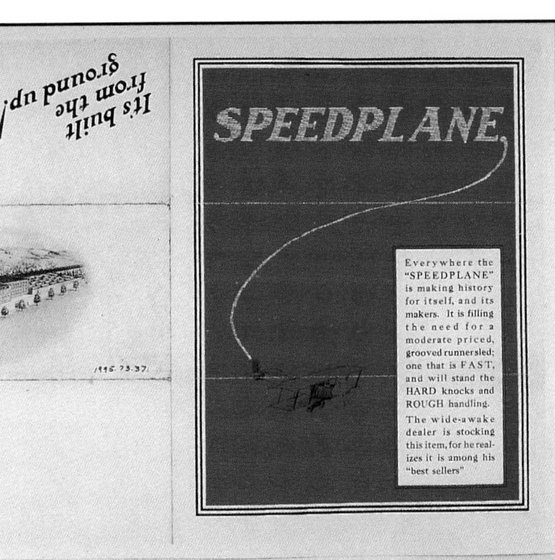

American Toy and Novelty Works brochure for Speedplane.

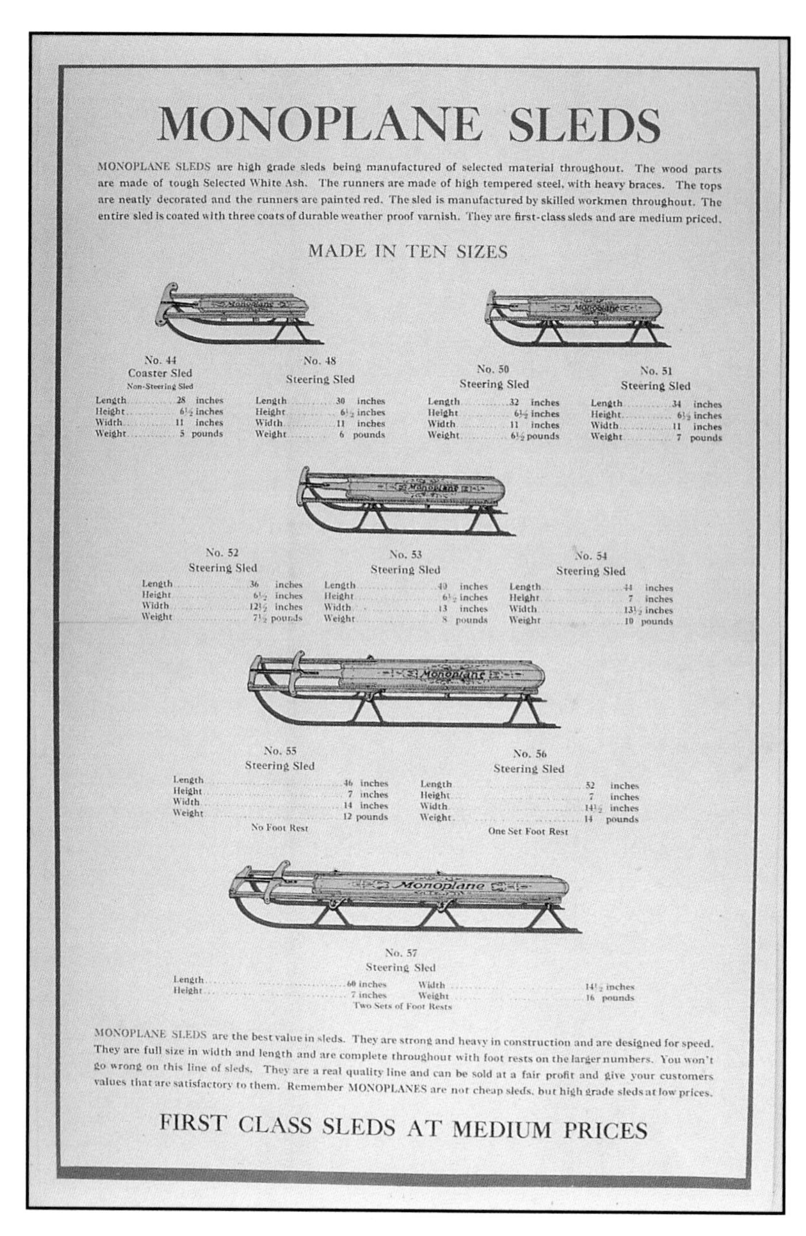

MONOPLANE SLEDS

MONOPLANE SLEDS are high grade sleds being manufactured of selected material throughout. The wood parts are made of tough Selected White Ash. The runners are made of high tempered steel, with heavy braces. The tops are neatly decorated and the runners are painted red. The sled is manufactured by skilled workmen throughout. The entire sled is coated with three coats of durable weather proof varnish. They are first-class sleds and are medium priced.

MADE IN TEN SIZES

No. 44
Coaster Sled
Non-Steering Sled

Length	28 inches
Height	6½ inches
Width	11 inches
Weight	5 pounds

No. 48
Steering Sled

Length	30 inches
Height	6½ inches
Width	11 inches
Weight	6 pounds

No. 50
Steering Sled

Length	32 inches
Height	6½ inches
Width	11 inches
Weight	6½ pounds

No. 51
Steering Sled

Length	34 inches
Height	6½ inches
Width	11 inches
Weight	7 pounds

No. 52
Steering Sled

Length	36 inches
Height	6½ inches
Width	12½ inches
Weight	7½ pounds

No. 53
Steering Sled

Length	40 inches
Height	6½ inches
Width	13 inches
Weight	8 pounds

No. 54
Steering Sled

Length	44 inches
Height	7 inches
Width	13½ inches
Weight	10 pounds

No. 55
Steering Sled

Length	46 inches
Height	7 inches
Width	14 inches
Weight	12 pounds

No Foot Rest

No. 56
Steering Sled

Length	52 inches
Height	7 inches
Width	14½ inches
Weight	14 pounds

One Set Foot Rest

No. 57
Steering Sled

Length	60 inches	Width	14½ inches
Height	7 inches	Weight	16 pounds

Two Sets of Foot Rests

MONOPLANE SLEDS are the best value in sleds. They are strong and heavy in construction and are designed for speed. They are full size in width and length and are complete throughout with foot rests on the larger numbers. You won't go wrong on this line of sleds. They are a real quality line and can be sold at a fair profit and give your customers values that are satisfactory to them. Remember MONOPLANES are not cheap sleds, but high grade sleds at low prices.

FIRST CLASS SLEDS AT MEDIUM PRICES

American Toy and Novelty Works brochure, c. 1920, with the Monoplane.

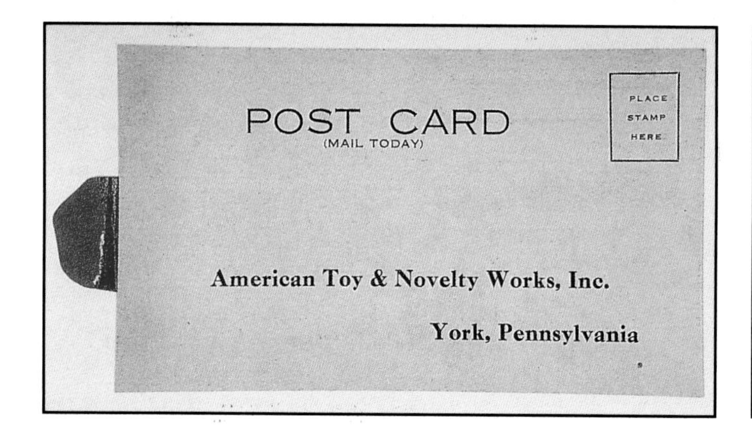

POST CARD
(MAIL TODAY)

PLACE
STAMP
HERE

American Toy & Novelty Works, Inc.

York, Pennsylvania

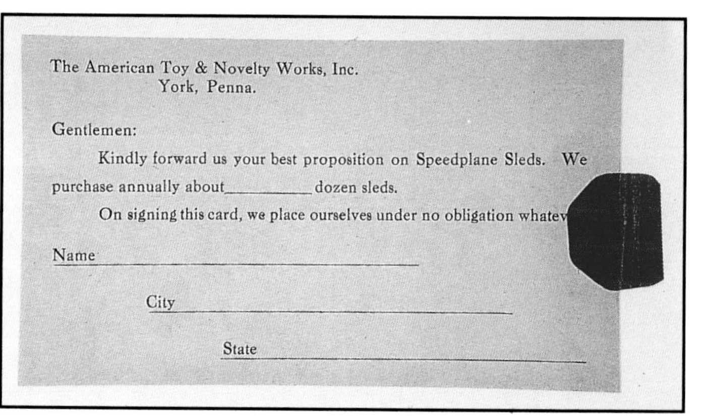

The American Toy & Novelty Works, Inc.
York, Penna.

Gentlemen:

Kindly forward us your best proposition on Speedplane Sleds. We purchase annually about__________ dozen sleds.

On signing this card, we place ourselves under no obligation whatev

Name

City

State

American Toy and Novelty Works mail back postcard for Speedplane, c. 1920s.

STROBEL-WILKEN COMPANY, INC.

IMPORTERS & MANUFACTURERS AGENTS

SNOW KING SLEDS
TOYS, DOLLS, NOVELTIES
AND HOUSE FURNISHING GOODS

33-35-37 EAST 17TH STREET
NEW YORK

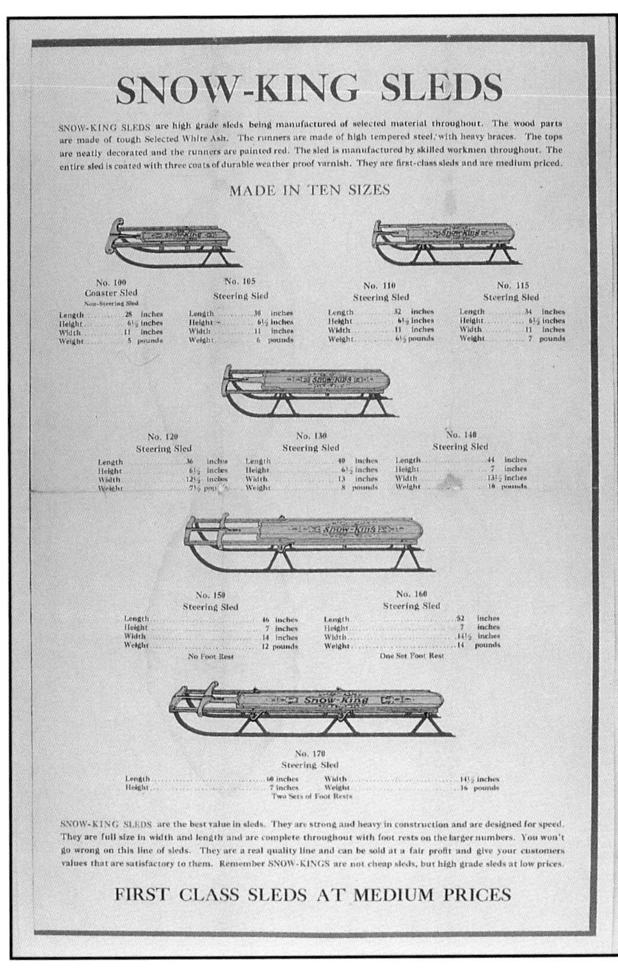
SNOW-KING SLEDS

SNOW-KING SLEDS are high grade sleds being manufactured of selected material throughout. The wood parts are made of tough Selected White Ash. The runners are made of high tempered steel, with heavy braces. The tops are neatly decorated and the runners are painted red. The sled is manufactured by skilled workmen throughout. The entire sled is coated with three coats of durable weather proof varnish. They are first-class sleds and are medium priced.

MADE IN TEN SIZES

No. 100
Coaster Sled
Non-Steering Sled
Length 28 inches
Height 6½ inches
Width 11 inches
Weight 5 pounds

No. 105
Steering Sled
Length 30 inches
Height 6½ inches
Width 11 inches
Weight 6 pounds

No. 110
Steering Sled
Length 32 inches
Height 6½ inches
Width 11 inches
Weight 6½ pounds

No. 115
Steering Sled
Length 34 inches
Height 6½ inches
Width 11 inches
Weight 7 pounds

No. 120
Steering Sled
Length 36 inches
Height 6½ inches
Width 12½ inches
Weight 7½ pounds

No. 130
Steering Sled
Length 40 inches
Height 6½ inches
Width 13 inches
Weight 8 pounds

No. 140
Steering Sled
Length 44 inches
Height 7 inches
Width 13½ inches
Weight 10 pounds

No. 150
Steering Sled
Length 46 inches
Height 7 inches
Width 14 inches
Weight 12 pounds
No Foot Rest

No. 160
Steering Sled
Length 52 inches
Height 7 inches
Width 14½ inches
Weight 14 pounds
One Set Foot Rest

No. 170
Steering Sled
Length 60 inches
Height 7 inches
Width 14½ inches
Weight 16 pounds
Two Sets of Foot Rests

SNOW-KING SLEDS are the best value in sleds. They are strong and heavy in construction and are designed for speed. They are full size in width and length and are complete throughout with foot rests on the larger numbers. You won't go wrong on this line of sleds. They are a real quality line and can be sold at a fair profit and give your customers values that are satisfactory to them. Remember SNOW-KINGS are not cheap sleds, but high grade sleds at low prices.

FIRST CLASS SLEDS AT MEDIUM PRICES

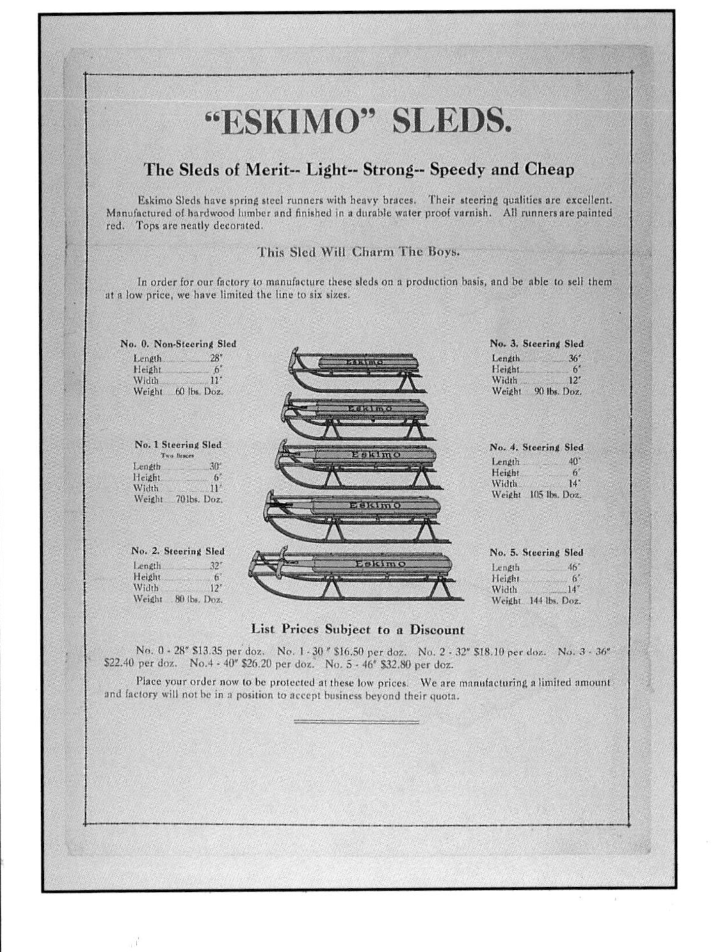
"ESKIMO" SLEDS.

The Sleds of Merit-- Light-- Strong-- Speedy and Cheap

Eskimo Sleds have spring steel runners with heavy braces. Their steering qualities are excellent. Manufactured of hardwood lumber and finished in a durable water proof varnish. All runners are painted red. Tops are neatly decorated.

This Sled Will Charm The Boys.

In order for our factory to manufacture these sleds on a production basis, and be able to sell them at a low price, we have limited the line to six sizes.

No. 0. Non-Steering Sled
Length 28"
Height 6"
Width 11"
Weight 60 lbs. Doz.

No. 1 Steering Sled
Two Braces
Length 30"
Height 6"
Width 11"
Weight 70lbs. Doz.

No. 2. Steering Sled
Length 32"
Height 6"
Width 12"
Weight 80 lbs. Doz.

No. 3. Steering Sled
Length 36"
Height 6"
Width 12"
Weight 90 lbs. Doz.

No. 4. Steering Sled
Length 40"
Height 6"
Width 14"
Weight 105 lbs. Doz.

No. 5. Steering Sled
Length 46"
Height 6"
Width 14"
Weight 144 lbs. Doz.

List Prices Subject to a Discount

No. 0 - 28" $13.35 per doz. No. 1 - 30" $16.50 per doz. No. 2 - 32" $18.10 per doz. No. 3 - 36" $22.40 per doz. No.4 - 40" $26.20 per doz. No. 5 - 46" $32.80 per doz.

Place your order now to be protected at these low prices. We are manufacturing a limited amount and factory will not be in a position to accept business beyond their quota.

American Toy and Novelty Works brochure from Strobel-Wilken Co., an importer, mentioning Snow King and Eskimo, c. 1920.

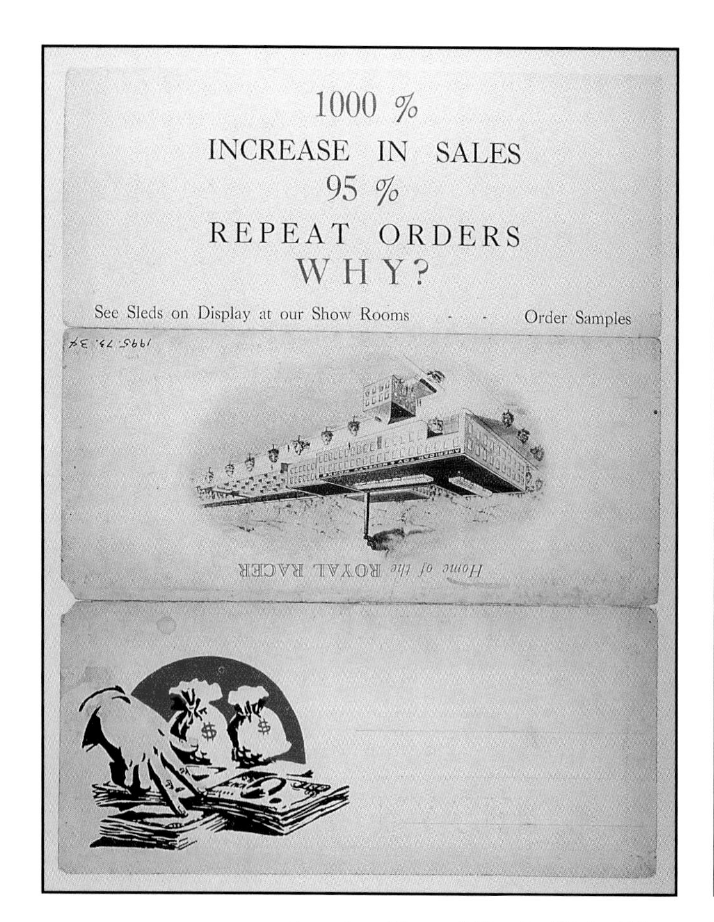

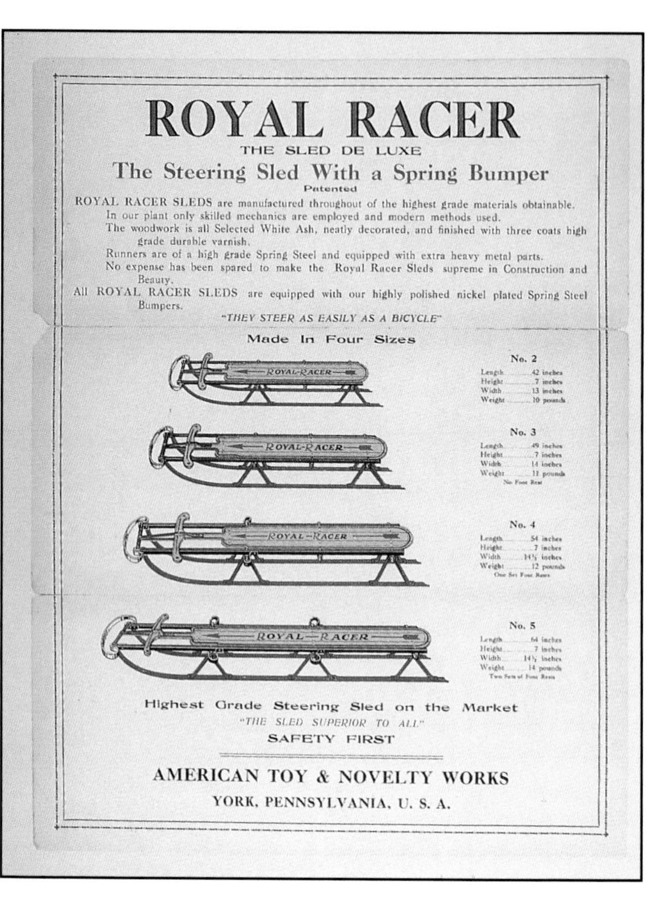

American Toy and Novelty Works "Home of the Royal Racer" brochure c. 1924.

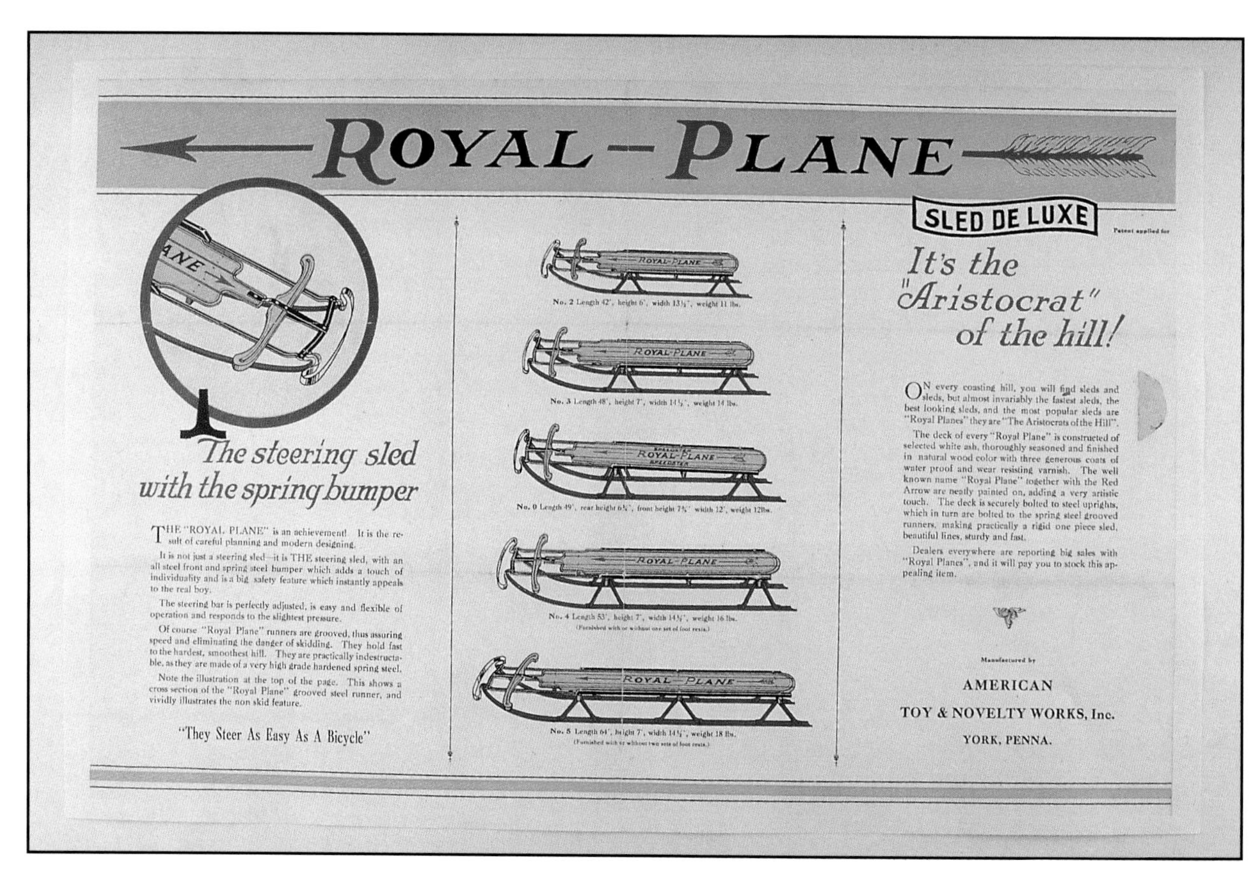

Royal Plane, c. 1925.

ISSUE Dec

Proof from

BOYS' LIFE

THE BOY SCOUTS' MAGAZINE

200 FIFTH AVENUE, NEW YORK

SPACE 106

Corrections for this advertisement cannot be made unless they reach our office before our closing date. Inside forms close on the 25th of the second month preceding date of issue; covers and color forms close on the 10th, two weeks earlier.

PROOFS AS INSERTED

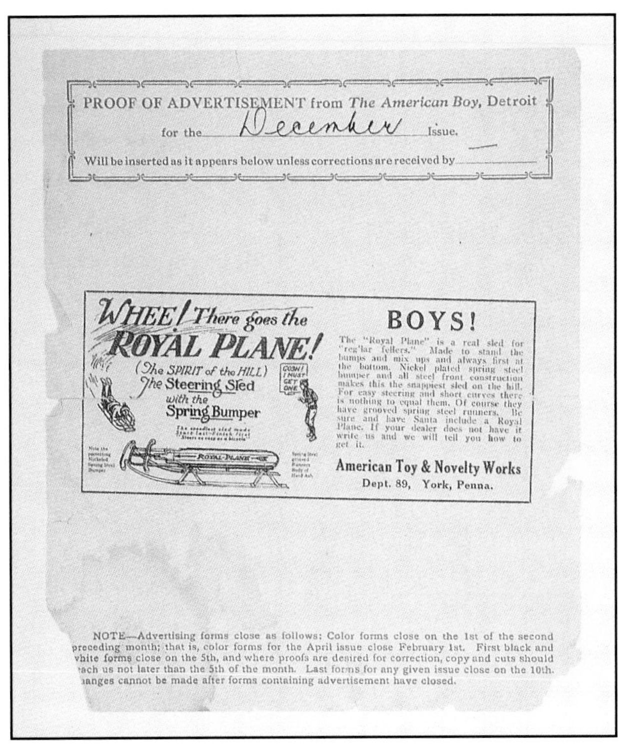

PROOF OF ADVERTISEMENT from *The American Boy*, Detroit

for the December Issue.

Will be inserted as it appears below unless corrections are received by

NOTE—Advertising forms close as follows: Color forms close on the 1st of the second preceding month; that is, color forms for the April issue close February 1st. First black and white forms close on the 5th, and where proofs are desired for correction, copy and cuts should reach us not later than the 5th of the month. Last forms for any given issue close on the 10th. Changes cannot be made after forms containing advertisement have closed.

American Toy and Novelty Works 1924 advertisement for Royal Plane. *Boys' Life* proof

American Toy and Novelty Works 1925 *Boys' Life* proof for Royal Plane advertisement.

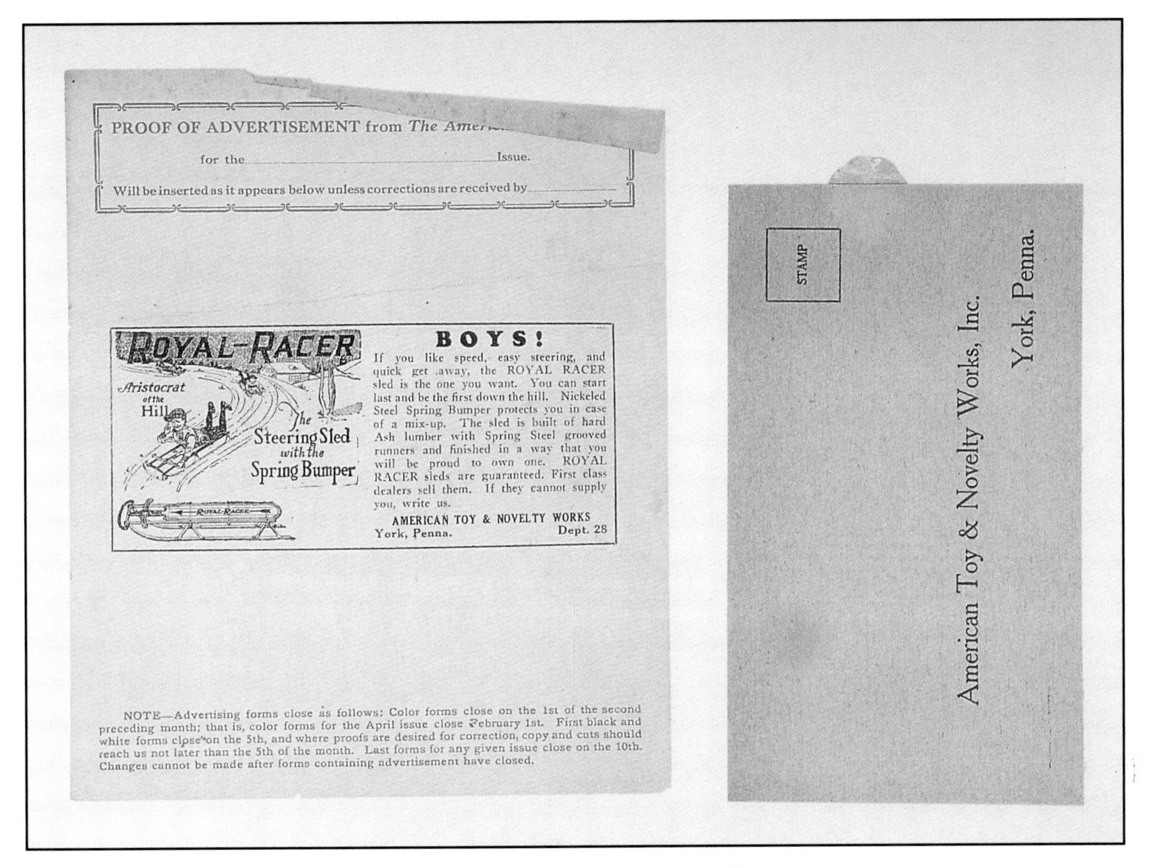

PROOF OF ADVERTISEMENT from *The Amer...*

for the Issue.

Will be inserted as it appears below unless corrections are received by

NOTE—Advertising forms close as follows: Color forms close on the 1st of the second preceding month; that is, color forms for the April issue close February 1st. First black and white forms close on the 5th, and where proofs are desired for correction, copy and cuts should reach us not later than the 5th of the month. Last forms for any given issue close on the 10th. Changes cannot be made after forms containing advertisement have closed.

American Toy and Novelty Works 1925 *Boys' Life* proof for Royal Racer advertisement.

Advertisement for the Royal-Racer, Manufactured by the American Toy & Novelty Works, York, Pennsylvania. c. 1925.

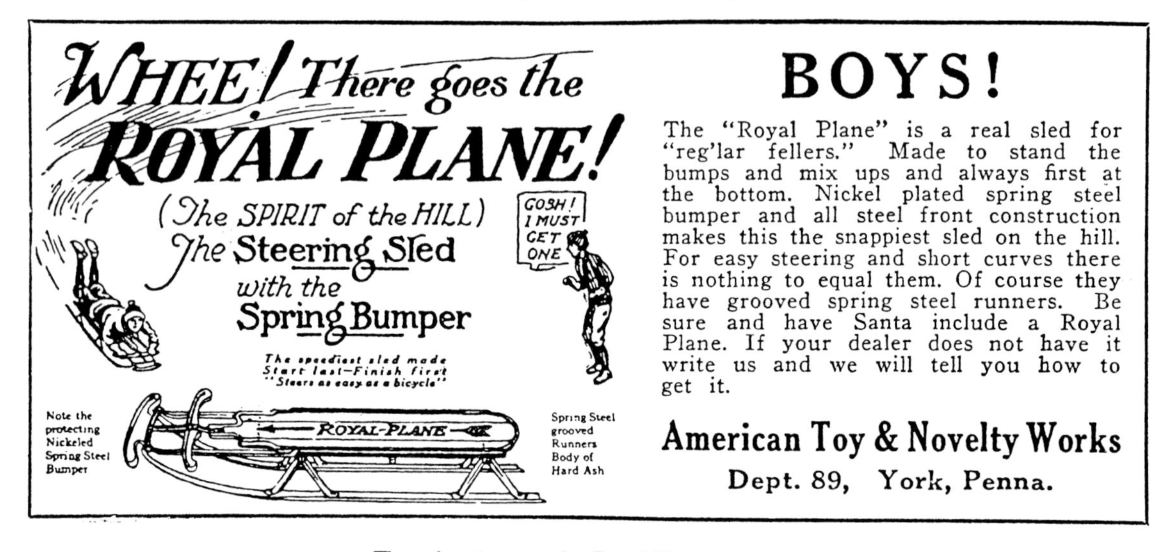

The advertisement for Royal Plane, as it ran.

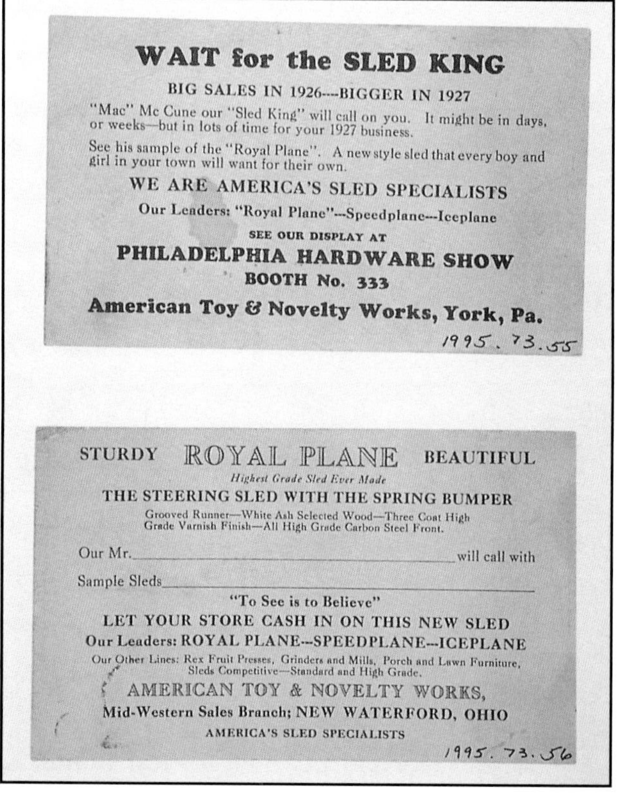

American Toy and Novelty Works postcard request for samples. It has a 1926 advertisement on the back.

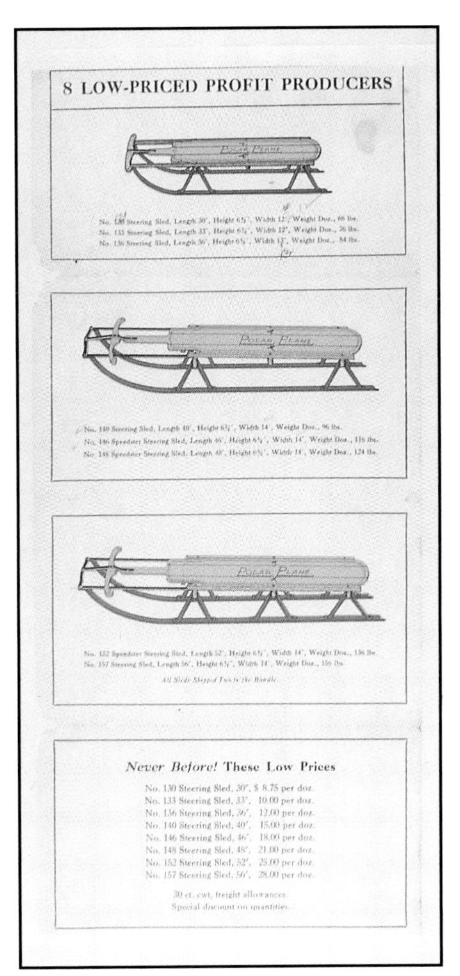

American Acme brochure, c. 1926, for Polar Plane.

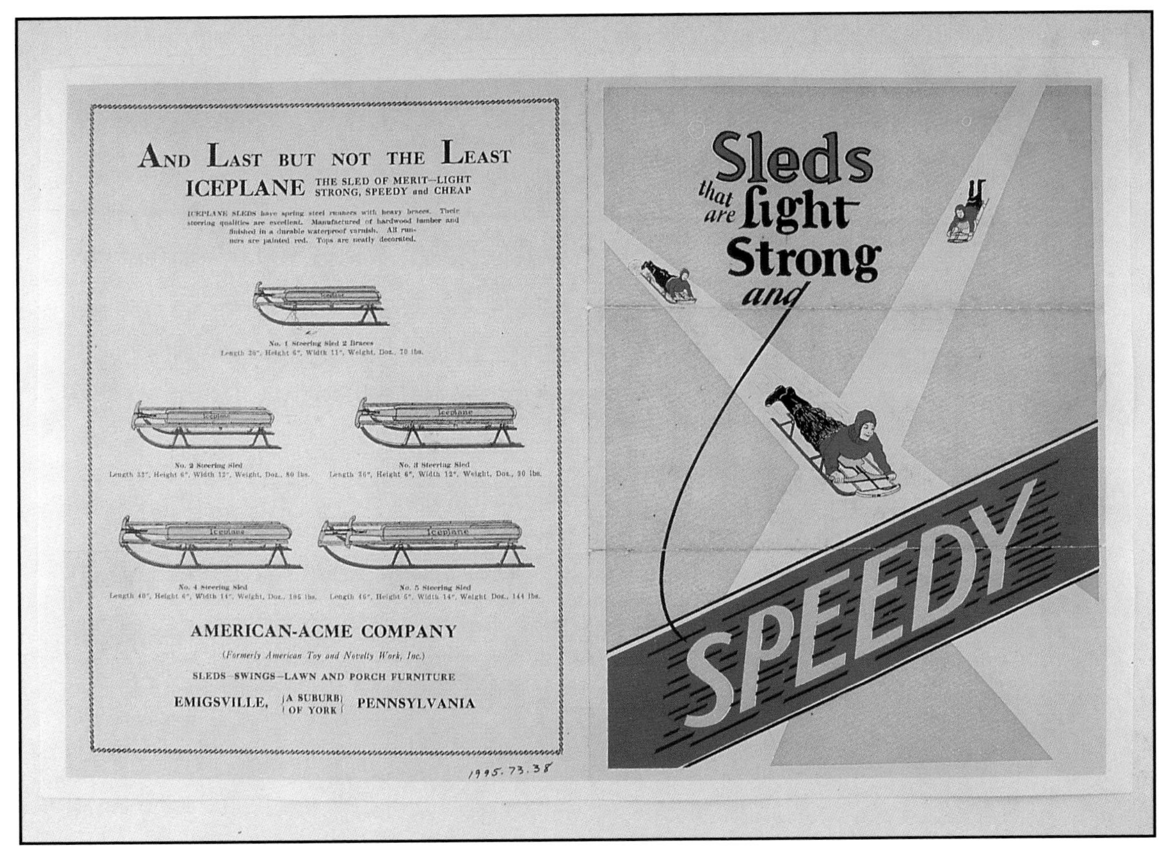

Speedy, c. 1927, featuring the Ice Plane.

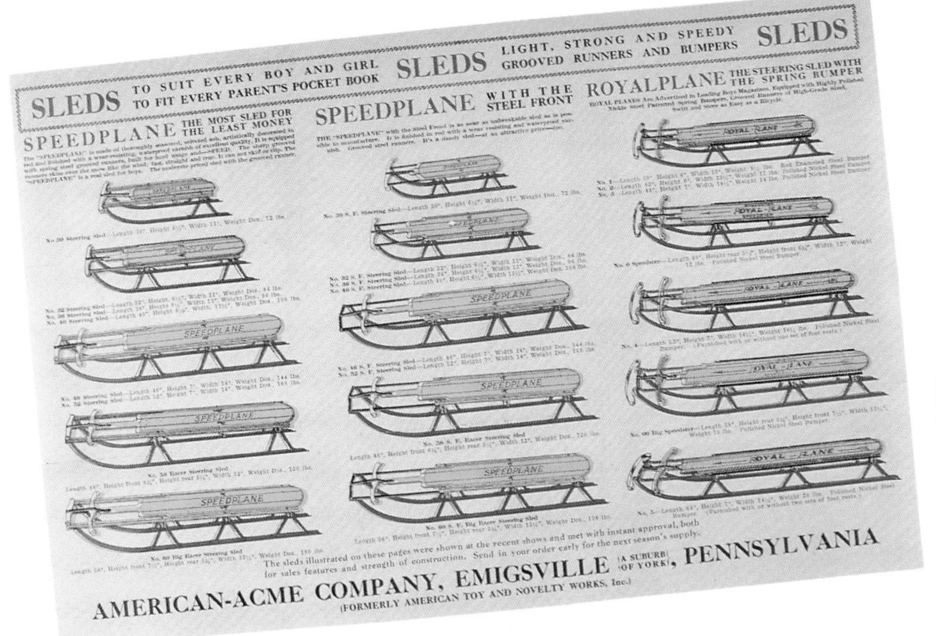

American Acme, c. 1927 with Speed Plane and Royal Plane with spring bumper.

"A Preview of Sleds." American Acme catalog, 1960.

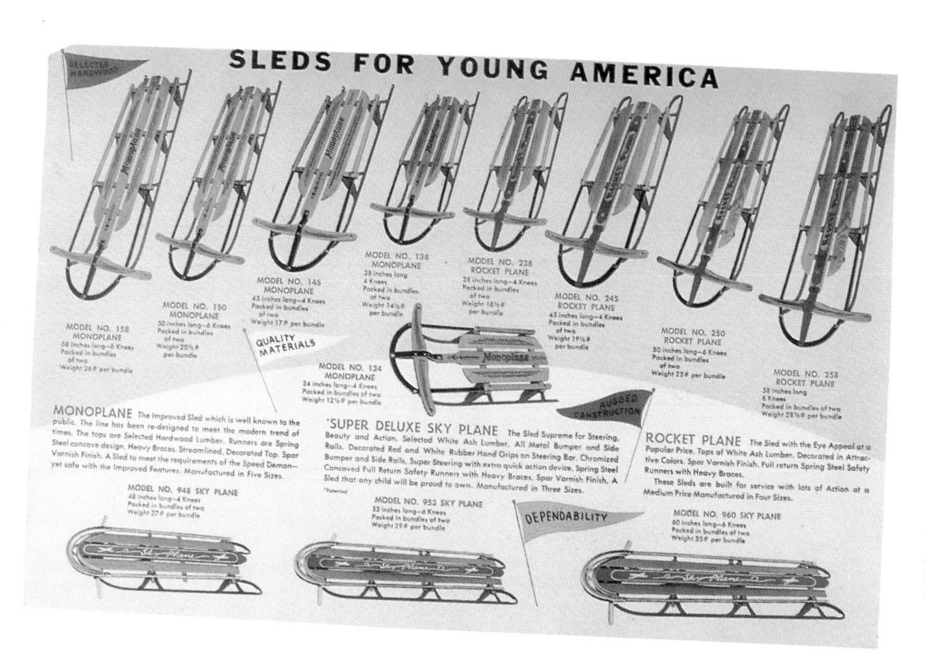

"Sleds for Young America" featuring the Monoplane, Rocket Plane and Sky Plane.

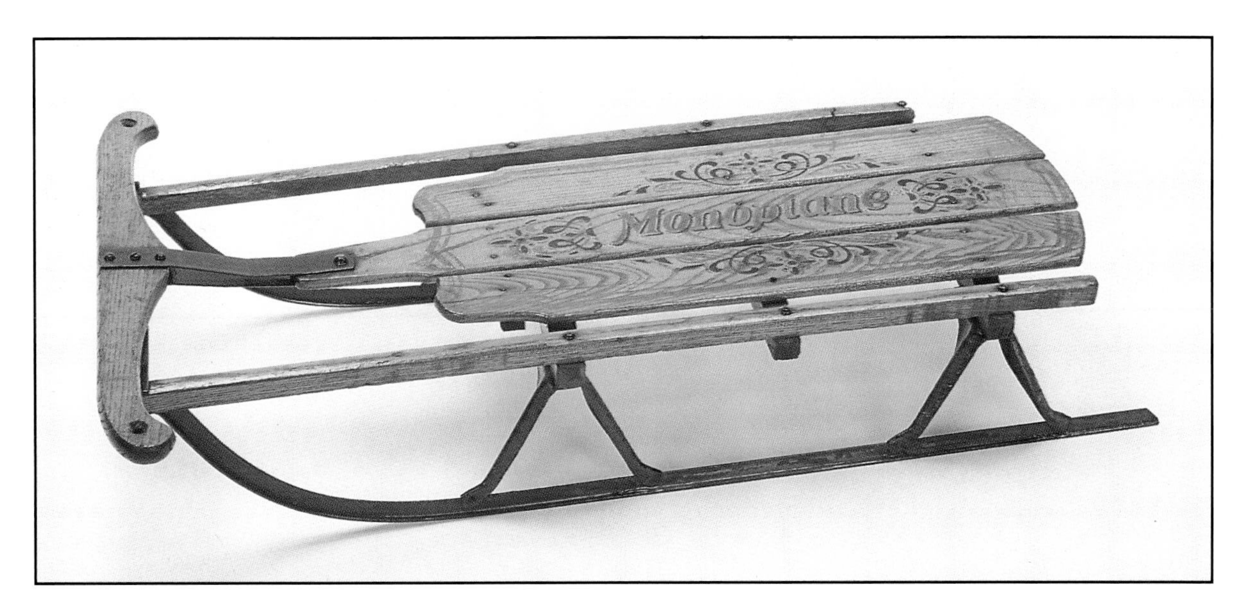

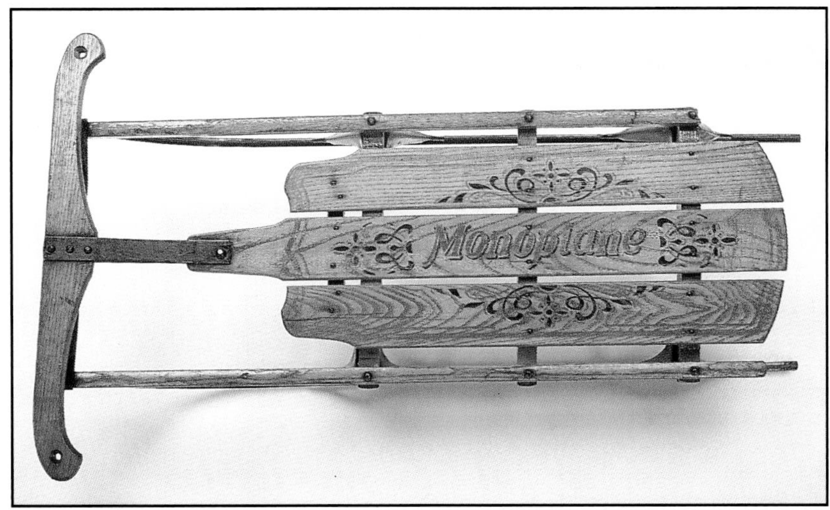

1917. Monoplane. #51, American Toy & Novelty Works. L: 34", W: 11", H: 6.5". $75-150.

Royal Sky Plane. Possibly by the American Acme Co., home of Royal Plane and Sky Plane. No advertisement has been found to confirm this. L: 48", W: 13", H: 7". $50-100. Continued on Top of Next Page.

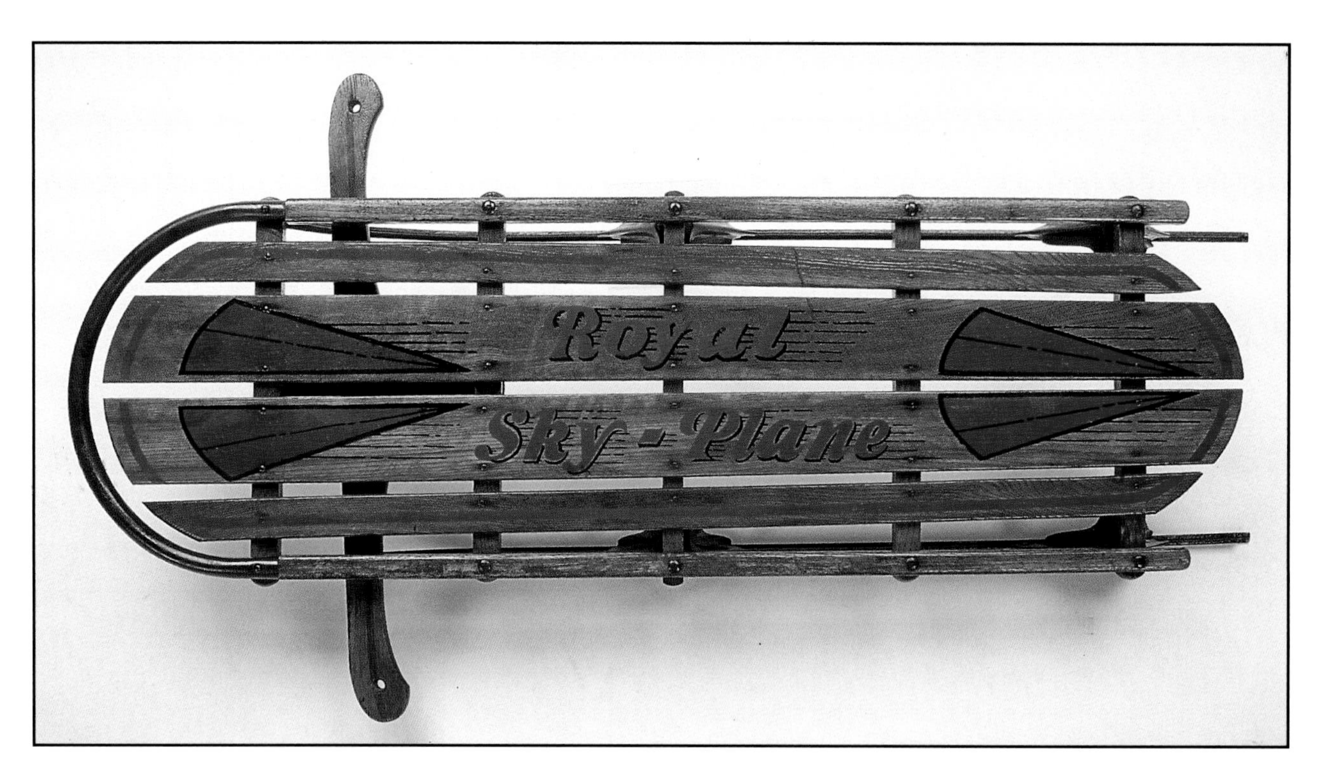

Continued from previous page.

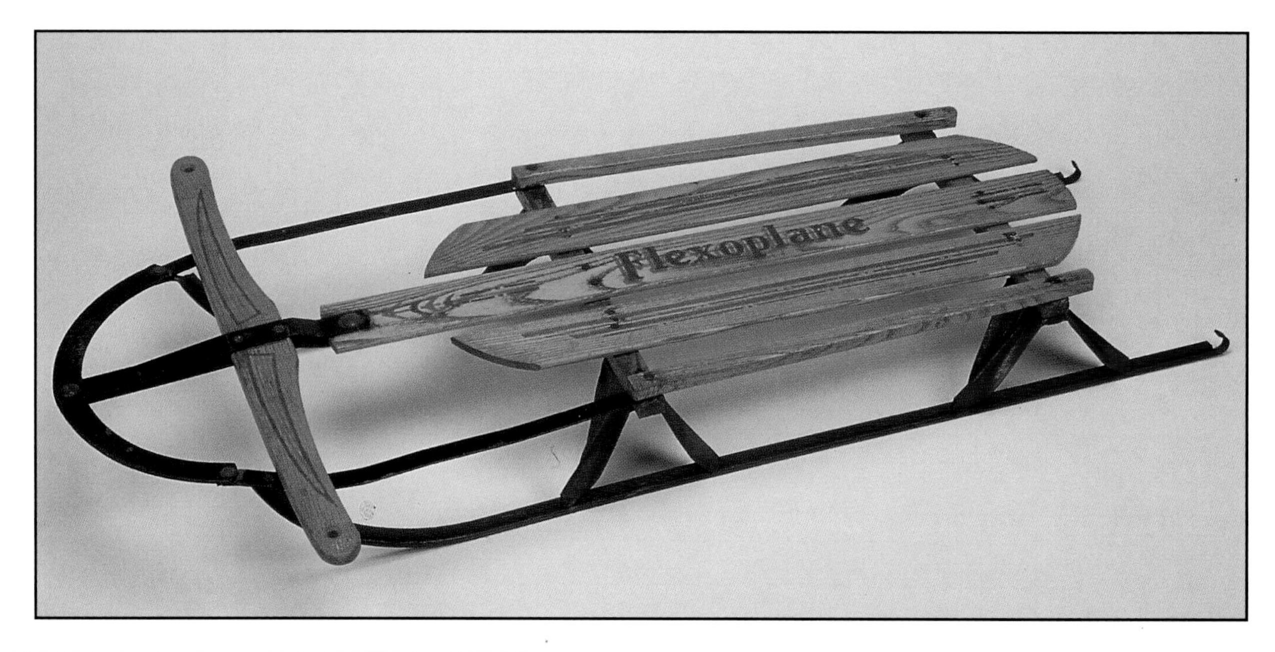

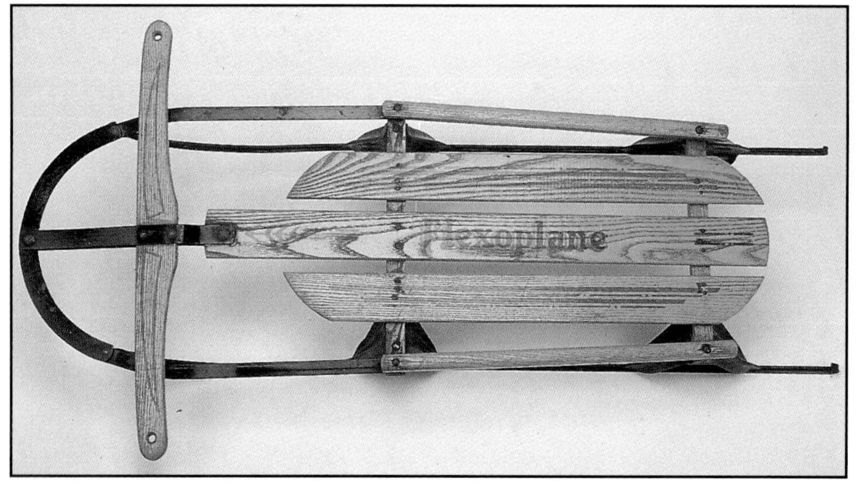

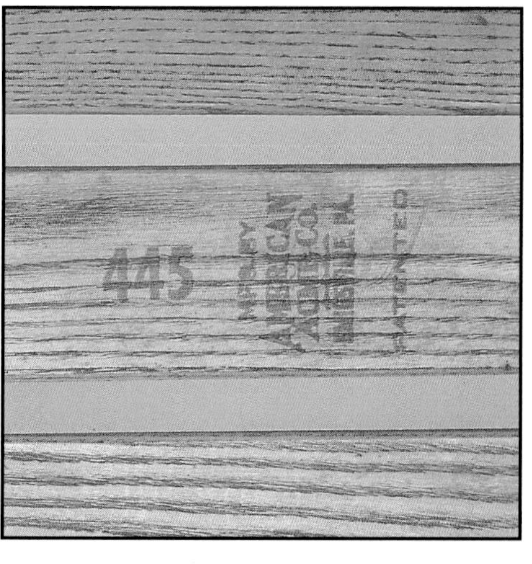

American Acme Flexoplane. $50-75.

Buffalo Sled Company - Auto Wheel Coaster

A Chronology

1904: Founded by John Snyder.
1910: Manufactured Gliderole, The Roller Sled.
1921: Became Auto Wheel Coaster Company, 95 Schenck Street North Tonawanda, New York. Makers of Fleetwing, Fleetwing Racer, Coaster King, Fleetwing Flash (Spring top) and Fleetwing Bobs. Sleds made of selected Northern Ash.
1964: Auto Wheel Coaster Company sold its last products and closed their doors due to bankruptcy.
1972: May 29 Fire destroyed building on Schenck Street

Advertising

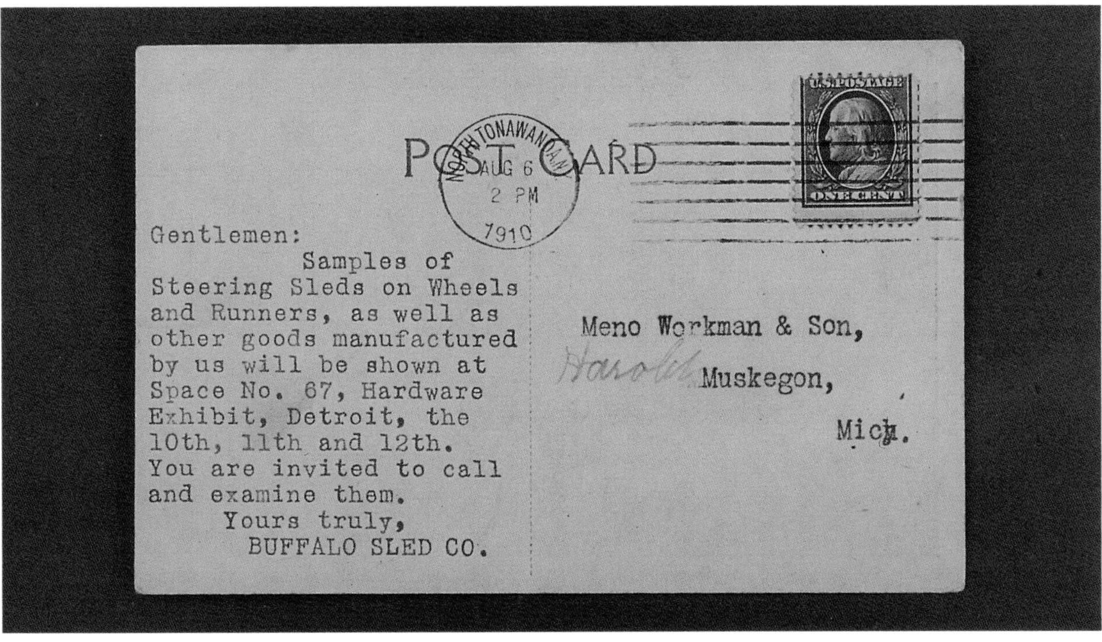

Postcard from the Buffalo Sled Company, 1910. 3.5" x 5.5". $50-100.

RICE LEADERS
OF THE WORLD ASSOCIATION

This Certifies that

THE BUFFALO SLED COMPANY
of North Tonawanda, New York, U.S.A.
Manufacturers of
THE AUTO WHEEL COASTER & CONVERTIBLE ROADSTER

Has joined in the work of this Association to co-operate in keeping before the public mind those high standards of "Business Principles" and "Merchandise Quality" which constitute the Association's

QUALIFICATIONS FOR MEMBERSHIP

HONOR A recognized reputation for fair and honorable business dealings

QUALITY An honest product of quality truthfully represented

STRENGTH A responsible and substantial financial standing

SERVICE A recognized reputation for conducting business in prompt and efficient manner

IN RECOGNITION OF WHICH this Association of eminent institutions has issued this

RECORD OF BUSINESS PRINCIPLES
as a tribute to this Member's recognized adherence to these high standards

This Member uses the Association Emblem as an evidence of co-operation in this work and also as a symbol of the spirit of integrity which governs its activities

IN WITNESS WHEREOF the Rice Leaders of the World Association has caused its official Seal to be affixed and attested by its Founder and President this 7th day of Feb A.D. 1920 at
358 Fifth Avenue New York U.S.A.
RICE LEADERS OF THE WORLD ASSOCIATION
Elwood E. Rice
FOUNDER AND PRESIDENT

Copy of a 1920 tribute to The Buffalo Sled Company. *Courtesy of the Historic Society of the Tonawandas.*

WHEE!

WHAT A BOB SLED

Here is something new in sliding down hill—steers like an auto, holds three kids or two grown ups, an goes like the wind. The most exciting thing on the hill, and boy can it break records! If you cannot get this Bob Sled from your Dealer, you can have it by writing to AUTO-WHEEL COASTER COMPANY, INC., North Tonawanda, N. Y., and sending $10.00. (If you live west of the Mississippi River, $12.00). If you want the thrill that comes once in a life time just ask Dad for the money, put it in an envelope and send it off to us as soon as possible, and the first thing you know you will be on your way down the steepest hill in the neighborhood. Lickity-Split. Send $10.00. No Express to pay.

AUTO-WHEEL COASTER COMPANY, INC.
North Tonawanda, New York

Boys' Life ad, 1934, Fleetwing bob

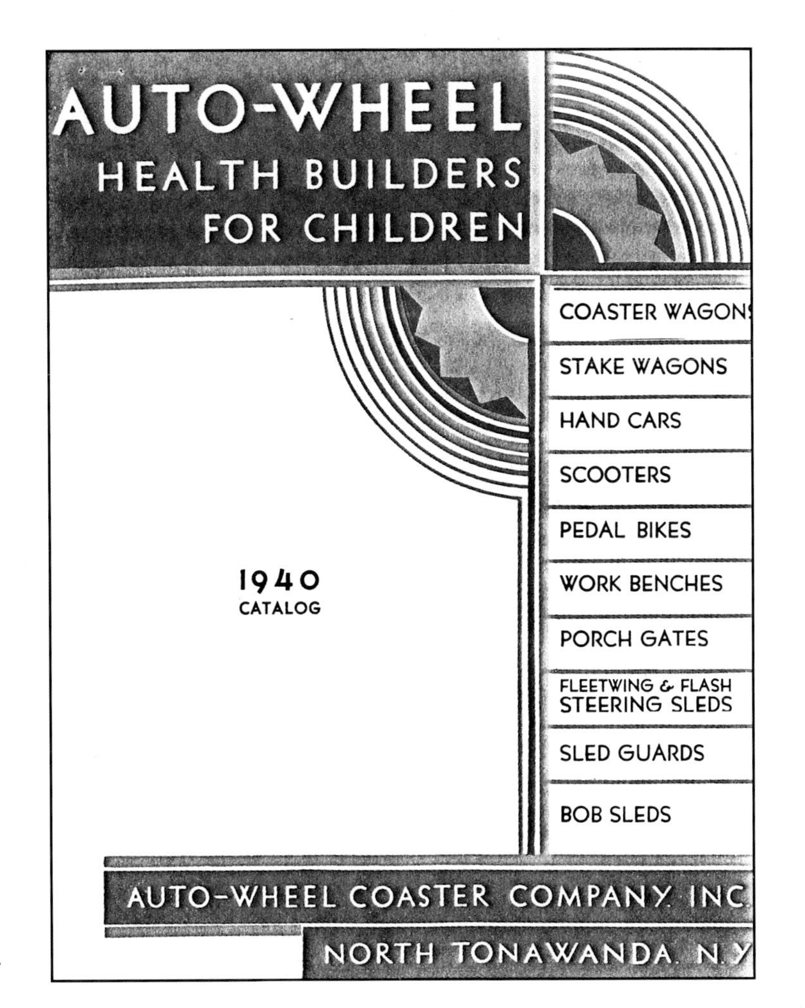

Auto-Wheel 1940 catalog.
Courtesy of the Historic Society of the Tonawandas.

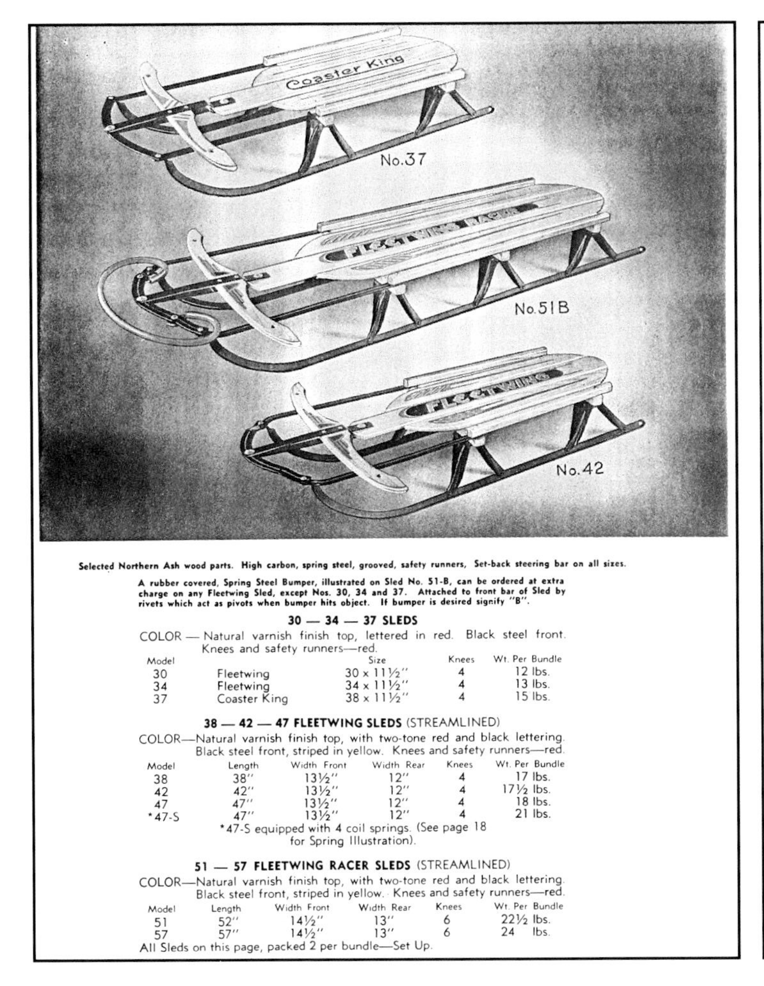

Selected Northern Ash wood parts. High carbon, spring steel, grooved, safety runners, Set-back steering bar on all sizes.

A rubber covered, Spring Steel Bumper, illustrated on Sled No. 51-B, can be ordered at extra charge on any Fleetwing Sled, except Nos. 30, 34 and 37. Attached to front bar of Sled by rivets which act as pivots when bumper hits object. If bumper is desired signify "B".

30 — 34 — 37 SLEDS

COLOR — Natural varnish finish top, lettered in red. Black steel front. Knees and safety runners—red.

Model		Size	Knees	Wt. Per Bundle
30	Fleetwing	30 x 11½"	4	12 lbs.
34	Fleetwing	34 x 11½"	4	13 lbs.
37	Coaster King	38 x 11½"	4	15 lbs.

38 — 42 — 47 FLEETWING SLEDS (STREAMLINED)

COLOR—Natural varnish finish top, with two-tone red and black lettering. Black steel front, striped in yellow. Knees and safety runners—red.

Model	Length	Width Front	Width Rear	Knees	Wt. Per Bundle
38	38"	13½"	12"	4	17 lbs.
42	42"	13½"	12"	4	17½ lbs.
47	47"	13½"	12"	4	18 lbs.
*47-S	47"	13½"	12"	4	21 lbs.

*47-S equipped with 4 coil springs. (See page 18 for Spring Illustration).

51 — 57 FLEETWING RACER SLEDS (STREAMLINED)

COLOR—Natural varnish finish top, with two-tone red and black lettering. Black steel front, striped in yellow. Knees and safety runners—red.

Model	Length	Width Front	Width Rear	Knees	Wt. Per Bundle
51	52"	14½"	13"	6	22½ lbs.
57	57"	14½"	13"	6	24 lbs.

All Sleds on this page, packed 2 per bundle—Set Up.

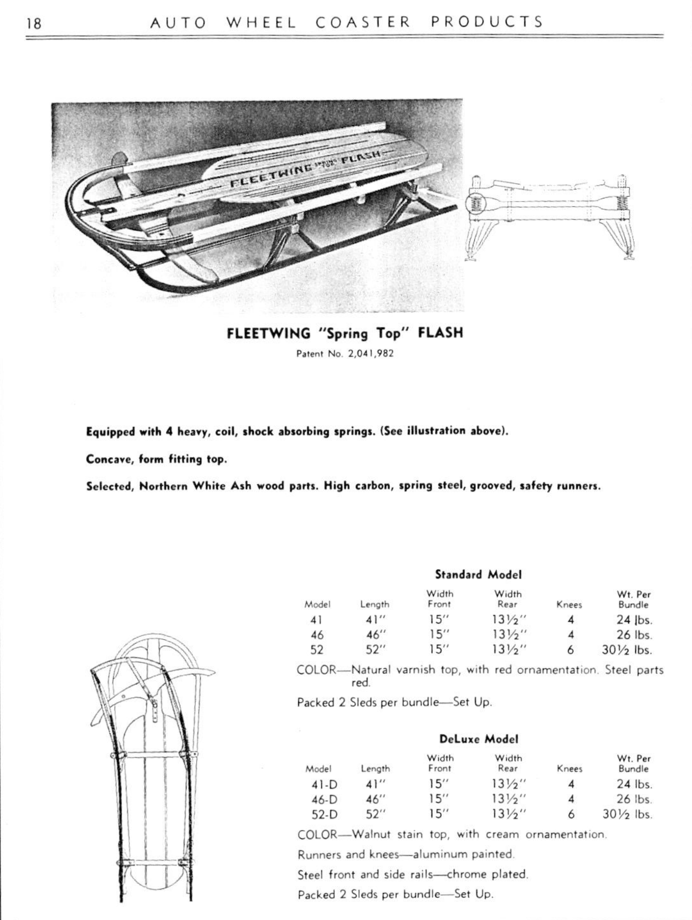

18 AUTO WHEEL COASTER PRODUCTS

FLEETWING "Spring Top" FLASH

Patent No. 2,041,982

Equipped with 4 heavy, coil, shock absorbing springs. (See illustration above).

Concave, form fitting top.

Selected, Northern White Ash wood parts. High carbon, spring steel, grooved, safety runners.

Standard Model

Model	Length	Width Front	Width Rear	Knees	Wt. Per Bundle
41	41"	15"	13½"	4	24 lbs.
46	46"	15"	13½"	4	26 lbs.
52	52"	15"	13½"	6	30½ lbs.

COLOR—Natural varnish top, with red ornamentation. Steel parts red.

Packed 2 Sleds per bundle—Set Up.

DeLuxe Model

Model	Length	Width Front	Width Rear	Knees	Wt. Per Bundle
41-D	41"	15"	13½"	4	24 lbs.
46-D	46"	15"	13½"	4	26 lbs.
52-D	52"	15"	13½"	6	30½ lbs.

COLOR—Walnut stain top, with cream ornamentation.

Runners and knees—aluminum painted.

Steel front and side rails—chrome plated.

Packed 2 Sleds per bundle—Set Up.

HEALTH BUILDERS FOR CHILDREN 19

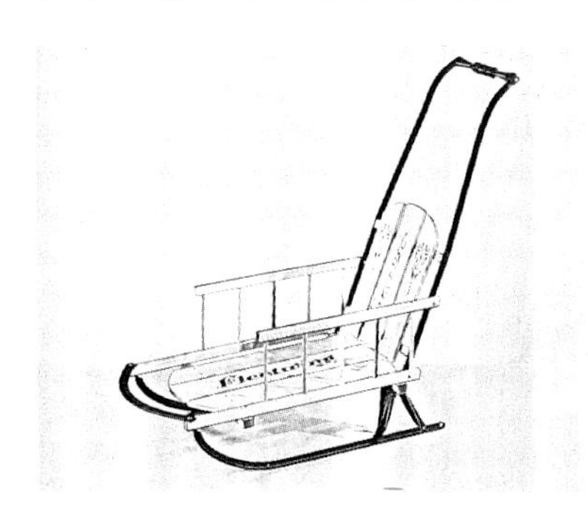

118 SLED STROLLER

Selected White Ash wood parts, natural varnish finish. Tee Steel runners, enameled bright red. Black steel handle. Black wood grip.

SIZE—Length — 38"; Width — 14"; Height of back to top of handle—33".

Inside height—12¼"; Height of back rest—14".

WEIGHT—112 lbs. per dozen. Packed 2 per bundle, with handle K.D.

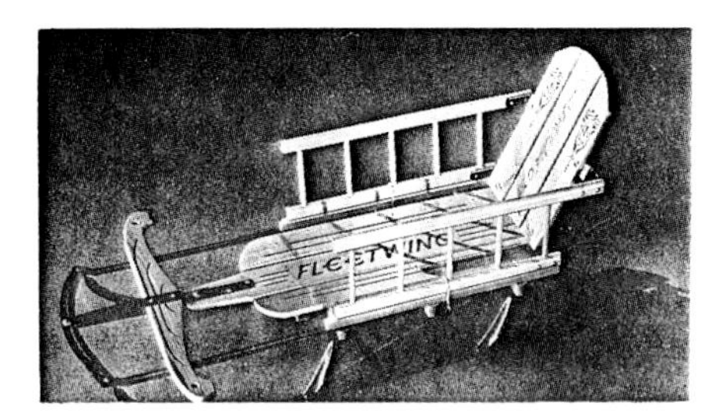

18 ADJUSTABLE SLED GUARD

Selected hard wood, natural varnish finish, ornamented in green and red.

SIZE—Sides — 15½" long, 7" high. Back 14" high.

Will fit all standard size Sleds. Fastened to Sled by 2 U-Clamps.

WEIGHT—4 lbs. each. Packed 1 per carton.

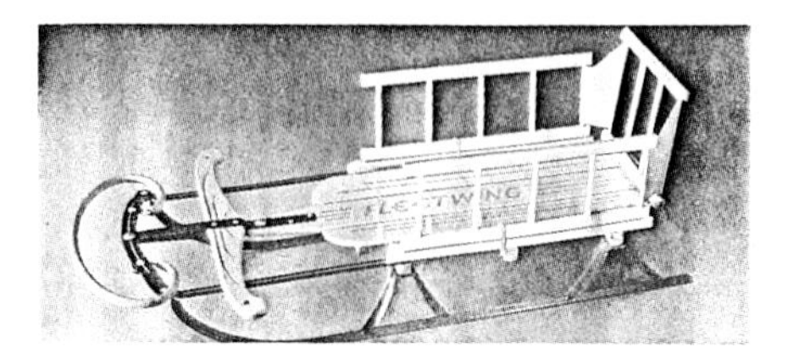

018 ADJUSTABLE SLED GUARD

Wood parts selected hard wood, natural varnish finish. Metal corner hinges enameled bright red.

SIZE—Sides — 18½" long, 7" high. Back—9½" high.

Will fit all standard size Sleds. Fastened to Sled by 2 U-Clamps.

WEIGHT—31 lbs. per carton. Packed 6 per carton.

20 AUTO WHEEL COASTER PRODUCTS

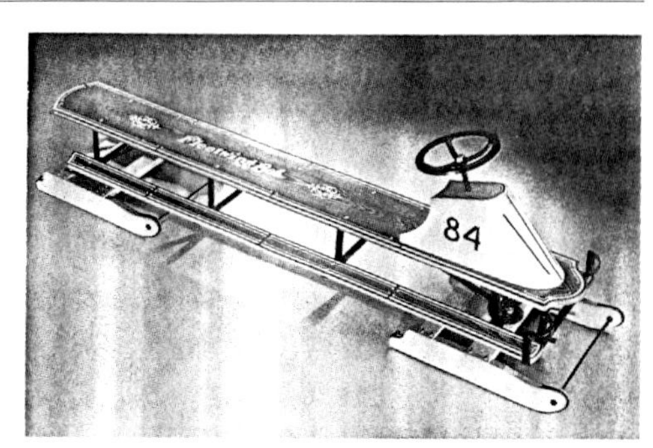

Selected Northern Ash Top and Foot Rests. "Knee Action", Hard Maple Bobs with ⅜" round steel runners. Sheet Steel hood.

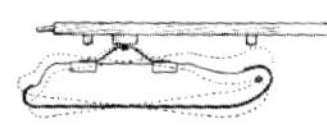

Knee Action

84 FLEETWING BOB SLED

SIZE — Length of top 80"; Width 16½"; Length overall 84".
COLOR — Top and foot rests stained green and varnished, with orange ornamentation and lettering. Orange hood and Bobs. Black steering wheel and gear.
WEIGHT — 74 lbs. each. Will hold 4 adults comfortably.

Packed in 2 cartons—easily assembled. Full assembly instructions with each Sled.

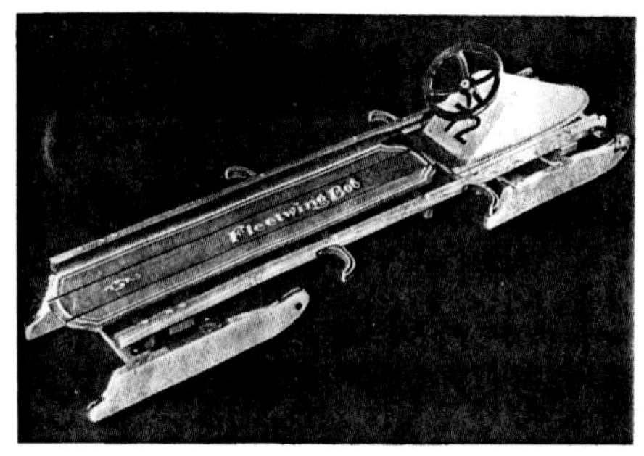

Selected Northern Ash Top. Steel foot rests. "Knee Action", Hard Maple Bobs with ⅜" round steel runners. Sheet Steel Hood.

Powerful Brake for Coasting Safety

The illustration below shows the wide saw tooth scraper blade attached to the rear bob. This blade, actuated by the hand lever, exerts a powerful downward pressure against the ice and snow. A safety feature that will be appreciated.

72 FLEETWING BOB SLED

SIZE —Length of top—63"; Width 14½"; Length overall 72".
COLOR — Top stained green, with orange lettering and striping. Hood and Bobs enameled orange. Black steering wheel and gear.
WEIGHT — 55 lbs. each. Will hold 3 adults comfortably.

Packed in 2 cartons—easily assembled.
Full assembly instructions with each Sled.

Sleds

C. 1940. Fleetwing Racer, Autowheel Coaster Co., Tonawanda, New York. L: 51.5" W: 12.5" H: 6". $25-50.

1940. Fleetwing Spring Top Flash. Auto Wheel Coaster Co. #41 as seen in 1940 catalog. L: 41" W: 15", H: 7.5". $50-75.

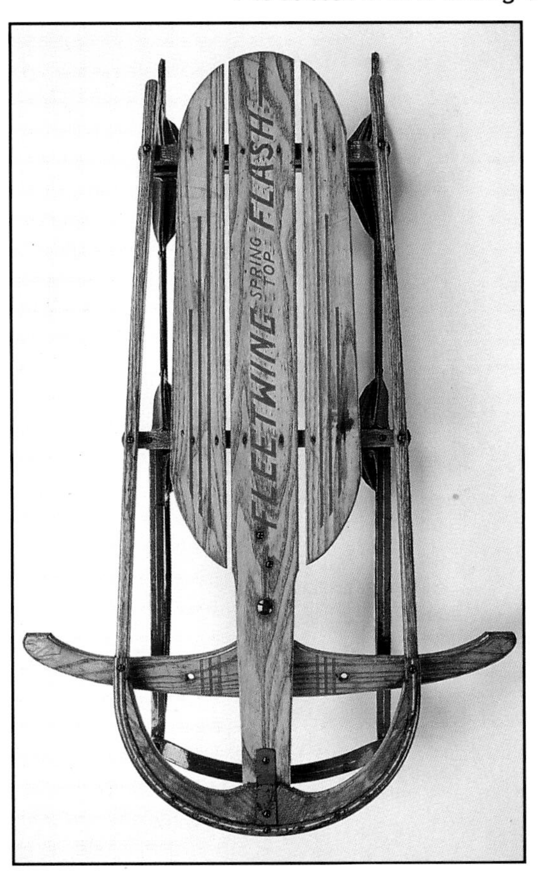

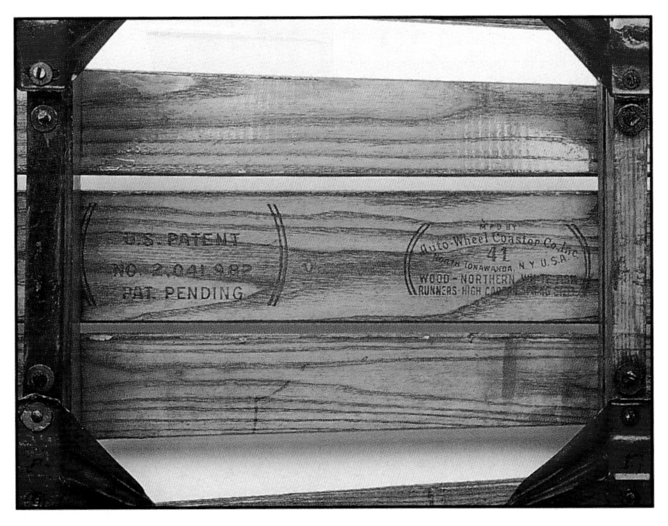

Ellington Turning Company, South Paris, Maine

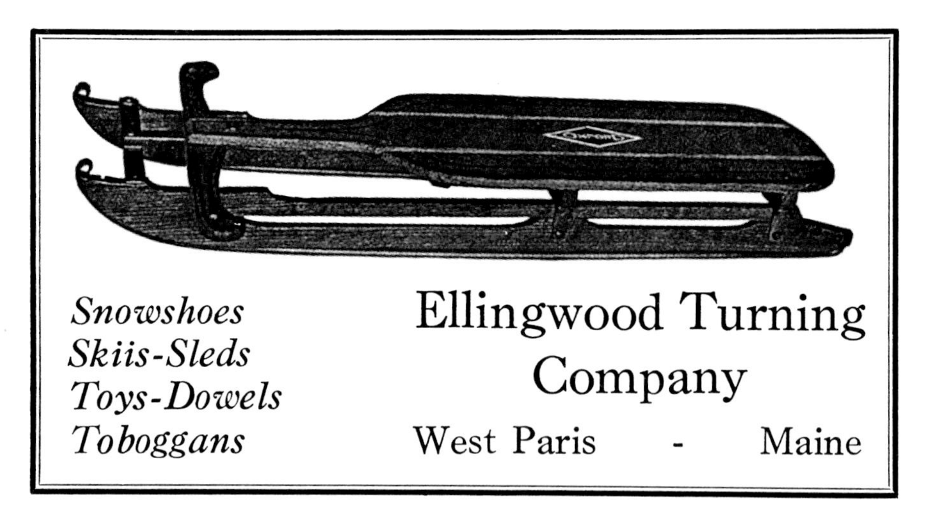

This advertisement ran in *Sun-Up, Maine's Own Magazine,* in March, 1927. The Ellingwood Turning Company was located in W. Paris, Maine. Their logo is the word "Oxford" in a diamond. Oxford is the county in which W. Paris is located.

Garton

The Garton Toy Company of Sheboygan, Wisconsin was founded by Eusebius B. Garton in 1879. Although famous for Coaster Wagons and sidewalk autos, I would be amiss not to mention Mr. Garton's contribution to the history of sled making.

Garton sled decks were made from a special weather resistant five-layered plywood imported from Finland. Flamboyant, colorful designs were silk screened on each sled deck. The brilliant red paint used was known in the trade as "Garton Red."

Almost all Garton Sleds were marketed with the Garton Trademark, but many retailers contracted with Garton to put their own designs and brand names over the Garton frame. For instance, the Coast to Coast Apollo Sled, the Gambles Sled, and Ace Hardware Sled were none other than The Garton Eskimo, Royal Racer, and the classiest of all, the Silver Streak.

Garton produced 8,500 sleds a week from June to Thanksgiving which was approximately 15% of its overall business.

In 1973 Garton Toys Company was sold to Monitor Corporation, a Milwaukee based finance and investment corporation. "Toy" was dropped from the corporate name.

Business operations were suspended in 1975, almost one hundred years after they began.

(Information was provided by Mead Public Library, IWF Service Sheboygan and found in *One Hundred Years of Sheboygan, 1846-1946* by Leberman)

A Chronology of Garton Toy Company

1890: Disastrous Fire
1929: Fire. Sleds manufactured. The wood was a layered plywood imported form Finland. Produced the Eskimo, Royal Racer, and Silver Streak, under their own name. These sleds were also marketed under different names: Coast to Coast Apollo - Eskimo frame; The Gambles - Royal Racer frame; Ace Hardware - Silver Streak frame. Famous in trade for brilliant red paint: "Garton Red."
1973: Company sold to Monitor Corporation. "Toy" dropped from name.
1975: Business operations suspended

Sleds

C. 1930. Garton Eskimo. L: 31", w: 13", h: 5.5". $15-25.

Hunt, Helm, and Ferris - Starline

Charles E. Hunt, Nathan B. Helm, and Henry L. Ferris formed a partnership in 1883 to manufacture and sell Mr. Ferris's patented "Starline" hay carrier. This was the beginning of a long line of patents from hay carriers to feeding systems and eventually to Starline Toys, including Star Steel Coaster Sleds and American Boy Bob Sleds.

A full line of Starline Steel Coaster sleds along with overland wagons, scooters, and pedal cars were developed. However, it was the invention of a unique barn door track and hanger device that gave the company their slogan and trademark: "Cannonball Beats 'Em All."

In 1931, Starline Toys became Starline Inc. Sometime in the years that followed toy manufacturing was suspended. Starline continued to manufacture farm equipment and in 1975, Starline became a part of the Farm Systems of the Chromalloy Farm and Industrial Equipment Company of Harvard, Illinois.

Chronology of Hunt, Helm, and Ferris

1883: Star Steel Coaster Sleds.
1924: Skeeboggan introduced, three thrills in one a surfboard, ski, and toboggan combined
1925: Cannon Ball Beats 'em All trademark. Skeebob and Skeeboggan advertised.
1931: Starline Toy becomes Starline Inc.
19??: Toy Manufacturing suspended
1975: Starline Inc. became part of the Farm Systems of the Chromalloy Farm and Industrial Equipment Company of Harvard, Illinois

Advertising

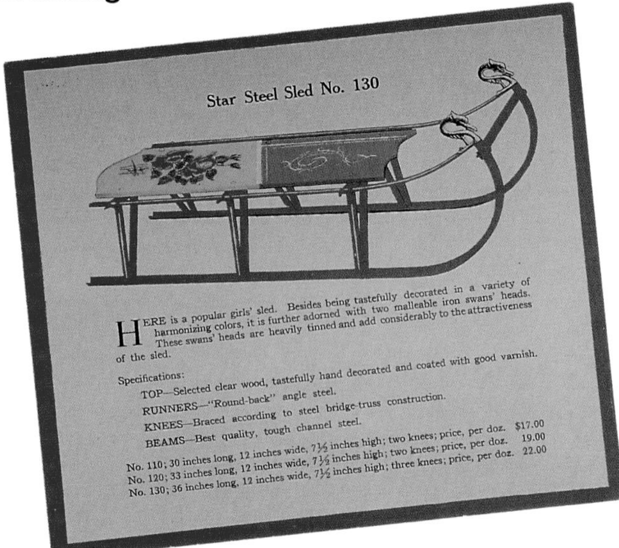

Star Steel Sled No. 130

HERE is a popular girls' sled. Besides being tastefully decorated in a variety of harmonizing colors, it is further adorned with two malleable iron swans' heads. These swans' heads are heavily tinned and add considerably to the attractiveness of the sled.

Specifications:

TOP—Selected clear wood, tastefully hand decorated and coated with good varnish.
RUNNERS—"Round-back" angle steel.
KNEES—Braced according to steel bridge-truss construction.
BEAMS—Best quality, tough channel steel.

No. 110; 30 inches long, 12 inches wide, 7½ inches high; two knees; price, per doz. $17.00
No. 120; 33 inches long, 12 inches wide, 7½ inches high; two knees; price, per doz. 19.00
No. 130; 36 inches long, 12 inches wide, 7½ inches high; three knees; price, per doz. 22.00

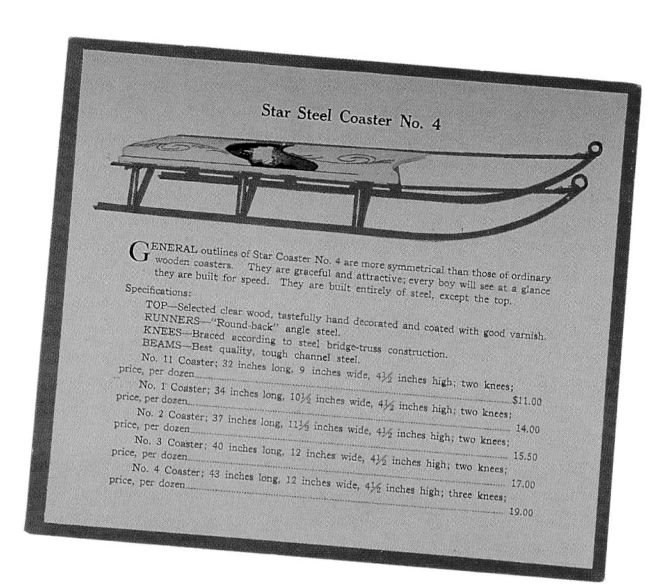

Star Steel Coaster No. 4

GENERAL outlines of Star Coaster No. 4 are more symmetrical than those of ordinary wooden coasters. They are graceful and attractive; every boy will see at a glance they are built for speed. They are built entirely of steel, except the top.

Specifications:

TOP—Selected clear wood, tastefully hand decorated and coated with good varnish.
RUNNERS—"Round-back" angle steel.
KNEES—Braced according to steel bridge-truss construction.
BEAMS—Best quality, tough channel steel.

No. 11 Coaster; 32 inches long, 9 inches wide, 4½ inches high; two knees; price, per dozen........$11.00
No. 1 Coaster; 34 inches long, 10½ inches wide, 4½ inches high; two knees; price, per dozen........14.00
No. 2 Coaster; 37 inches long, 11½ inches wide, 4½ inches high; two knees; price, per dozen........15.50
No. 3 Coaster; 40 inches long, 12 inches wide, 4½ inches high; two knees; price, per dozen........17.00
No. 4 Coaster; 43 inches long, 12 inches wide, 4½ inches high; three knees; price, per dozen........19.00

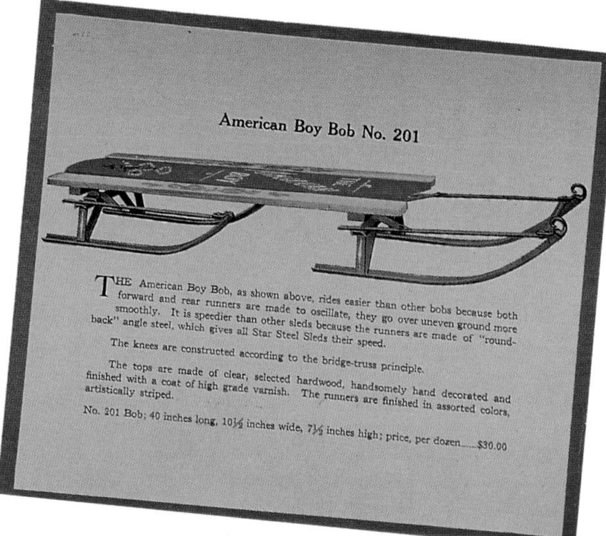

American Boy Bob No. 201

THE American Boy Bob, as shown above, rides easier than other bobs because both forward and rear runners are made to oscillate, they go over uneven ground more smoothly. It is speedier than other sleds because the runners are made of "round-back" angle steel, which gives all Star Steel Sleds their speed.

The knees are constructed according to the bridge-truss principle.

The tops are made of clear, selected hardwood, handsomely hand decorated and finished with a coat of high grade varnish. The runners are finished in assorted colors, artistically striped.

No. 201 Bob; 40 inches long, 10½ inches wide, 7½ inches high; price, per dozen......$30.00

Pages of a catalog of Hunt-Helm-Ferris & Co. featuring their Starline Sleds. *Courtesy of the Greater Harvard Area Historical Society.*

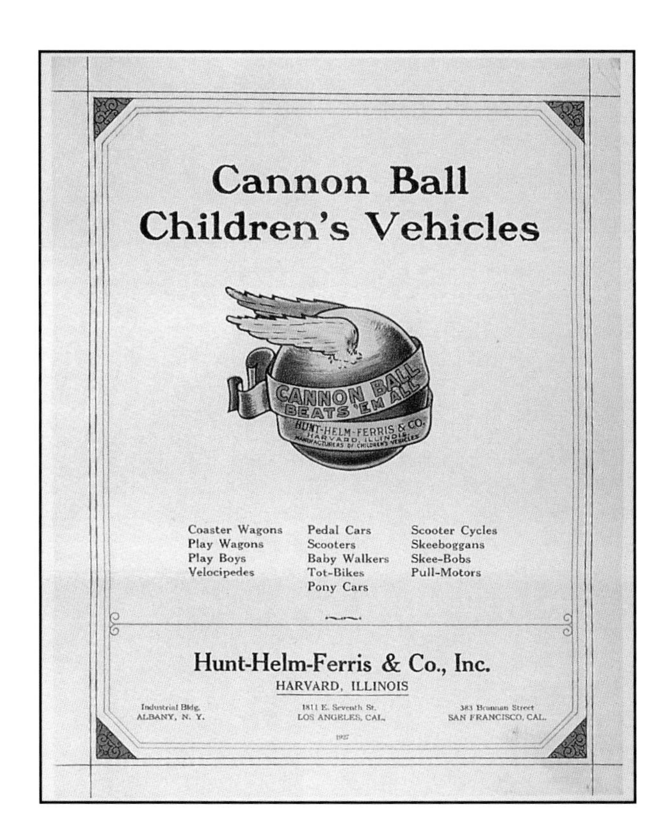

Cannon Ball Children's Vehicles

Coaster Wagons
Play Wagons
Play Boys
Velocipedes

Pedal Cars
Scooters
Baby Walkers
Tot-Bikes
Pony Cars

Scooter Cycles
Skeeboggans
Skee-Bobs
Pull-Motors

Hunt-Helm-Ferris & Co., Inc.
HARVARD, ILLINOIS

Industrial Bldg. ALBANY, N. Y. — 1811 E. Seventh St. LOS ANGELES, CAL. — 383 Brannan Street SAN FRANCISCO, CAL.

1927

HUNT-HELM-FERRIS & CO. Inc.

Cannon Ball Skee-Bob

THE No. 95 Skee-Bob has a heavy, strong channel steel backbone and is surmounted by a heavy, hardwood running board covered with a rubber mat to prevent slipping.

The runners are pivoted so that they conform to uneven parts over which the Skee-Bob runs.

The steering post, of solid, cold rolled steel, is ¾ inch in diameter and is held firmly in a carefully fitted steel bearing so that wear is practically eliminated.

When the guide runner is turned to the right or left, it is turned up on edge and the sharp edge of this hollow ground runner takes hold in a way that makes the Skee-Bob easy to steer at all speeds.

The runners are broad and will not cut down through packed snow.

The channel steel backbone is enameled in rich orange. The runners are orange with black trim. The running board and clear, selected hardwood handle are finished in natural color with two coats of clear, tough varnish. Footrest included as shown with No. 96.

No. 95 Skee-Bob, shipping weight, 18 lbs. each.

Skee-Bobs are packed one in a carton.

Cannon Ball Skee-Bob

NO. 96 Cannon Ball Skee-Bob has 4 broad runners connected in pairs. Each runner is pivoted on an axle so that it passes easily over any rough or uneven spots in its path.

The rear runners are free to pivot on the axle, but are locked in line with the body of the Skee-Bob which is steered by turning the front pair of runners.

The heavy channel steel backbone runs from front to back.

On this is mounted a heavy hardwood running board. This running board is covered with a corrugated rubber mat to prevent slipping.

The malleable iron foot rest shown in the illustration is for the convenience of those who would rather ride sitting down than to ride standing up. The position of this foot rest can be adjusted to the size of the rider.

No. 96 Skee-Bob, shipping weight, 20 lbs. each.

Skee-Bobs are packed one in a carton.

31

HUNT-HELM-FERRIS & CO. Inc.

Cannon Ball Skee-Bob

THE No. 95 Skee-Bob has a heavy, strong channel steel backbone and is surmounted by a heavy, hardwood running board covered with a rubber mat to prevent slipping.

The runners are pivoted so that they conform to uneven parts over which the Skee-Bob runs.

The steering post, of solid, cold rolled steel, is ¾ inch in diameter and is held firmly in a carefully fitted steel bearing so that wear is practically eliminated.

When the guide runner is turned to the right or left, it is turned up on edge and the sharp edge of this hollow ground runner takes hold in a way that makes the Skee-Bob easy to steer at all speeds.

The runners are broad and will not cut down through packed snow.

The channel steel backbone is enameled in rich orange. The runners are orange with black trim. The running board and clear, selected hardwood handle are finished in natural color with two coats of clear, tough varnish. Footrest included as shown with No. 96.

No. 95 Skee-Bob, shipping weight, 18 lbs. each.

Skee-Bobs are packed one in a carton.

Cannon Ball Skee-Bob

NO. 96 Cannon Ball Skee-Bob has 4 broad runners connected in pairs. Each runner is pivoted on an axle so that it passes easily over any rough or uneven spots in its path.

The rear runners are free to pivot on the axle, but are locked in line with the body of the Skee-Bob which is steered by turning the front pair of runners.

The heavy channel steel backbone runs from front to back.

On this is mounted a heavy hardwood running board. This running board is covered with a corrugated rubber mat to prevent slipping.

The malleable iron foot rest shown in the illustration is for the convenience of those who would rather ride sitting down than to ride standing up. The position of this foot rest can be adjusted to the size of the rider.

No. 96 Skee-Bob, shipping weight, 20 lbs. each.

Skee-Bobs are packed one in a carton.

31

HUNT-HELM-FERRIS & CO. Inc.

Cannon Ball Skeeboggan

THE No. 97 Skeeboggan is brand new. Nothing like it on the market before. In appearance and operation it is something like a ski and it is something like a toboggan but it combines the delightful thrill of skiing and coasting with that of riding a surf board.

But the sensation you get from the Skeeboggan because of its spring steel construction and the peculiar way in which it is handled is much more exciting as it glides noiselessly over the snow.

Anybody who lives near a snow-covered hill can enjoy Skeebogganing. It is a new winter sport that knows no age limit and there is no limit to its fun and thrills.

The Skeeboggan is shoved off from the top of the hill with one foot. Then both feet are brought together with the heels at the rear end of the rubber mat. The rider grasps the handles with both hands and leans back.

Steering is done by "warping." A slight sway of the body to either side will steer the Skeeboggan in that direction.

The Skeeboggan will go as fast and as far as any sled and the ride—standing up—is a lot more exhilarating.

Skeebogganing induces active red-blooded folks to stay out of doors in the crisp, fresh winter air and enjoy healthy fun filled with a new kind of thrill—a thrill vastly different from any ever experienced before.

Really, there is no way to explain the "kick" you get from using the Skeeboggan. No words exactly describe it.

The body of the Skeeboggan is made of high carbon spring steel which is practically indestructible. It is 8½ in. wide, 3 ft. long and weighs 9½ lbs. Rounded runners are formed into it and run the full length of each side. To the body is fastened a rubber topped platform. The rider stands on this platform, which keeps his shoes from marring the attractive finish, which is a deep, rich orange enamel trimmed in green.

The tongue is of clear grain, selected hardwood, measuring 1 in. x 1¼ in. The pressed steel bracket to which it is bolted is securely fastened to the body of the Skeeboggan with four heavy rivets. The handle is ⅝ in. in diameter and extends 3½ in. from each side of the tongue.

There's only one Skeeboggan—and nothing else like it.

(Pat. Applied For)
No. 97

The Biggest Thrill on the Hill

Pat. applied for

Packed six in a box, weighing 69 lbs.

30

Catalog cover and pages of the Cannon Ball line of the Hunt-Helm-Ferris & Co., Inc., Harvard, Illinois. 1920s. *Courtesy the Greater Harvard Area Historical Society.*

CANNON BALL
BEATS 'EM ALL

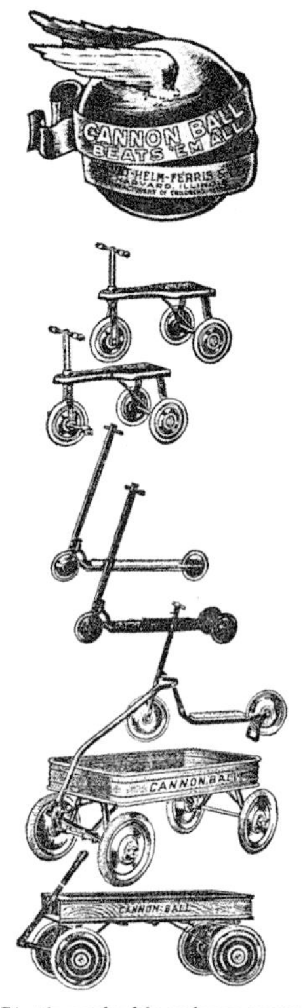

Saturday Evening Post Advertising

Attractive Posters

Window Cut-Outs

Movie Slides

Natural Attractiveness

Distinctive orange finish with green trim

Sturdy construction

— all are working together to help dealers sell Cannon Ball Wheel Goods on a profitable basis.

Cannon Ball Coaster Wagons in 3 sizes. Junior Wagons in 3 sizes. All steel wagons, Pedal Cars, Play Boys, Scooters and Skee-Skooters, Skeeboggans and Skee-Bobs are now on hand for prompt delivery.

Single or double column newspaper cuts of any illustration on this page, mailed on request. Life size cut-out of Cannon Ball Boy, fits large size wagon, sent with order for $50.00 or more.

Phone, wire, or mail your fill-in orders to nearest branch for quick service and prompt delivery.

Winter Thrillers and Heavy Sellers

A long winter is predicted with plenty of cold weather and snow; just the kind of weather to invite the use of Cannon Ball Skeeboggans and Skee-Bobs and help their sales. Cannon Ball Skeeboggan is made of heavy spring steel with runners formed in the sides. Finished in brilliant orange, green trim with rubber covered platform. Hinged handle folds flat — easily carried in car or bus.

CANNON BALL SKEE-BOBS

Cannon Ball Skee-Bob, No. 1146, has channel steel frame with hardwood rubber-covered running board. Easy to steer by turning handle. Can be quickly stopped by a full quarter turn. Operator sits on rubber topped wood platform.

No. 1145 Cannon Ball Skee-Bob. Same as No. 1146 except for a single instead of double front runner.

HUNT-HELM-FERRIS & CO.

Harvard, Illinois

Industrial Building Albany, New York	383 Brannan Street San Francisco, Cal.
22nd and Arch Streets Philadelphia, Pa.	422 Stinson Boulevard Minneapolis, Minn.
1811 E. 7th Street Los Angeles, Cal.	1322-1330 W. 13th St. Kansas City, Mo.

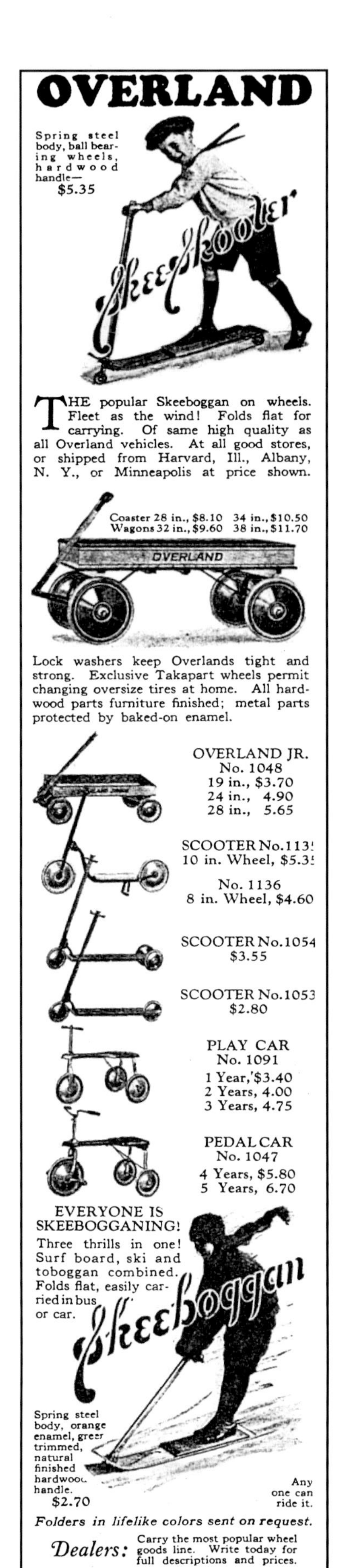

Above:
1925 advertisement for Cannon Ball.

Right:
September, 1924 advertisement for Hunt, Helm, Ferris & Co.'s Skeeboggan.

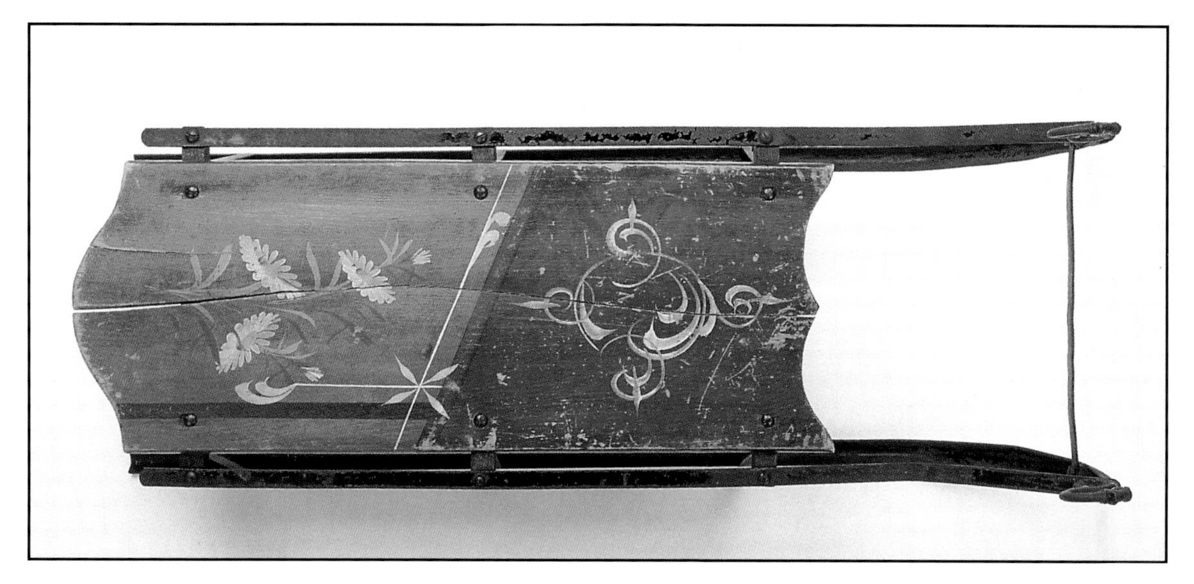

Star. Metal frame. L: 36.5", W: 12", H: 7.5". *Courtesy of Lou and Carole Scudillo.* $375-950.

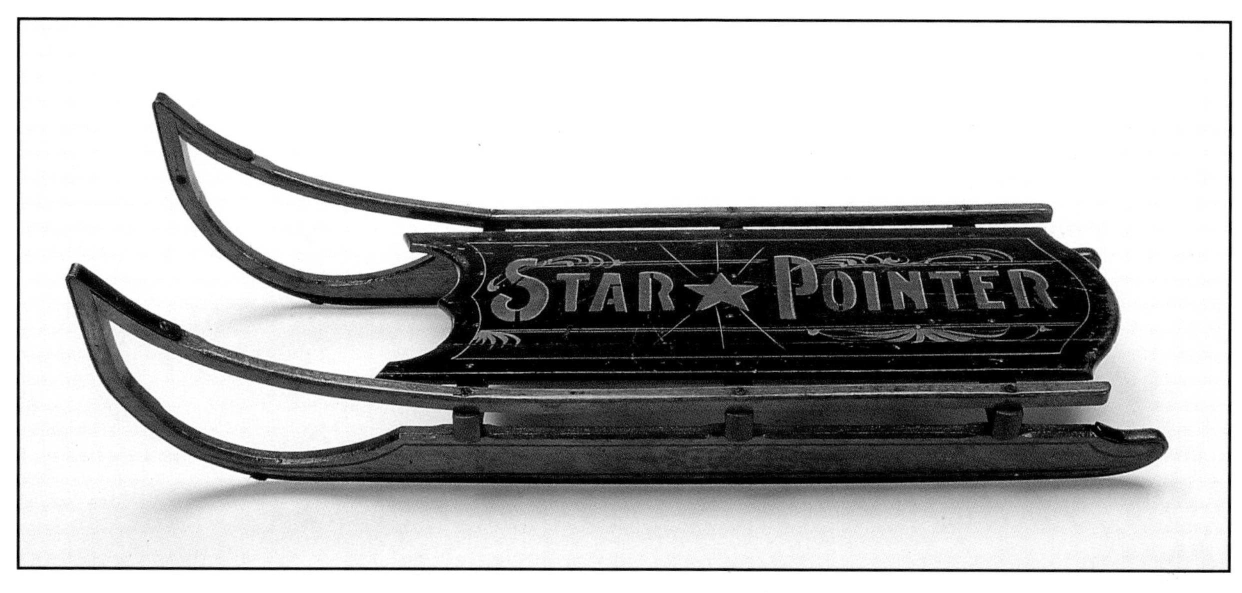

Star Pointer. L: 39" W: 11" H: 4.75". *Courtesy of Lou and Carole Scudillo.* $850-1950.

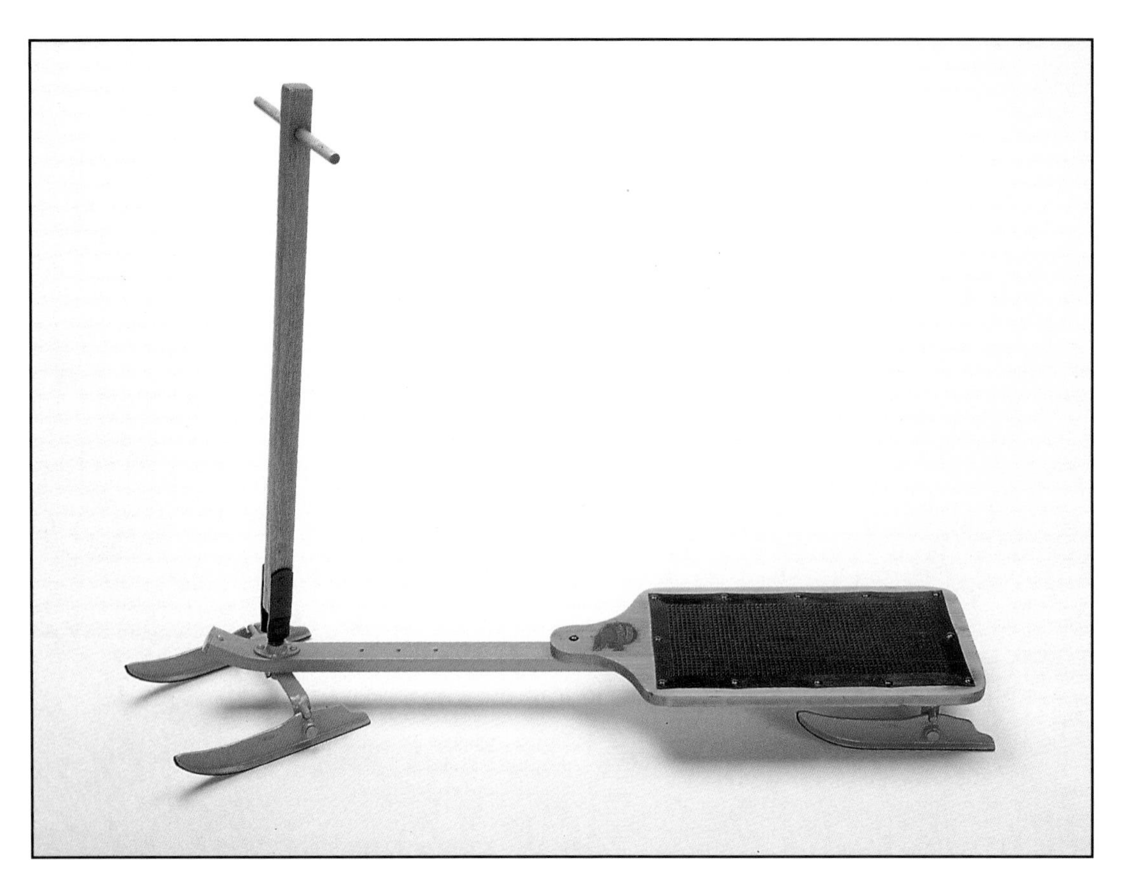

1925. Hunt, Helm & Ferris Double Runner Skee-Bob #96. $350-500.

Icecycle, Contoocook, New Hampshire

A photograph of the Ski Sled, by the Icecycle Company, Contoocook, New Hampshire. The people are the brothers of John E. Bean, the inventor. They are Emmett and Harold. The larger version is for adults and was the only one made. It is a single runner model. The other has a double runner.

These advertisements are from the winter of 1916-1917. Without a clear patent, the company only manufactured for one season.

Kalamazoo Sled Company

The Kalamazoo Sled Company was founded by H.P. Kauffer, Horace B. Peck, and W.E. Kidder after purchasing the stock of the Columbia Sled Company in 1894.

The company manufactured Champion Sleds at its Kalamazoo, Michigan plant under the name of Kalamazoo Sled Company until 1962. At this time the name was changed to Kalamazoo Sled and Toy Company after purchasing a St. Louis, Missouri toy manufacturing firm.

In 1968 another change occurred. The Gladding Corporation, a leisure time products manufacturer purchased the Kalamazoo Sled and Toy Company. They continued to operate the company until 1972 when they abandoned the Kalamazoo plant and moved their sled operations to a corporately owned toboggan plant near Portland, Maine.

Gladding owned forests near the Portland plant and economically it made good sense to combine both sled and toboggan manufacturing under one roof.

Gladding also acquired the American Flyer brand and affixed both names a top of their sleds.

On April 15, 1974, fire destroyed the old abandoned Kalamazoo Sled and Toy Company building. For almost 80 years of sled making Kalamazooans had fond memories of their very first Champion Sled—as the ad reads: "The Sled equipped with Presto Magic Steering." (Thanks to the Kalamazoo Public Library, Reference Department, Local History for this background information)

A Chronology of the Kalamazoo Sled Company

1894-1961: Kalamazoo Sled Company
1962: St. Louis, Missouri toy manufacturing firm purchased the company changing the name to Kalamazoo Toy and Sled Company.
1968: Gladding Corporation purchased company
1972: Sled making moved to Portland, Maine. Gladding American Flyer
1974: Fire destroyed abandoned Sled and Toy Company building in Kalamazoo

FELLOWS—what does the word Champion mean?
The winner of course! Whether with autos, airplanes, boxing gloves or a football — the "Champion" always means the best—the winner. That's why we call our sleds CHAMPION.

They Win—they have exclusive features which make it possible for you to go around curves without dragging your foot—you go faster, safer, and don't skid. You steer better, go farther.

PRESTO Magic Curve Steering Control

That's the secret of how far and how fast they go. And champions are the only sleds which have this patented method of definite control. Let any fellow try to follow right in your tracks!

Champions are made from the best materials money can buy.

Finest spring steel runners, extra strong braces. Special rivet construction gives you finger tip steering control. Built for sturdy service with the same care used in automobile construction. All corners rounded to avoid splinters that damage hands and clothes. Fine quality glistening finish. CHAMPIONS ARE BEAUTIES.

Be sure to get the true and only *CHAMPION*. There is no "just as good." Buy only sleds equipped with the Presto Magic Steering Control. Sure to please and much better value for your money. If your dealer can't supply you, write direct and we will send you folder.

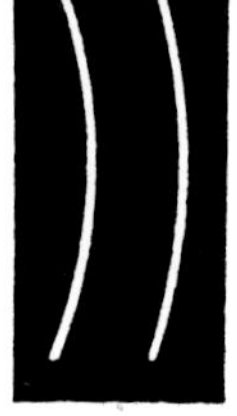

NEW WAY
Diagram shows how Presto curves the runners full length instead of only half way. Top does not move or side rails bend. A slight pull on the steering bar bends both runners in a correct arc, eliminating the braking effect that slows up other sleds.

THE KALAMAZOO SLED COMPANY
862 Third Street Kalamazoo, Michigan

December, 1934 advertisement for Champion sleds in *Boys' Life*. 1934

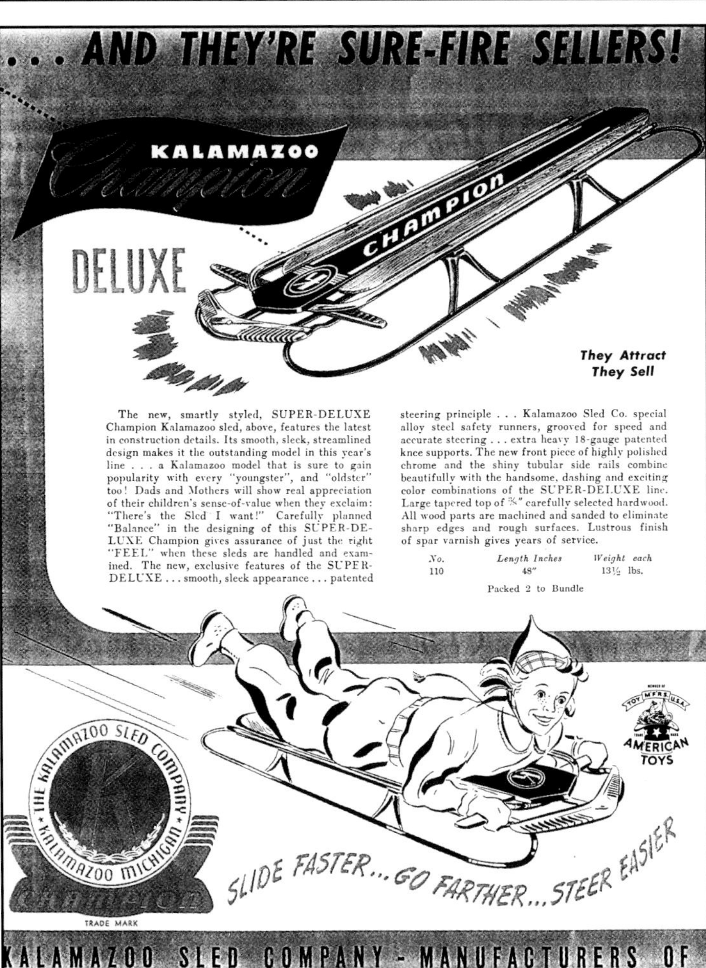

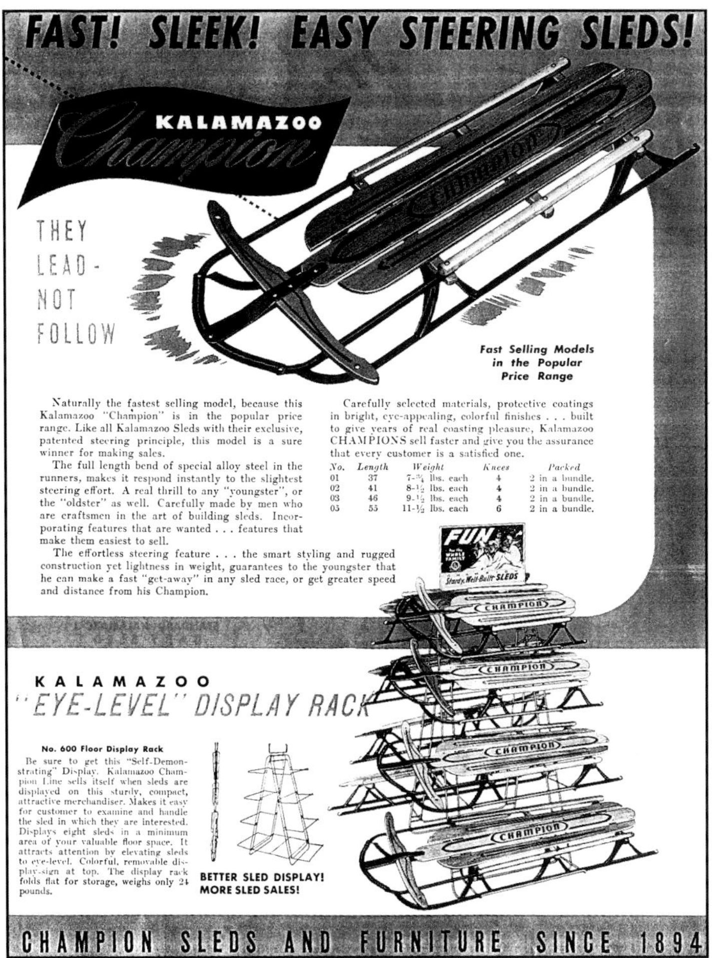

Kalamazoo Champion Sleds catalog,
Courtesy of the Kalamazoo Public Library.

Paris Manufacturing Company

The Paris Manufacturing Company had modest beginnings. In 1861 a man named Henry Franklin Morton, starting with a hobby, began what would become the largest and longest operating sled company in American History.

Mr. Morton was born in Corina, Penobscot County, Maine in 1839. Early on he was plagued with a serious eye condition that would alter his life plans of attending Harvard in pursuit of a teaching career.

Living in West Sumner, Maine with his wife Lucilla in 1861, he started making rakes and non-steerable sleds as a hobby to make extra money. He assembled fifty sleds that his wife Lucilla hand painted in the family kitchen and they were successful in selling them. The following year, Mr. Morton formed a stock company and hired four employees. The company was called West Sumner Manufacturing Company. A diverse line of sleds and toboggans were manufactured and shipped to Maine and Boston Markets. Word of his magnificent hand painted sleds traveled and he was invited to move his company to Paris Hill where he set up production under the name of Paris Hill Manufacturing Company. As the business grew, so did the line of products. Go-carts, wagons, wheelbarrows, step ladders, ironing boards, children's desks, furniture and more were added to keep the factory running all year long.

By the early 1880's, the use of oxen and horses to transport products to the nearest freight depot, which was three miles away, became more and more inefficient. Seeking a better location, the company relocated to South Paris, Maine where they set up shop next to the railroad on Western Ave. The company became known as the Paris Manufacturing Company (PMC). It was on this location that the company operated until its closure in 1989.

It was Samuel Leeds Allen who invented the first steerable sled, but it was Henry Franklin Morton who started it all by producing the first sleds and introducing ski making in this country.

In the early 1890's, Finnish immigrants living in West Paris were making their own skis. Mr. Morton capitalized on their skills by having them make skis at home and deliver them to the South Paris plant. PMC became the first commercial manufacturer of skis in this country, supplying skis to the U.S. Army during World War I and World War II. It produced skis until 1965.

Paris dominated the sled business for over fifty years with their eloquently painted motifs of scenery, animals, birds, flowers, and ships. Frequently copied but never equaled, their competitors began to slowly disappear from the market. In 1912, metal runners replaced wood and a steering sled was introduced by Paris. The steerable sled was called the "Speedway." The next change occurred around 1920 when an all steel frame was adopted.

By 1928, Flexible Flyer® was on top of the world with Admiral Byrd, but Paris was over the top of the world with the Famous Arctic Explorers Perry and MacMillan in 1914. Sledges were specially made as gifts for the Eskimo children, another first for PMC. As the years progressed other models were introduced: the Flying Cloud, Speedster, Snobird, and Speed Flex to name a few. Despite new models, sled sales began to decline as plastic products and saucers began to take their share of the market, forcing PMC to seek outside revenue. In 1970, the Gladding

PMC catalog c. 1915. *Courtesy of Henry R. Morton.*

Corporation purchased PMC and operated under the name of Gladding-Paris until 1978 when Gladding sold out. Again the company was under Maine ownership with Henry R. Morton as Vice President of Sales. He was the great-grandson of the founder Henry F. Morton. Mr. Morton continued manufacturing sleds under the Paris name and in 1986 released a reproduction of the original Snow King, an oak clipper. Paris sought bankruptcy protection in 1987 and in 1989 the company closed its doors, ending more than a century of sled making.

A fire in 1990 destroyed the original plant. Only a smoke stack remains marking the original site of PMC. Gone are the artisans who once produced one of Maine's most recognizable treasures.

Henry F. Morton would be proud to know that his business spanned five generations of Mortons and who knows how many more to come?

Since 1991, Torpedo Inc., a Canadian company and leading producer of sleds and toboggans to the North American market, has set up its United States sales and distribution center at the site of the former Paris Manufacturing Company.

A Chronology of the Paris Manufacturing Company

1861: Henry F. Morton built and sold fifty sleds hand painted by his wife, Lucilla.

1862: Stock company formed, West Sumner Manufacturing Company.

1869: Relocated to Paris Hill and renamed Paris Hill Manufacturing Company.

1876: Awarded first prize at the Philadelphia World's Fair.

1882: Moved to South Paris and became Paris Manufacturing Company.

1886: Fire destroyed new plant.

1887: Plant rebuilt and reopened.

circa 1890s: Norwegian Ski production.

1893: Colombian Clipper introduced, Chicago World's Fair model.

1894: Black Beauty introduced

1912: Speedway, first steerable sled

circa 1920: All steel frame introduced on Speedway.

1970: Paris Manufacturing Company sold to Gladding Corporation. Sleds bore the name of Gladding-Paris.

1978: Gladding sold, Morton Family in control. Company is called Paris Manufacturing.

1986: Snow King reproduction issued.

1989: Foreclosure.

February 6, 1990: Fire destroys original plant, smoke stack left as monument.

1991: Torpedo Inc., a Canadian company that produces sleds and toboggans, sets up a United States distribution center on the site of the former PMC.

1996: Henry R. Morton, grandson of Henry F. Morton, is Vice President of marketing overseas sales of sleds, toboggans, and snow related items for United States distribution.

Advertising

PMC catalog c. 1915.
Courtesy of Henry R. Morton.

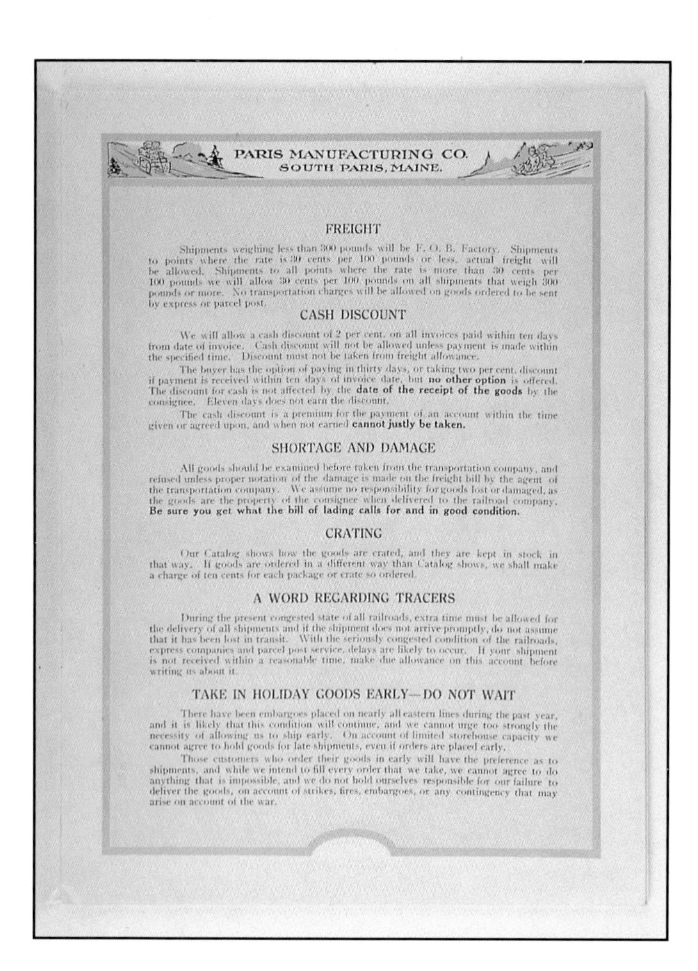

PARIS MANUFACTURING CO.
SOUTH PARIS, MAINE.

FREIGHT

Shipments weighing less than 300 pounds will be F. O. B. Factory. Shipments to points where the rate is 30 cents per 100 pounds or less, actual freight will be allowed. Shipments to all points where the rate is more than 30 cents per 100 pounds we will allow 30 cents per 100 pounds on all shipments that weigh 300 pounds or more. No transportation charges will be allowed on goods ordered to be sent by express or parcel post.

CASH DISCOUNT

We will allow a cash discount of 2 per cent. on all invoices paid within ten days from date of invoice. Cash discount will not be allowed unless payment is made within the specified time. Discount must not be taken from freight allowance.

The buyer has the option of paying in thirty days, or taking two per cent. discount if payment is received within ten days of invoice date, but **no other option** is offered. The discount for cash is not affected by the **date of the receipt of the goods** by the consignee. Eleven days does not earn the discount.

The cash discount is a premium for the payment of an account within the time given or agreed upon, and when not earned **cannot justly be taken.**

SHORTAGE AND DAMAGE

All goods should be examined before taken from the transportation company, and refused unless proper notation of the damage is made on the freight bill by the agent of the transportation company. We assume no responsibility for goods lost or damaged, as the goods are the property of the consignee when delivered to the railroad company. **Be sure you get what the bill of lading calls for and in good condition.**

CRATING

Our Catalog shows how the goods are crated, and they are kept in stock in that way. If goods are ordered in a different way than Catalog shows, we shall make a charge of ten cents for each package or crate so ordered.

A WORD REGARDING TRACERS

During the present congested state of all railroads, extra time must be allowed for the delivery of all shipments and if the shipment does not arrive promptly, do not assume that it has been lost in transit. With the seriously congested condition of the railroads, express companies and parcel post service, delays are likely to occur. If your shipment is not received within a reasonable time, make due allowance on this account before writing us about it.

TAKE IN HOLIDAY GOODS EARLY—DO NOT WAIT

There have been embargoes placed on nearly all eastern lines during the past year, and it is likely that this condition will continue, and we cannot urge too strongly the necessity of allowing us to ship early. On account of limited storehouse capacity we cannot agree to hold goods for late shipments, even if orders are placed early.

Those customers who order their goods in early will have the preference as to shipments, and while we intend to fill every order that we take, we cannot agree to do anything that is impossible, and we do not hold ourselves responsible for our failure to deliver the goods, on account of strikes, fires, embargoes, or any contingency that may arise on account of the war.

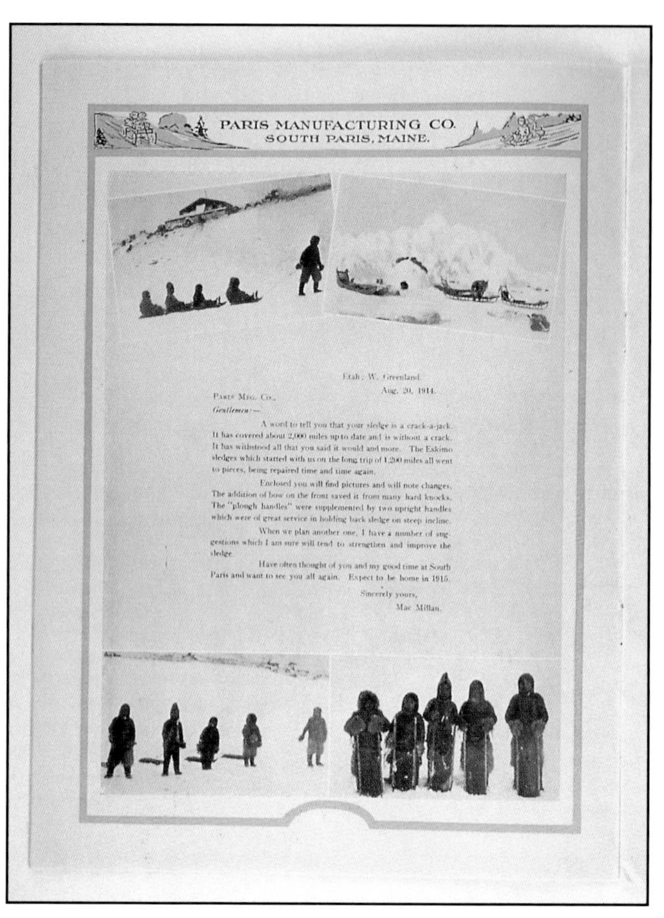

PARIS MANUFACTURING CO.
SOUTH PARIS, MAINE.

Etah, W. Greenland.
Aug. 20, 1914.

PARIS MFG. CO.,

Gentlemen:—

A word to tell you that your sledge is a crack-a-jack. It has covered about 2,000 miles up to date and is without a crack. It has withstood all that you said it would and more. The Eskimo sledges which started with us on the long trip of 1,200 miles all went to pieces, being repaired time and time again.

Enclosed you will find pictures and will note changes. The addition of bow on the front saved it from many hard knocks. The "plough handles" were supplemented by two upright handles which were of great service in holding back sledge on steep incline.

When we plan another one, I have a number of suggestions which I am sure will tend to strengthen and improve the sledge.

Have often thought of you and my good time at South Paris and want to see you all again. Expect to be home in 1915.

Sincerely yours,
Mac Millan.

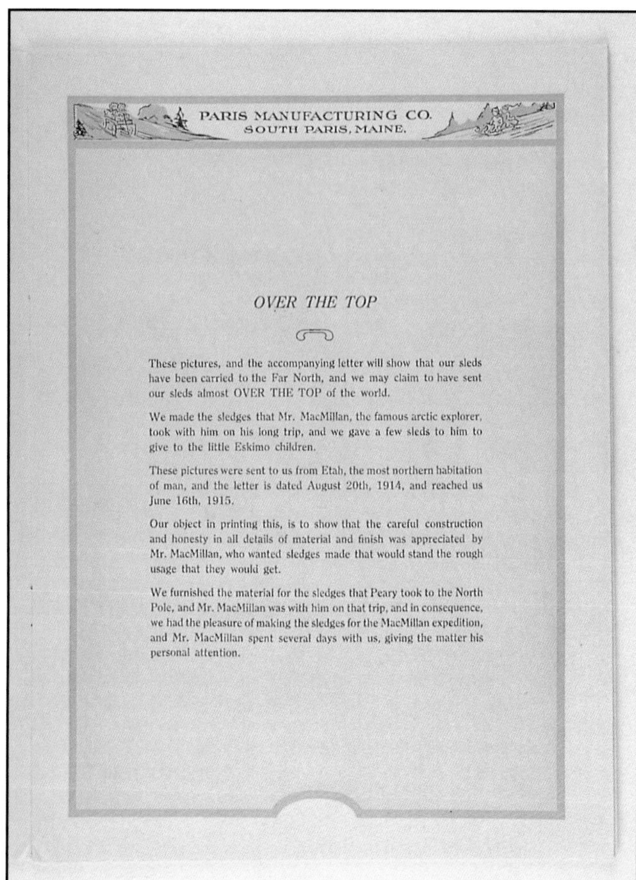

PARIS MANUFACTURING CO.
SOUTH PARIS, MAINE.

OVER THE TOP

These pictures, and the accompanying letter will show that our sleds have been carried to the Far North, and we may claim to have sent our sleds almost OVER THE TOP of the world.

We made the sledges that Mr. MacMillan, the famous arctic explorer, took with him on his long trip, and we gave a few sleds to him to give to the little Eskimo children.

These pictures were sent to us from Etah, the most northern habitation of man, and the letter is dated August 20th, 1914, and reached us June 16th, 1915.

Our object in printing this, is to show that the careful construction and honesty in all details of material and finish was appreciated by Mr. MacMillan, who wanted sledges made that would stand the rough usage that they would get.

We furnished the material for the sledges that Peary took to the North Pole, and Mr. MacMillan was with him on that trip, and in consequence, we had the pleasure of making the sledges for the MacMillan expedition, and Mr. MacMillan spent several days with us, giving the matter his personal attention.

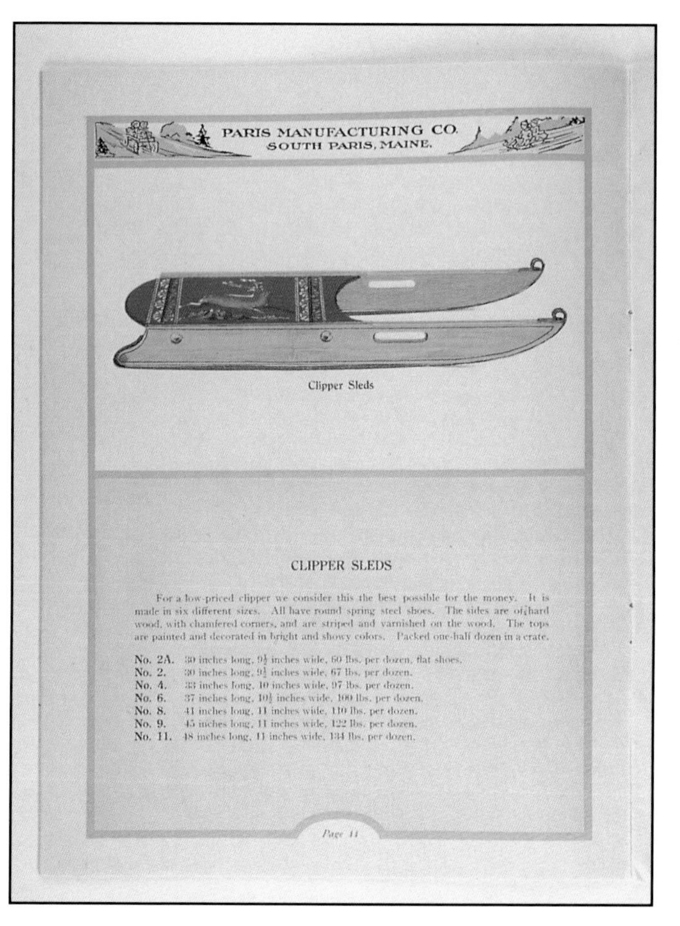

PARIS MANUFACTURING CO.
SOUTH PARIS, MAINE.

Clipper Sleds

CLIPPER SLEDS

For a low-priced clipper we consider this the best possible for the money. It is made in six different sizes. All have round spring steel shoes. The sides are of hard wood, with chamfered corners, and are striped and varnished on the wood. The tops are painted and decorated in bright and showy colors. Packed one-half dozen in a crate.

No. 2A. 30 inches long, $9\frac{1}{2}$ inches wide, 60 lbs. per dozen, flat shoes.
No. 2. 30 inches long, $9\frac{1}{2}$ inches wide, 67 lbs. per dozen.
No. 4. 33 inches long, 10 inches wide, 97 lbs. per dozen.
No. 6. 37 inches long, $10\frac{1}{2}$ inches wide, 100 lbs. per dozen.
No. 8. 41 inches long, 11 inches wide, 110 lbs. per dozen.
No. 9. 45 inches long, 11 inches wide, 122 lbs. per dozen.
No. 11. 48 inches long, 11 inches wide, 134 lbs. per dozen.

Page 11

PMC catalog c. 1915. *Courtesy of Henry R. Morton.*

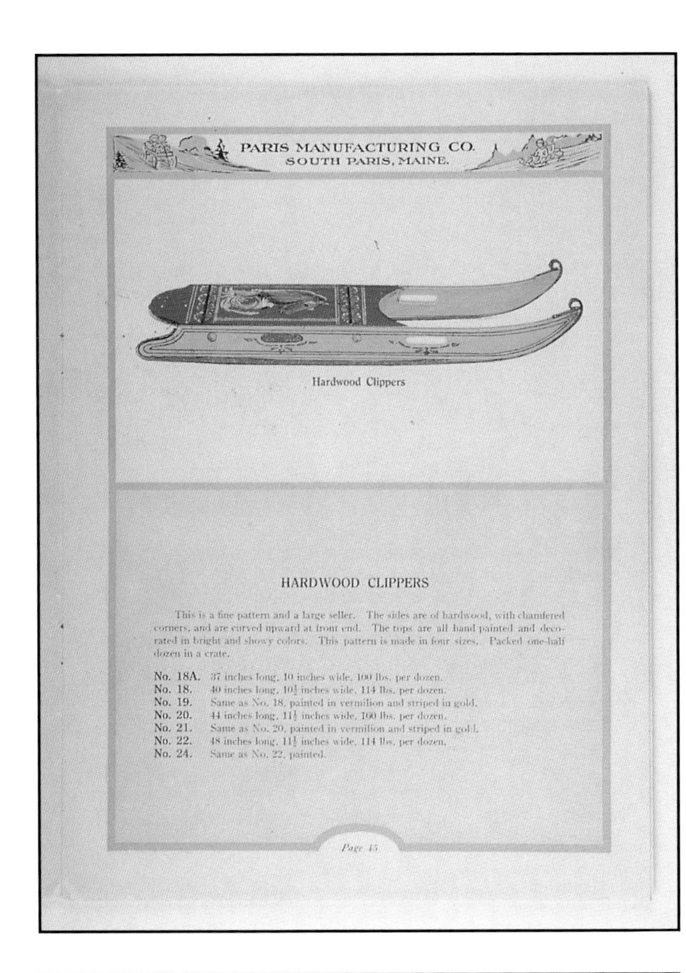
PARIS MANUFACTURING CO.
SOUTH PARIS, MAINE.

Hardwood Clippers

HARDWOOD CLIPPERS

This is a fine pattern and a large seller. The sides are of hardwood, with chamfered corners, and are curved upward at front end. The tops are all hand painted and decorated in bright and showy colors. This pattern is made in four sizes. Packed one-half dozen in a crate.

No. 18A. 37 inches long, 10 inches wide, 100 lbs. per dozen.
No. 18. 40 inches long, $10\frac{1}{2}$ inches wide, 114 lbs. per dozen.
No. 19. Same as No. 18, painted in vermilion and striped in gold.
No. 20. 44 inches long, $11\frac{1}{2}$ inches wide, 160 lbs. per dozen.
No. 21. Same as No. 20, painted in vermilion and striped in gold.
No. 22. 48 inches long, $11\frac{1}{2}$ inches wide, 114 lbs. per dozen.
No. 24. Same as No. 22, painted.

Page 45

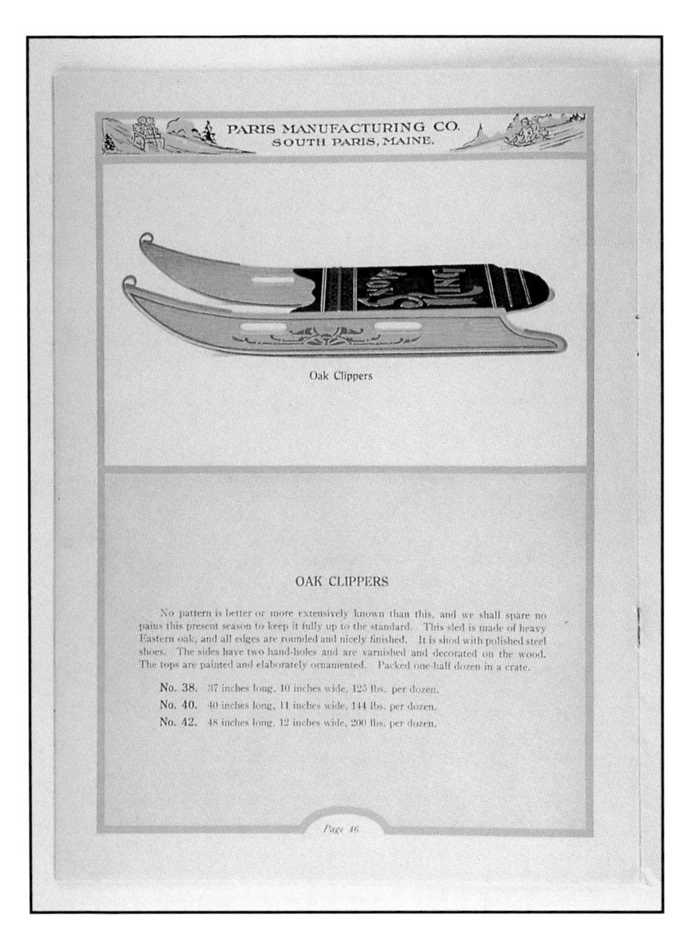
PARIS MANUFACTURING CO.
SOUTH PARIS, MAINE.

Oak Clippers

OAK CLIPPERS

No pattern is better or more extensively known than this, and we shall spare no pains this present season to keep it fully up to the standard. This sled is made of heavy Eastern oak, and all edges are rounded and nicely finished. It is shod with polished steel shoes. The sides have two hand-holes and are varnished and decorated on the wood. The tops are painted and elaborately ornamented. Packed one-half dozen in a crate.

No. 38. 37 inches long, 10 inches wide, 125 lbs. per dozen.
No. 40. 40 inches long, 11 inches wide, 144 lbs. per dozen.
No. 42. 48 inches long, 12 inches wide, 200 lbs. per dozen.

Page 46

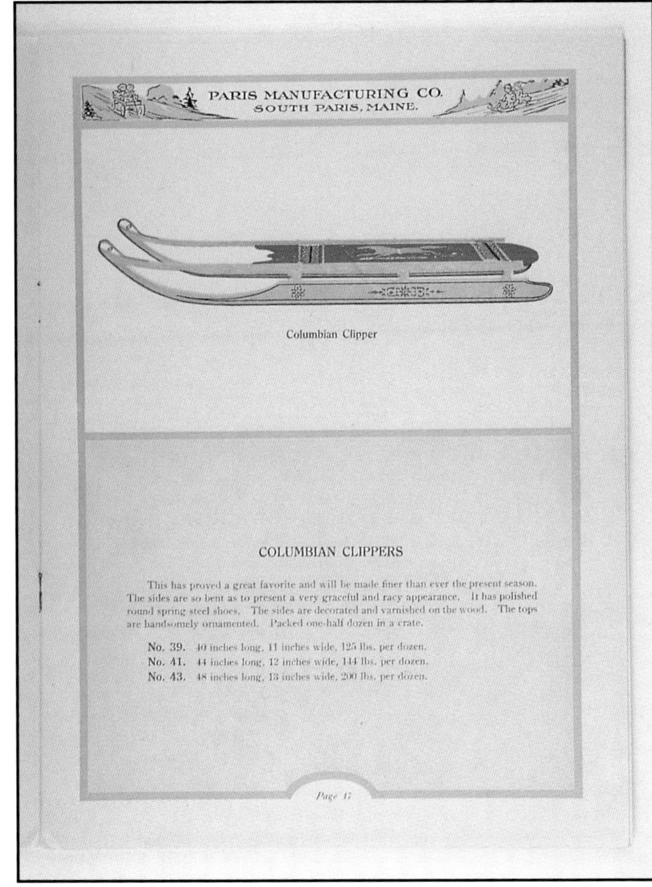
PARIS MANUFACTURING CO.
SOUTH PARIS, MAINE.

Columbian Clipper

COLUMBIAN CLIPPERS

This has proved a great favorite and will be made finer than ever the present season. The sides are so bent as to present a very graceful and racy appearance. It has polished round spring steel shoes. The sides are decorated and varnished on the wood. The tops are handsomely ornamented. Packed one-half dozen in a crate.

No. 39. 40 inches long, 11 inches wide, 125 lbs. per dozen.
No. 41. 44 inches long, 12 inches wide, 144 lbs. per dozen.
No. 43. 48 inches long, 13 inches wide, 200 lbs. per dozen.

Page 47

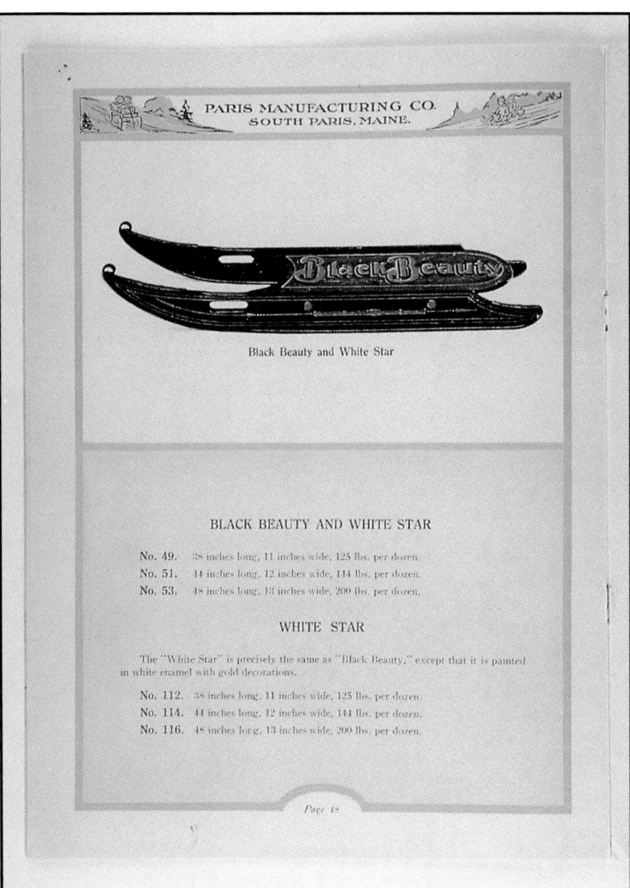
PARIS MANUFACTURING CO.
SOUTH PARIS, MAINE.

Black Beauty and White Star

BLACK BEAUTY AND WHITE STAR

No. 49. 38 inches long, 11 inches wide, 125 lbs. per dozen.
No. 51. 44 inches long, 12 inches wide, 144 lbs. per dozen.
No. 53. 48 inches long, 13 inches wide, 200 lbs. per dozen.

WHITE STAR

The "White Star" is precisely the same as "Black Beauty," except that it is painted in white enamel with gold decorations.

No. 112. 38 inches long, 11 inches wide, 125 lbs. per dozen.
No. 114. 44 inches long, 12 inches wide, 144 lbs. per dozen.
No. 116. 48 inches long, 13 inches wide, 200 lbs. per dozen.

Page 48

PMC catalog c. 1915. *Courtesy of Henry R. Morton.*

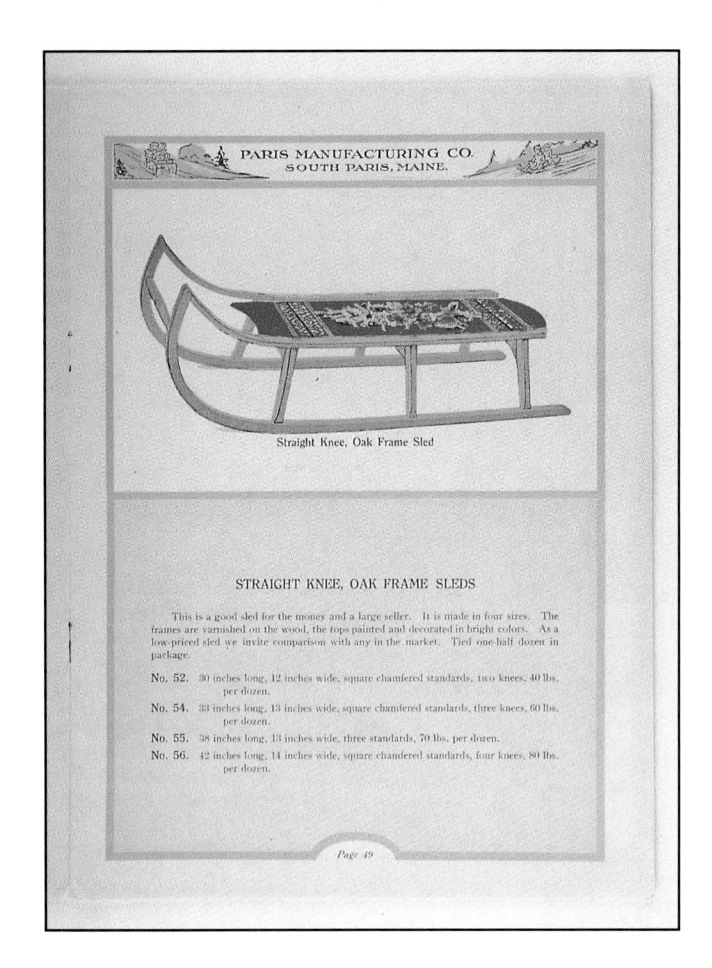

PARIS MANUFACTURING CO.
SOUTH PARIS, MAINE.

Straight Knee, Oak Frame Sled

STRAIGHT KNEE, OAK FRAME SLEDS

This is a good sled for the money and a large seller. It is made in four sizes. The frames are varnished on the wood, the tops painted and decorated in bright colors. As a low-priced sled we invite comparison with any in the market. Tied one-half dozen in package.

No. 52. 30 inches long, 12 inches wide, square chamfered standards, two knees, 40 lbs. per dozen.

No. 54. 33 inches long, 13 inches wide, square chamfered standards, three knees, 60 lbs. per dozen.

No. 55. 38 inches long, 13 inches wide, three standards, 70 lbs. per dozen.

No. 56. 42 inches long, 14 inches wide, square chamfered standards, four knees, 80 lbs. per dozen.

Page 49

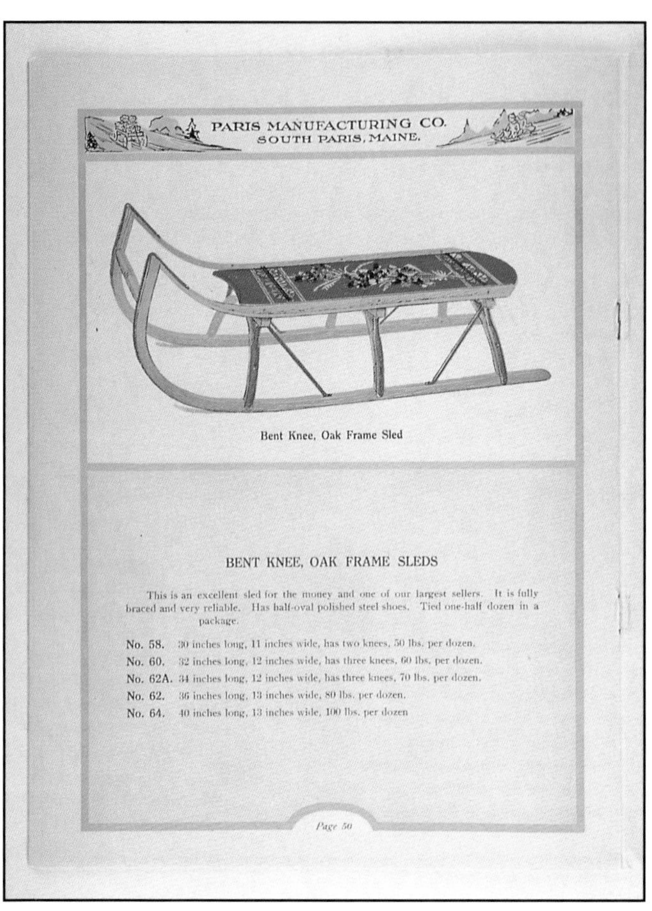

PARIS MANUFACTURING CO.
SOUTH PARIS, MAINE.

Bent Knee, Oak Frame Sled

BENT KNEE, OAK FRAME SLEDS

This is an excellent sled for the money and one of our largest sellers. It is fully braced and very reliable. Has half-oval polished steel shoes. Tied one-half dozen in a package.

No. 58. 30 inches long, 11 inches wide, has two knees, 50 lbs. per dozen.

No. 60. 32 inches long, 12 inches wide, has three knees, 60 lbs. per dozen.

No. 62A. 34 inches long, 12 inches wide, has three knees, 70 lbs. per dozen.

No. 62. 36 inches long, 13 inches wide, 80 lbs. per dozen.

No. 64. 40 inches long, 13 inches wide, 100 lbs. per dozen

Page 50

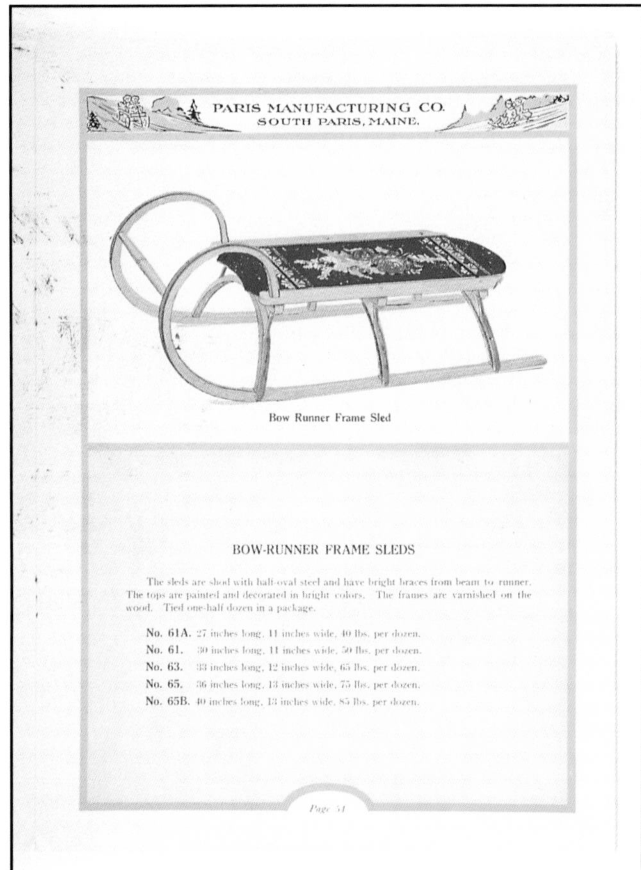

PARIS MANUFACTURING CO.
SOUTH PARIS, MAINE.

Bow Runner Frame Sled

BOW-RUNNER FRAME SLEDS

The sleds are shod with half-oval steel and have bright braces from beam to runner. The tops are painted and decorated in bright colors. The frames are varnished on the wood. Tied one-half dozen in a package.

No. 61A. 27 inches long, 11 inches wide, 40 lbs. per dozen.

No. 61. 30 inches long, 11 inches wide, 50 lbs. per dozen.

No. 63. 33 inches long, 12 inches wide, 65 lbs. per dozen.

No. 65. 36 inches long, 13 inches wide, 75 lbs. per dozen.

No. 65B. 40 inches long, 13 inches wide, 85 lbs. per dozen.

Page 51

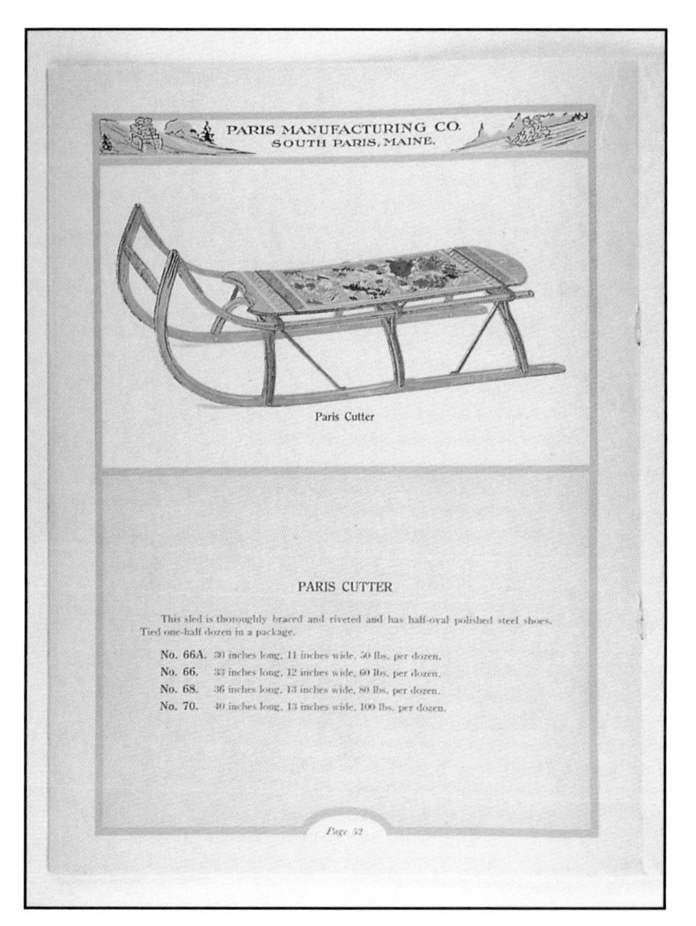

PARIS MANUFACTURING CO.
SOUTH PARIS, MAINE.

Paris Cutter

PARIS CUTTER

This sled is thoroughly braced and riveted and has half-oval polished steel shoes. Tied one-half dozen in a package.

No. 66A. 30 inches long, 11 inches wide, 50 lbs. per dozen.

No. 66. 33 inches long, 12 inches wide, 60 lbs. per dozen.

No. 68. 36 inches long, 13 inches wide, 80 lbs. per dozen.

No. 70. 40 inches long, 13 inches wide, 100 lbs. per dozen.

Page 52

PMC catalog c. 1915. *Courtesy of Henry R. Morton.*

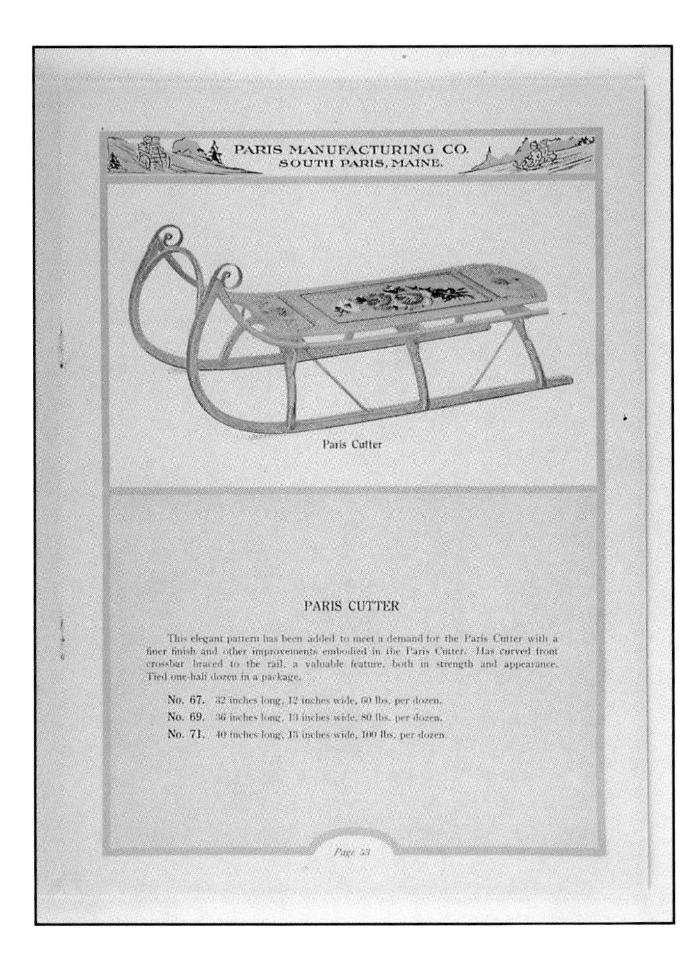
PARIS MANUFACTURING CO.
SOUTH PARIS, MAINE.

Paris Cutter

PARIS CUTTER

This elegant pattern has been added to meet a demand for the Paris Cutter with a finer finish and other improvements embodied in the Paris Cutter. Has curved front crossbar braced to the rail, a valuable feature, both in strength and appearance. Tied one-half dozen in a package.

No. 67. 32 inches long, 12 inches wide, 60 lbs. per dozen.
No. 69. 36 inches long, 13 inches wide, 80 lbs. per dozen.
No. 71. 40 inches long, 13 inches wide, 100 lbs. per dozen.

Page 53

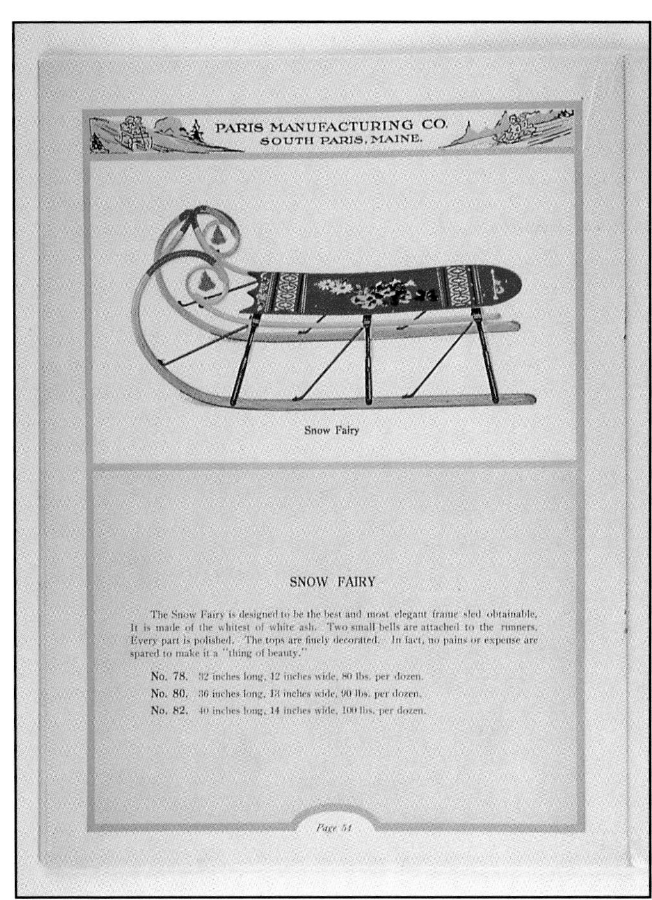
PARIS MANUFACTURING CO.
SOUTH PARIS, MAINE.

Snow Fairy

SNOW FAIRY

The Snow Fairy is designed to be the best and most elegant frame sled obtainable. It is made of the whitest of white ash. Two small bells are attached to the runners. Every part is polished. The tops are finely decorated. In fact, no pains or expense are spared to make it a "thing of beauty."

No. 78. 32 inches long, 12 inches wide, 80 lbs. per dozen.
No. 80. 36 inches long, 13 inches wide, 90 lbs. per dozen.
No. 82. 40 inches long, 14 inches wide, 100 lbs. per dozen.

Page 54

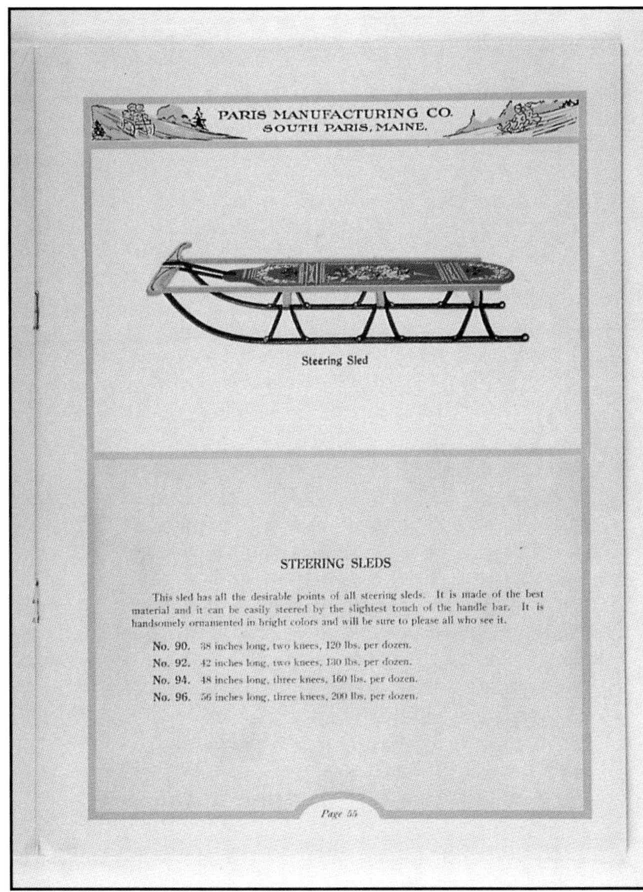
PARIS MANUFACTURING CO.
SOUTH PARIS, MAINE.

Steering Sled

STEERING SLEDS

This sled has all the desirable points of all steering sleds. It is made of the best material and it can be easily steered by the slightest touch of the handle bar. It is handsomely ornamented in bright colors and will be sure to please all who see it.

No. 90. 38 inches long, two knees, 120 lbs. per dozen.
No. 92. 42 inches long, two knees, 130 lbs. per dozen.
No. 94. 48 inches long, three knees, 160 lbs. per dozen.
No. 96. 56 inches long, three knees, 200 lbs. per dozen.

Page 55

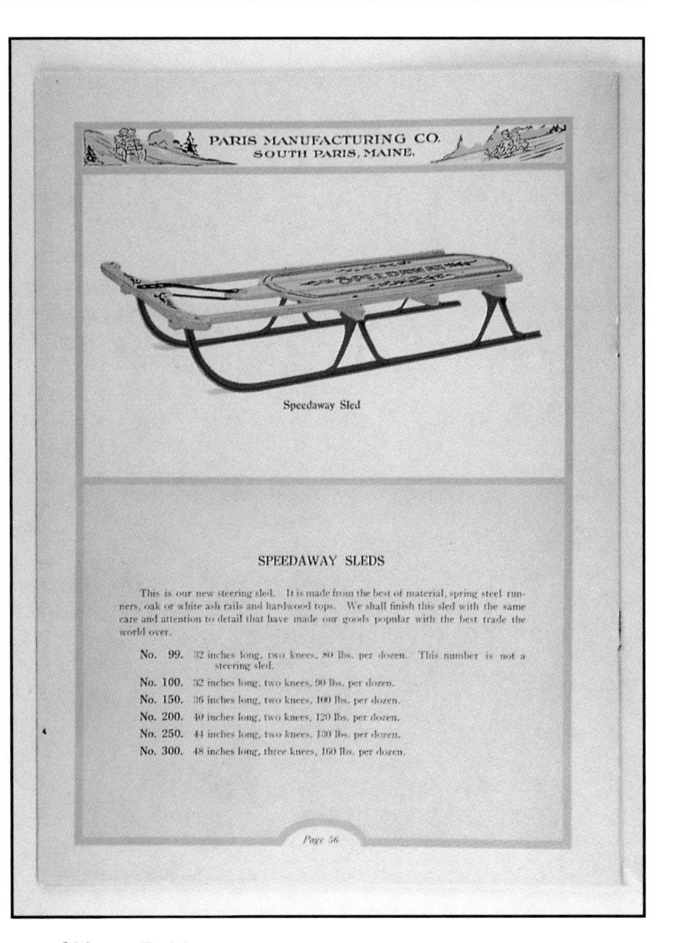
PARIS MANUFACTURING CO.
SOUTH PARIS, MAINE.

Speedaway Sled

SPEEDAWAY SLEDS

This is our new steering sled. It is made from the best of material, spring steel runners, oak or white ash rails and hardwood tops. We shall finish this sled with the same care and attention to detail that have made our goods popular with the best trade the world over.

No. 99. 32 inches long, two knees, 80 lbs. per dozen. This number is not a steering sled.
No. 100. 32 inches long, two knees, 90 lbs. per dozen.
No. 150. 36 inches long, two knees, 100 lbs. per dozen.
No. 200. 40 inches long, two knees, 120 lbs. per dozen.
No. 250. 44 inches long, two knees, 130 lbs. per dozen.
No. 300. 48 inches long, three knees, 160 lbs. per dozen.

Page 56

PMC catalog c. 1915. *Courtesy of Henry R. Morton.*

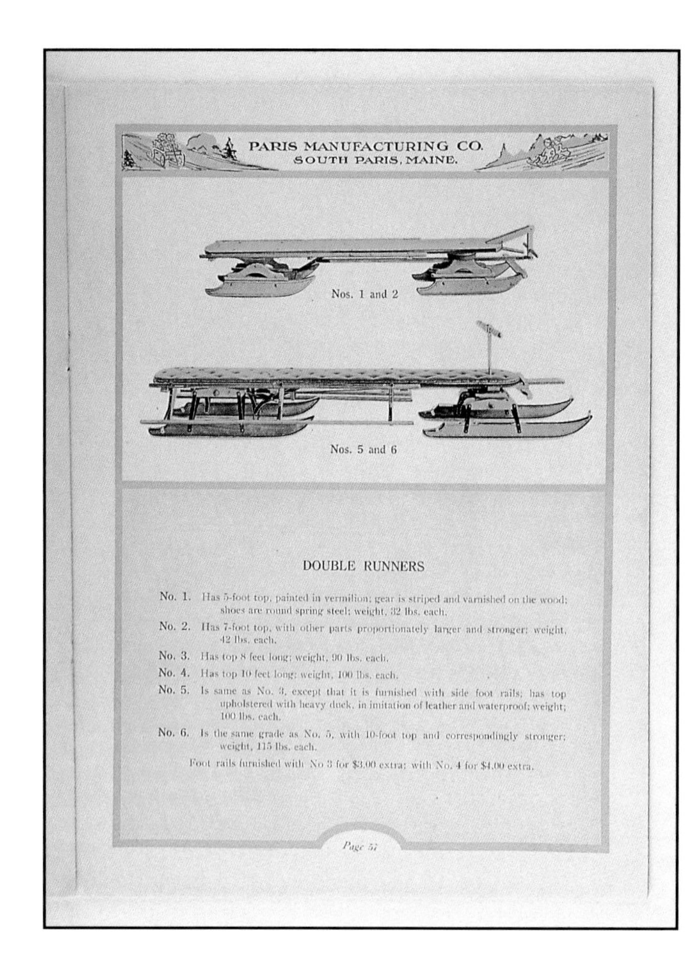

PARIS MANUFACTURING CO.
SOUTH PARIS, MAINE.

Nos. 1 and 2

Nos. 5 and 6

DOUBLE RUNNERS

No. 1. Has 5-foot top, painted in vermilion; gear is striped and varnished on the wood; shoes are round spring steel; weight, 32 lbs. each.

No. 2. Has 7-foot top, with other parts proportionately larger and stronger; weight, 42 lbs. each.

No. 3. Has top 8 feet long; weight, 90 lbs. each.

No. 4. Has top 10 feet long; weight, 100 lbs. each.

No. 5. Is same as No. 3, except that it is furnished with side foot rails; has top upholstered with heavy duck, in imitation of leather and waterproof; weight; 100 lbs. each.

No. 6. Is the same grade as No. 5, with 10-foot top and correspondingly stronger; weight, 115 lbs. each.

Foot rails furnished with No 3 for $3.00 extra; with No. 4 for $4.00 extra.

Page 57

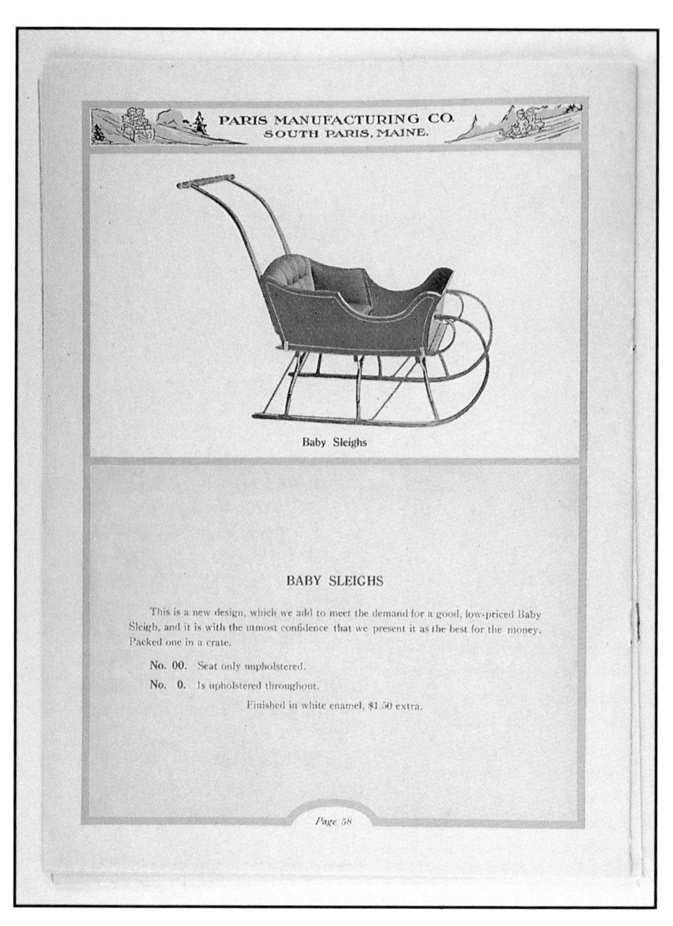

PARIS MANUFACTURING CO.
SOUTH PARIS, MAINE.

Baby Sleighs

BABY SLEIGHS

This is a new design, which we add to meet the demand for a good, low-priced Baby Sleigh, and it is with the utmost confidence that we present it as the best for the money. Packed one in a crate.

No. 00. Seat only upholstered.

No. 0. Is upholstered throughout.

Finished in white enamel, $1.50 extra.

Page 58

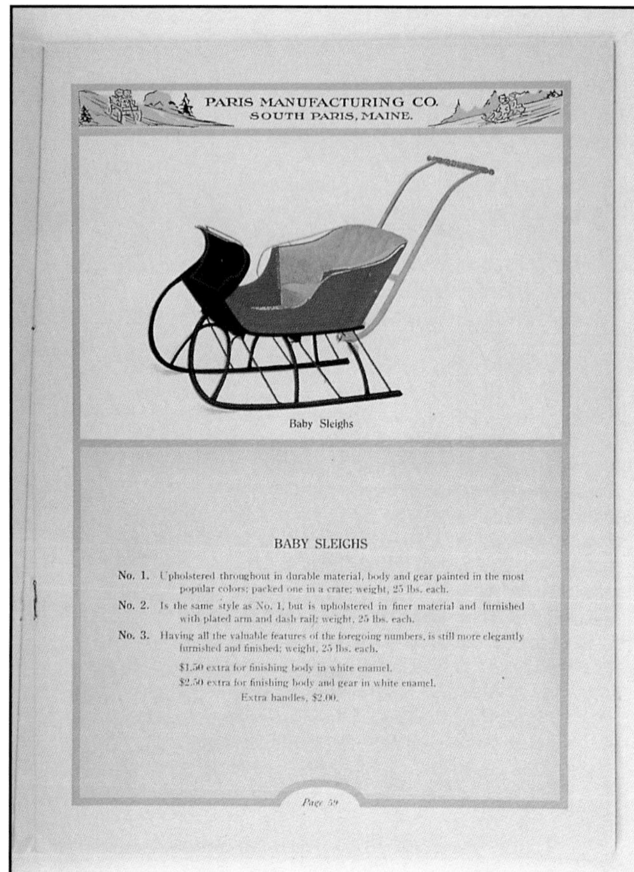

PARIS MANUFACTURING CO.
SOUTH PARIS, MAINE.

Baby Sleighs

BABY SLEIGHS

No. 1. Upholstered throughout in durable material, body and gear painted in the most popular colors; packed one in a crate; weight, 25 lbs. each.

No. 2. Is the same style as No. 1, but is upholstered in finer material and furnished with plated arm and dash rail; weight, 25 lbs. each.

No. 3. Having all the valuable features of the foregoing numbers, is still more elegantly furnished and finished; weight, 25 lbs. each.

$1.50 extra for finishing body in white enamel.
$2.50 extra for finishing body and gear in white enamel.
Extra handles, $2.00.

Page 59

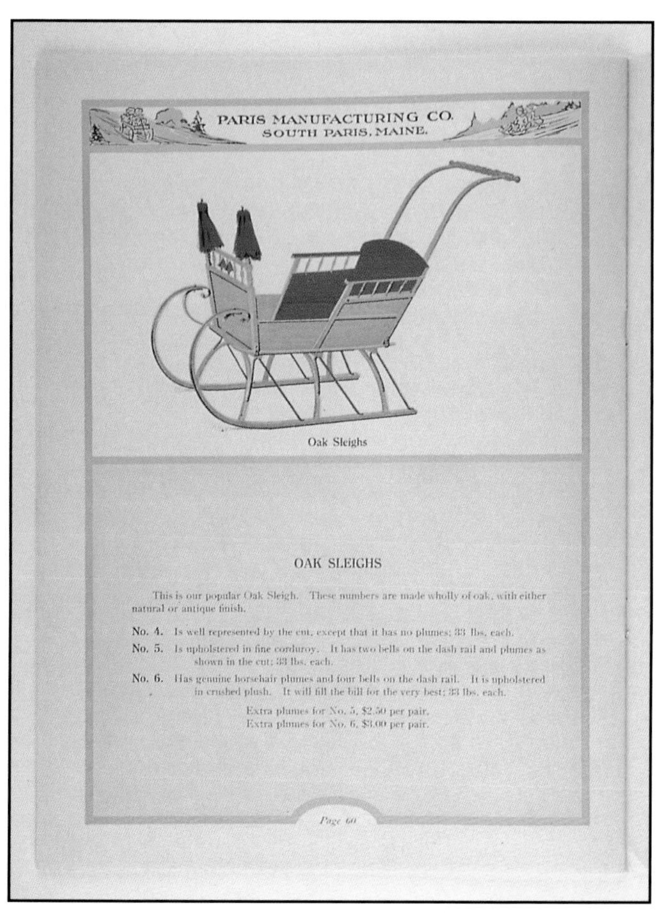

PARIS MANUFACTURING CO.
SOUTH PARIS, MAINE.

Oak Sleighs

OAK SLEIGHS

This is our popular Oak Sleigh. These numbers are made wholly of oak, with either natural or antique finish.

No. 4. Is well represented by the cut, except that it has no plumes; 33 lbs. each.

No. 5. Is upholstered in fine corduroy. It has two bells on the dash rail and plumes as shown in the cut; 33 lbs. each.

No. 6. Has genuine horsehair plumes and four bells on the dash rail. It is upholstered in crushed plush. It will fill the bill for the very best; 33 lbs. each.

Extra plumes for No. 5, $2.50 per pair.
Extra plumes for No. 6, $3.00 per pair.

Page 60

PMC catalog c. 1915. *Courtesy of Henry R. Morton.*

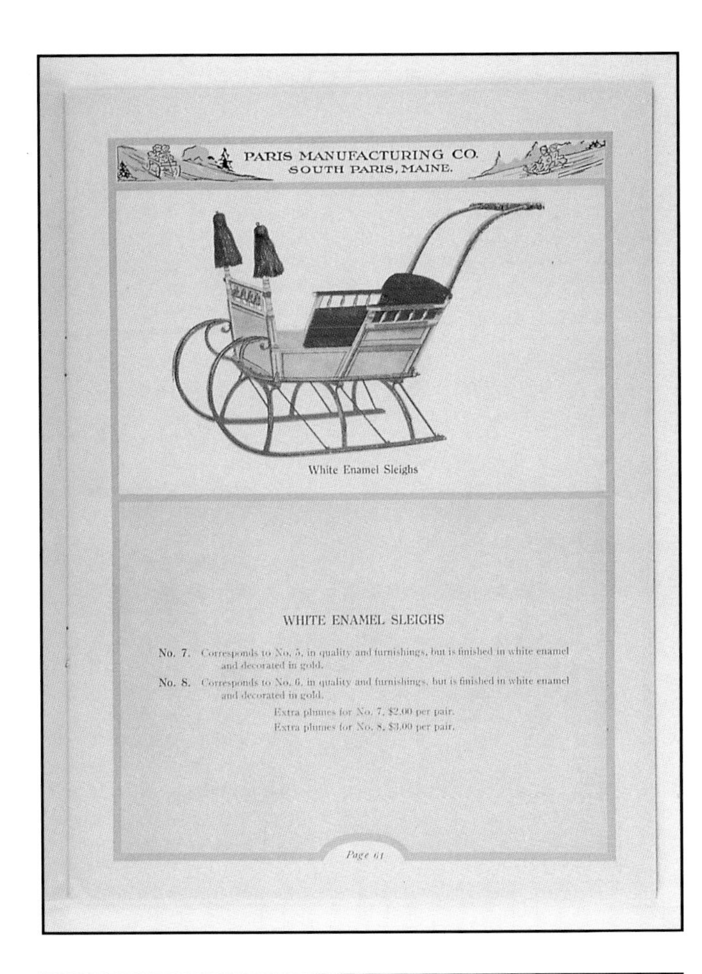

PARIS MANUFACTURING CO.
SOUTH PARIS, MAINE.

White Enamel Sleighs

WHITE ENAMEL SLEIGHS

No. 7. Corresponds to No. 5, in quality and furnishings, but is finished in white enamel and decorated in gold.

No. 8. Corresponds to No. 6, in quality and furnishings, but is finished in white enamel and decorated in gold.

Extra plumes for No. 7, $2.00 per pair.
Extra plumes for No. 8, $3.00 per pair.

Page 61

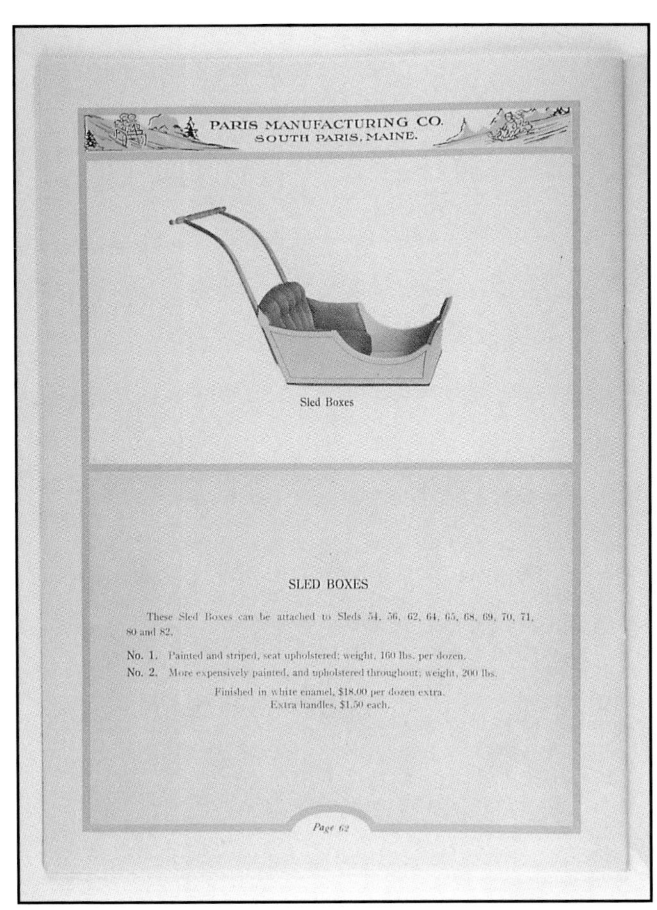

PARIS MANUFACTURING CO.
SOUTH PARIS, MAINE.

Sled Boxes

SLED BOXES

These Sled Boxes can be attached to Sleds 54, 56, 62, 64, 65, 68, 69, 70, 71, 80 and 82.

No. 1. Painted and striped, seat upholstered; weight, 160 lbs. per dozen.

No. 2. More expensively painted, and upholstered throughout; weight, 200 lbs.

Finished in white enamel, $18.00 per dozen extra.
Extra handles, $1.50 each.

Page 62

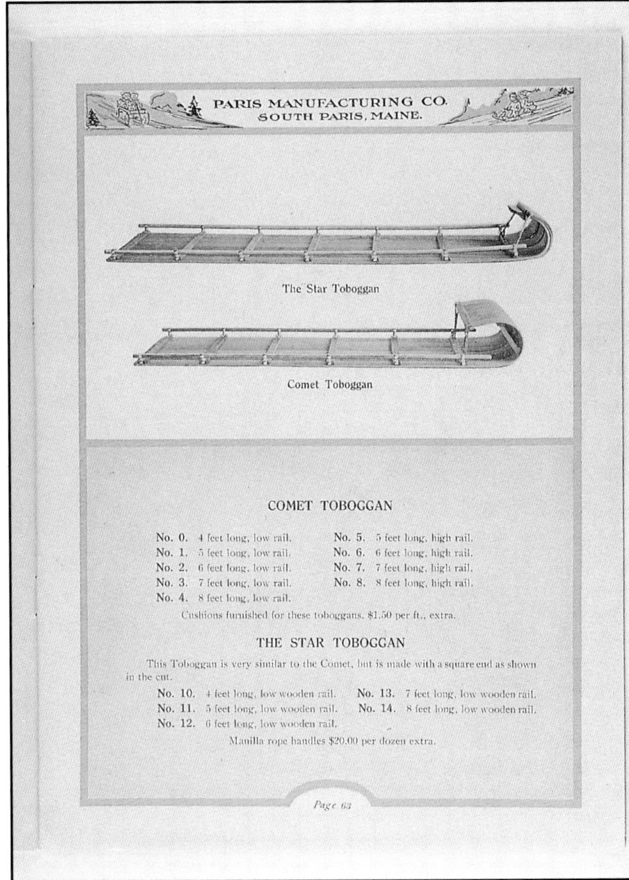

PARIS MANUFACTURING CO.
SOUTH PARIS, MAINE.

The Star Toboggan

Comet Toboggan

COMET TOBOGGAN

No. 0. 4 feet long, low rail.
No. 1. 5 feet long, low rail.
No. 2. 6 feet long, low rail.
No. 3. 7 feet long, low rail.
No. 4. 8 feet long, low rail.
No. 5. 5 feet long, high rail.
No. 6. 6 feet long, high rail.
No. 7. 7 feet long, high rail.
No. 8. 8 feet long, high rail.

Cushions furnished for these toboggans, $1.50 per ft., extra.

THE STAR TOBOGGAN

This Toboggan is very similar to the Comet, but is made with a square end as shown in the cut.

No. 10. 4 feet long, low wooden rail.
No. 11. 5 feet long, low wooden rail.
No. 12. 6 feet long, low wooden rail.
No. 13. 7 feet long, low wooden rail.
No. 14. 8 feet long, low wooden rail.

Manilla rope handles $20.00 per dozen extra.

Page 63

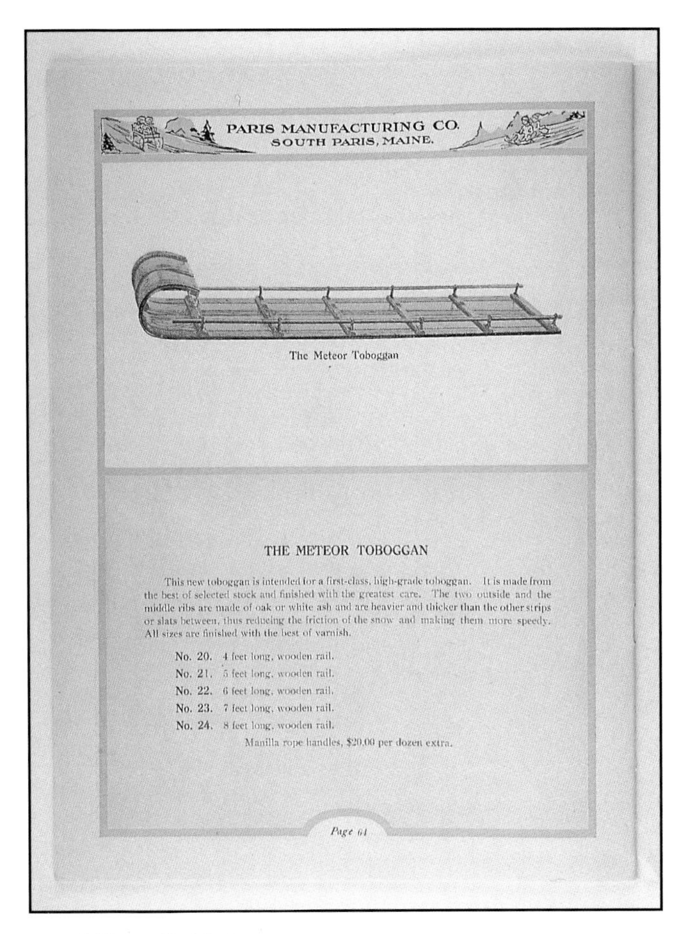

PARIS MANUFACTURING CO.
SOUTH PARIS, MAINE.

The Meteor Toboggan

THE METEOR TOBOGGAN

This new toboggan is intended for a first-class, high-grade toboggan. It is made from the best of selected stock and finished with the greatest care. The two outside and the middle ribs are made of oak or white ash and are heavier and thicker than the other strips or slats between, thus reducing the friction of the snow and making them more speedy. All sizes are finished with the best of varnish.

No. 20. 4 feet long, wooden rail.
No. 21. 5 feet long, wooden rail.
No. 22. 6 feet long, wooden rail.
No. 23. 7 feet long, wooden rail.
No. 24. 8 feet long, wooden rail.

Manilla rope handles, $20.00 per dozen extra.

Page 64

PMC catalog c. 1915. *Courtesy of Henry R. Morton.*

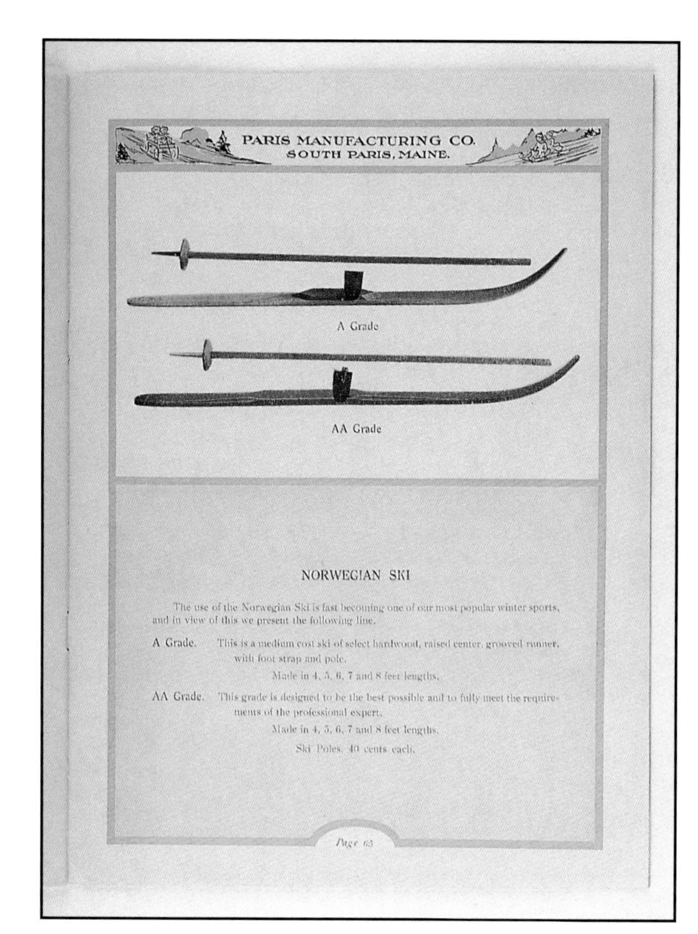

PARIS MANUFACTURING CO.
SOUTH PARIS, MAINE.

A Grade

AA Grade

NORWEGIAN SKI

The use of the Norwegian Ski is fast becoming one of our most popular winter sports, and in view of this we present the following line.

A Grade. This is a medium cost ski of select hardwood, raised center, grooved runner, with foot strap and pole.

Made in 4, 5, 6, 7 and 8 feet lengths.

AA Grade. This grade is designed to be the best possible and to fully meet the requirements of the professional expert.

Made in 4, 5, 6, 7 and 8 feet lengths.

Ski Poles, 40 cents each.

Page 63

PARIS MANUFACTURING CO.
SOUTH PARIS, MAINE.

These pictures show some of the logs being hauled to our saw mill by our steam log hauler.
We operate our own lumbering, and get out the greater port of our lumber ourselves.
We have a large number of men in the woods in the winter, and our lumbering operations are quite a part of our business.
One of our lumber camps is shown also, which is a typical view of a lumber camp in the Maine woods.
We carry a very large amount of native lumber on hand at all times.
The picture shows a load of 38,700 feet of hardwood lumber being hauled at one load, which was hauled over a seven mile road, from where we were cutting, to our saw mill.

PMC catalog c. 1915. *Courtesy of Henry R. Morton.*

Sleds

The following excerpt from a Eugene Field poem captures some of the joy of the clipper sled.

Jest 'Fore Christmas

Got a clipper sled, an' when us kids goes out to slide,
'Long comes the grocery cart, an' we all hook a ride!
But sometimes when the grocery man is worrited an' cross,
He reaches at us with his whip, an' larrups up his hoss,
An' then I laff an' holler, "Oh, ye never teched *me!*"
But jest 'fore Christmas I'm as good as I kin be!

Early Paris Cutter, c. 1870, in mint, never used condition. Owned by Henry R. Morton, painted by William P. Morton. 39" x 13" x 18".

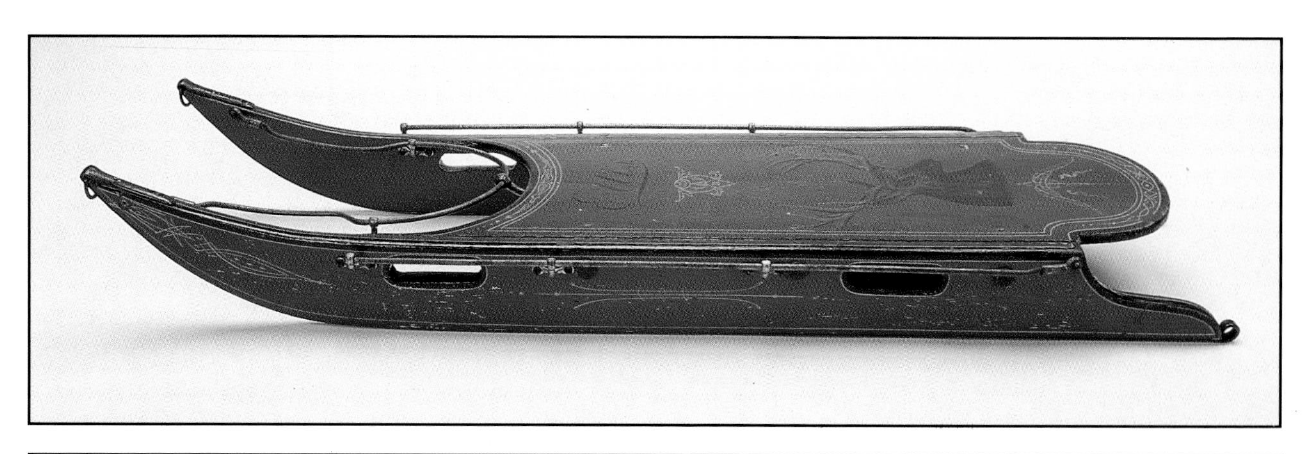

Ellis Reindeer Clipper Sled. L: 48", H: 4.25", W: 13.75". Maker unknown. *Courtesy of Lou and Carole Scudillo.* $500-1250.

Paris Cutter (as on postcard). L: 34", W: 14", H: 7". *Courtesy of Lou and Carole Scudillo.* $1000-1950.
Continued on Next Page.

Paris Cutter (as on postcard). L: 34", W: 14", H: 7". *Courtesy of Lou and Carole Scudillo.* $1000-1950.

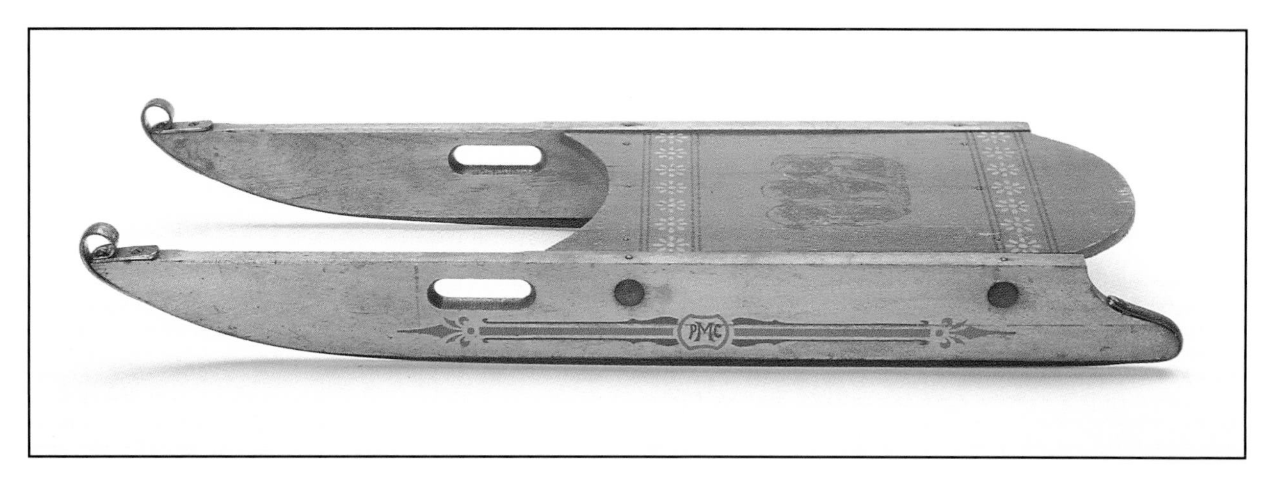

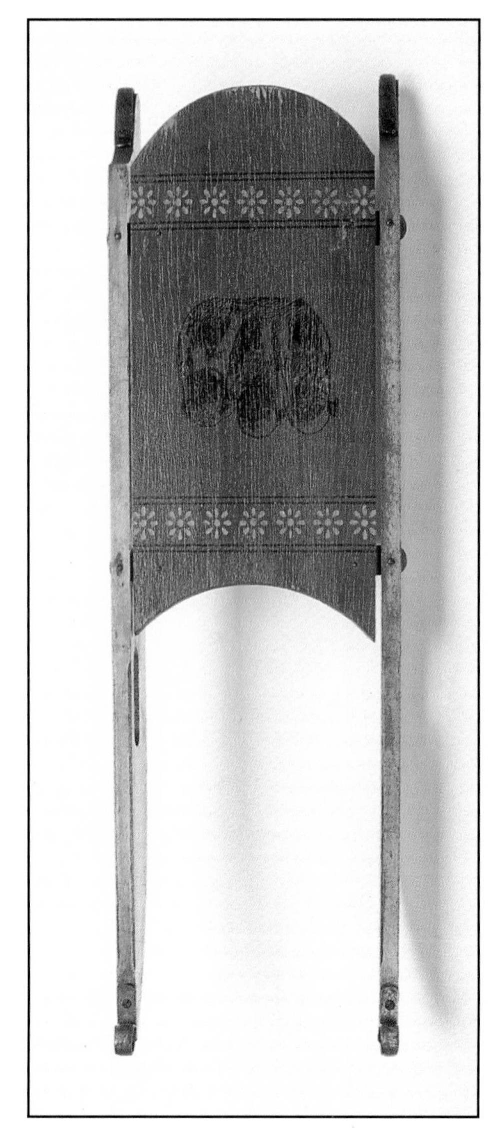

PMC Clipper. L: 32", W: 10", H: 4". *Courtesy of Lou and Carole Scudillo*. $295-550.

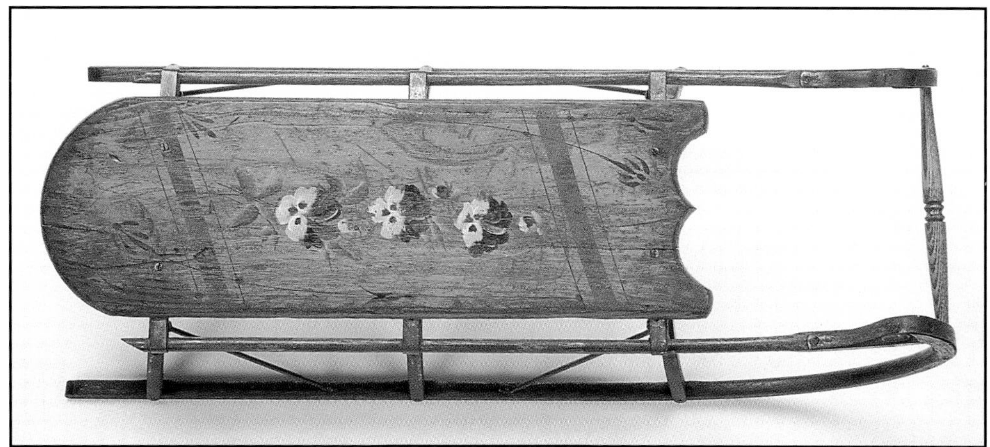

Paris Pansy. L: 36", W: 11" H: 6.5". *Courtesy of Lou and Carole Scudillo.* $450-950.

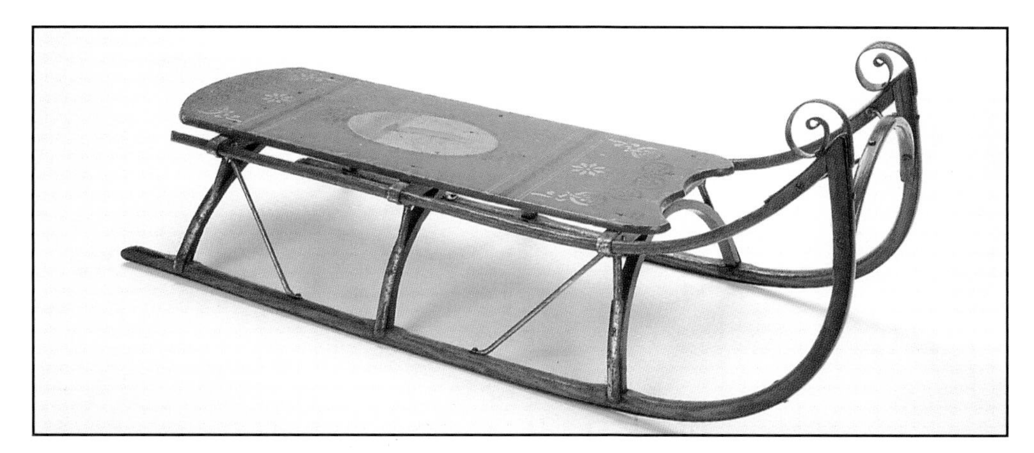

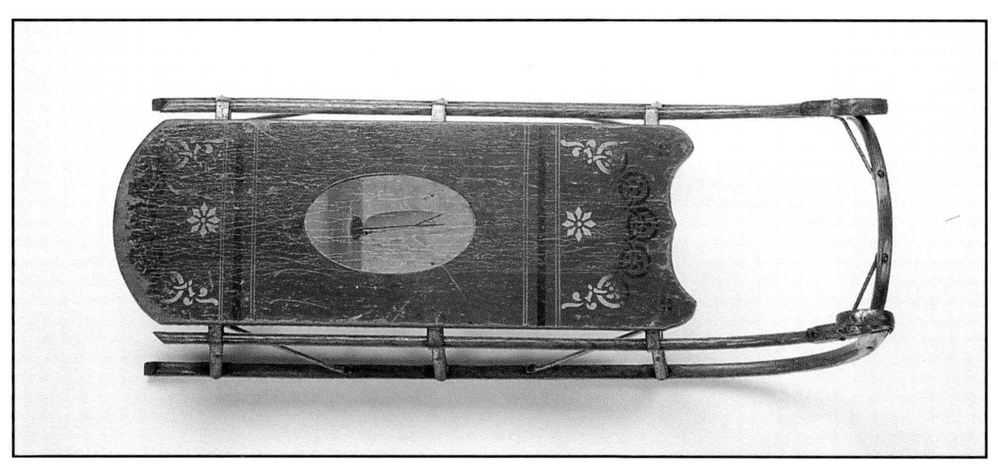

Sail Boat. Dated 1922, Ardsley, New York. Maker unknown, but possibly Paris.
L: 35.5", W: 11", H: 6.5". *Courtesy of Lou and Carole Scudillo.* $650-1450.

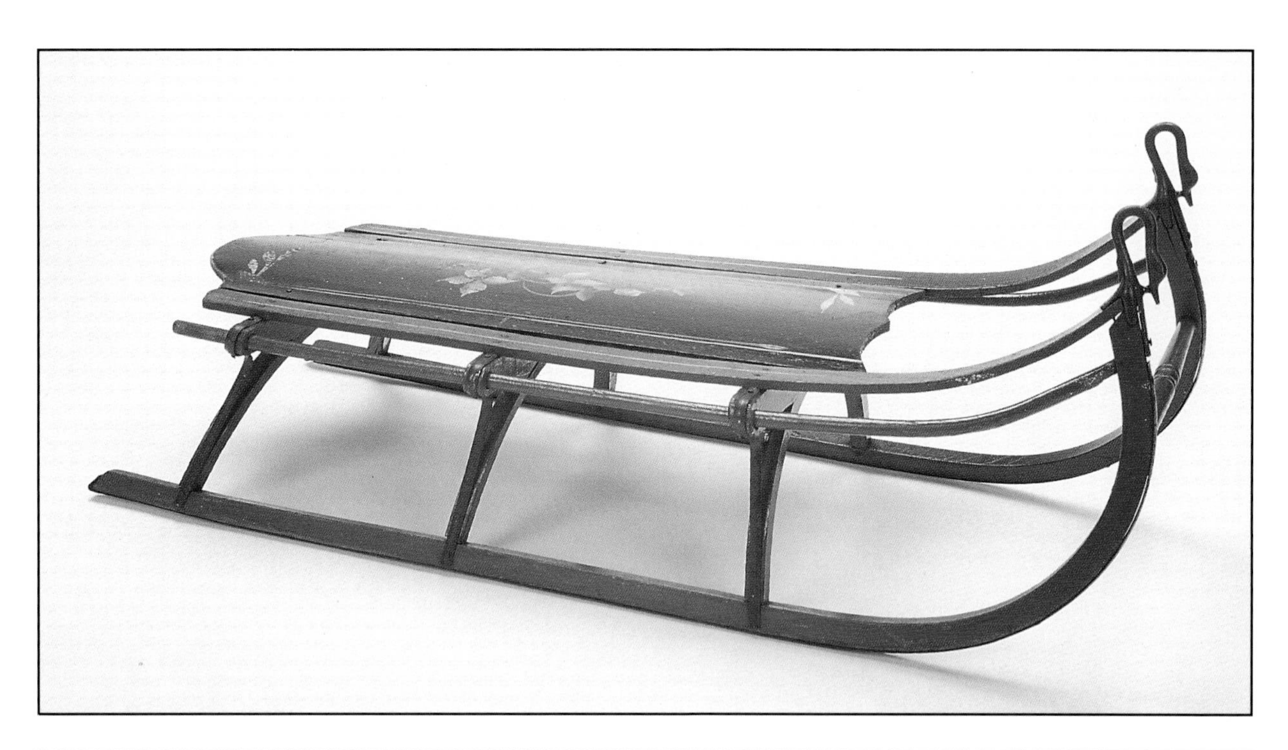

Swan (gooseneck). L: 36", W: 13", H: 7". *Courtesy of Lou and Carole Scudillo*. $450-950.

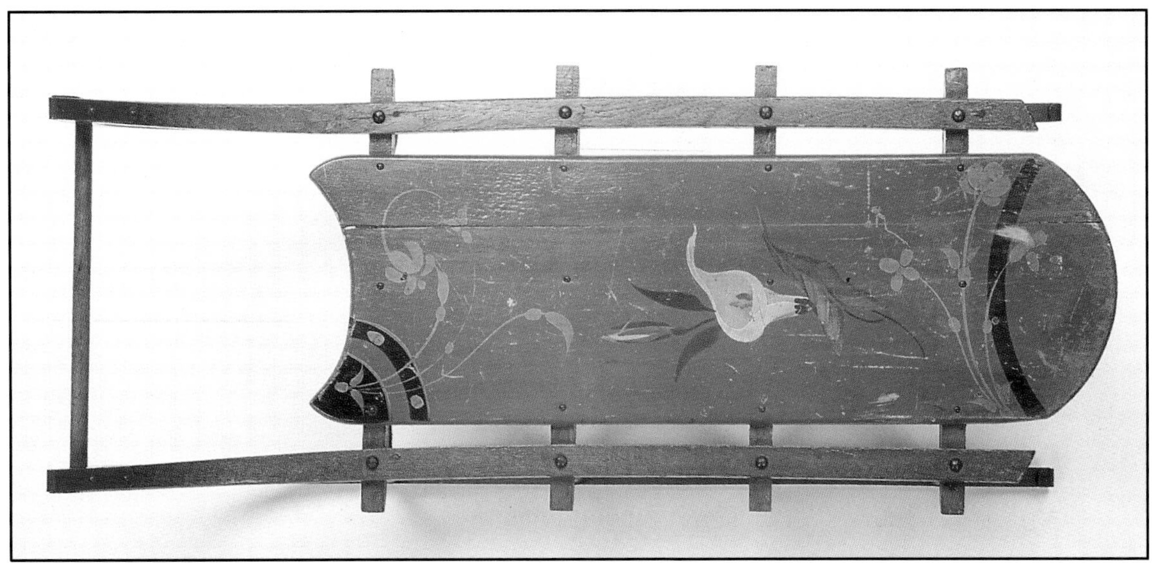

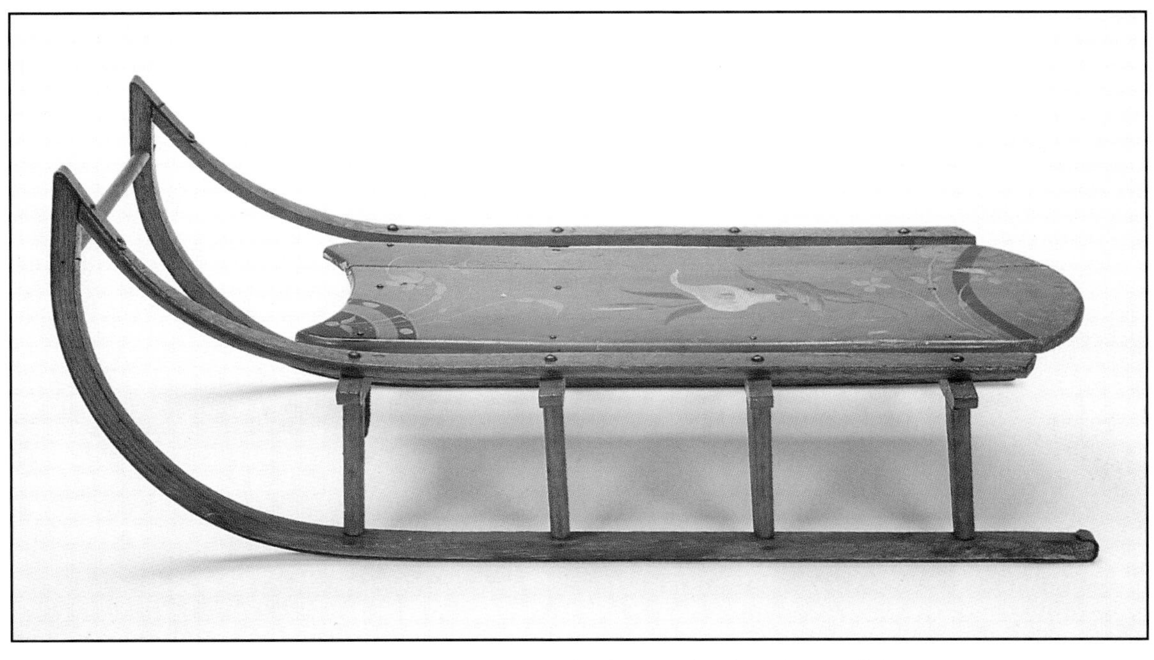

Calla Lily. L: 34", W: 12", H: 6.5". Note the four bent knees. Maker unknown. *Courtesy of Lou and Carole Scudillo.* $500-895.

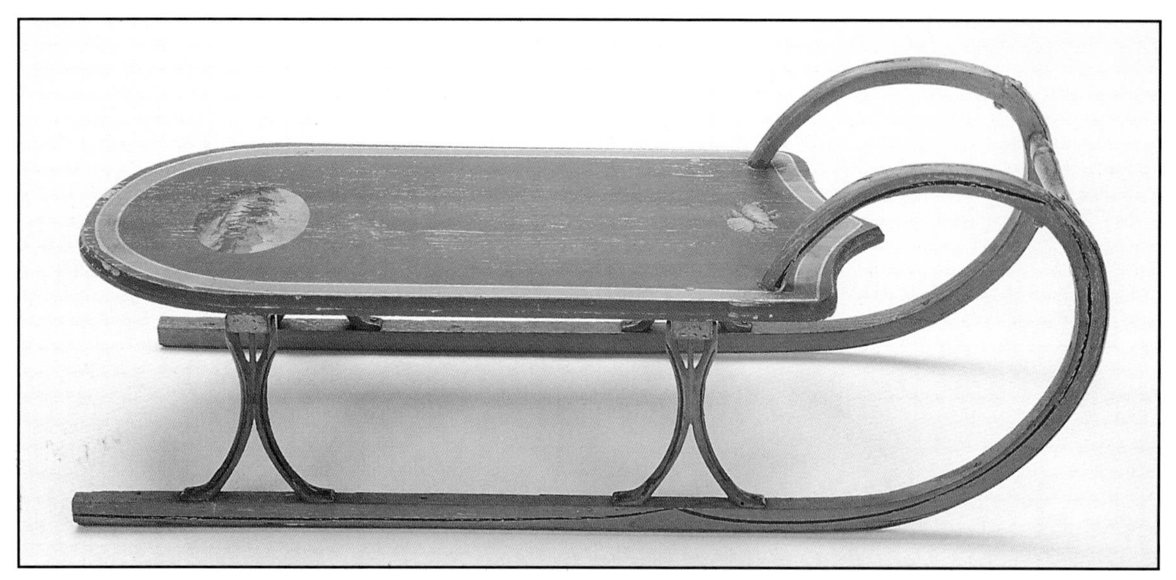

Sea shells & Camels. Typical bow runner sled. L: 31", W: 11.5", H: 7.5". Maker unknown. *Courtesy of Lou and Carole Scudillo.* $350-750.

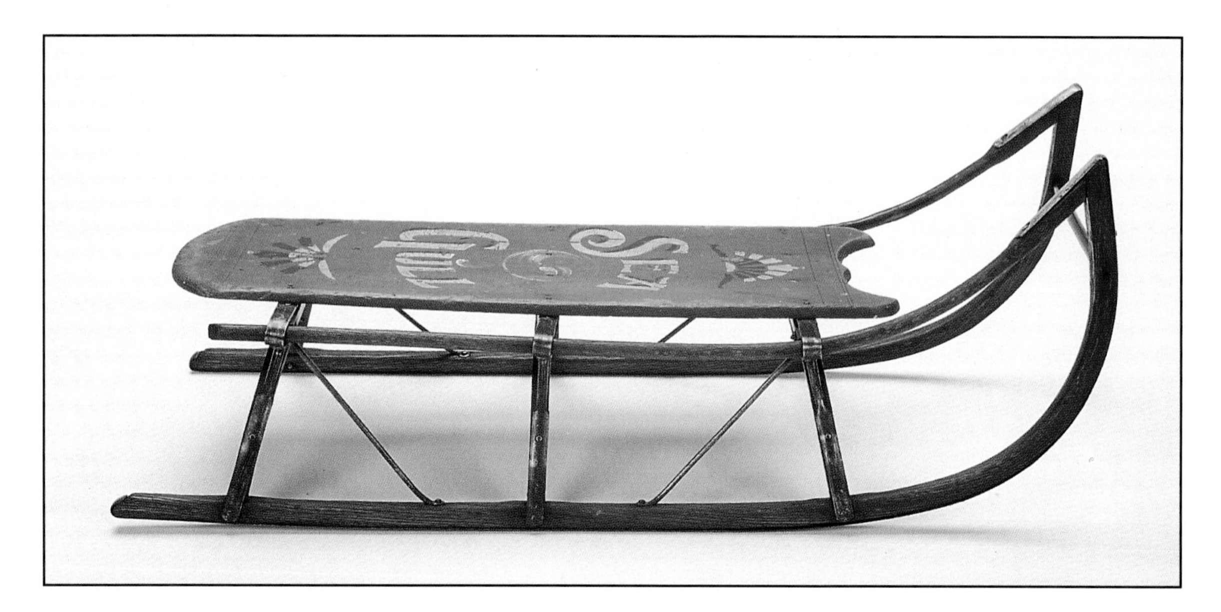

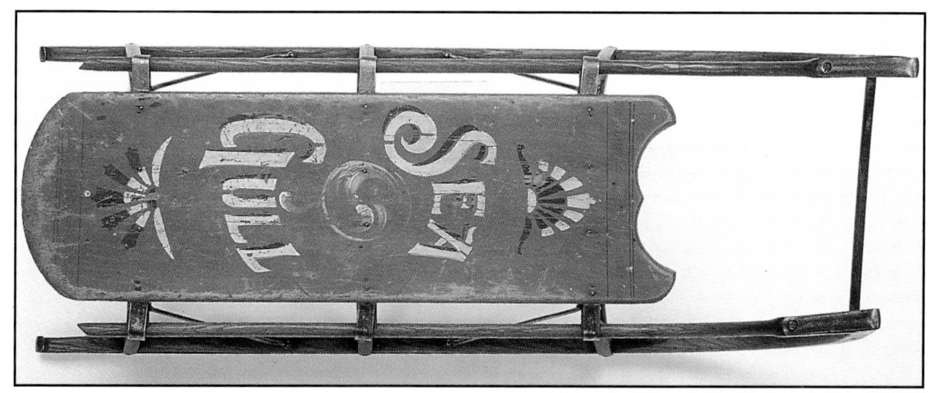

Sea Gull. L: 31", W: 10", H: 6". *Courtesy of Lou and Carole Scudillo.* $450-850.

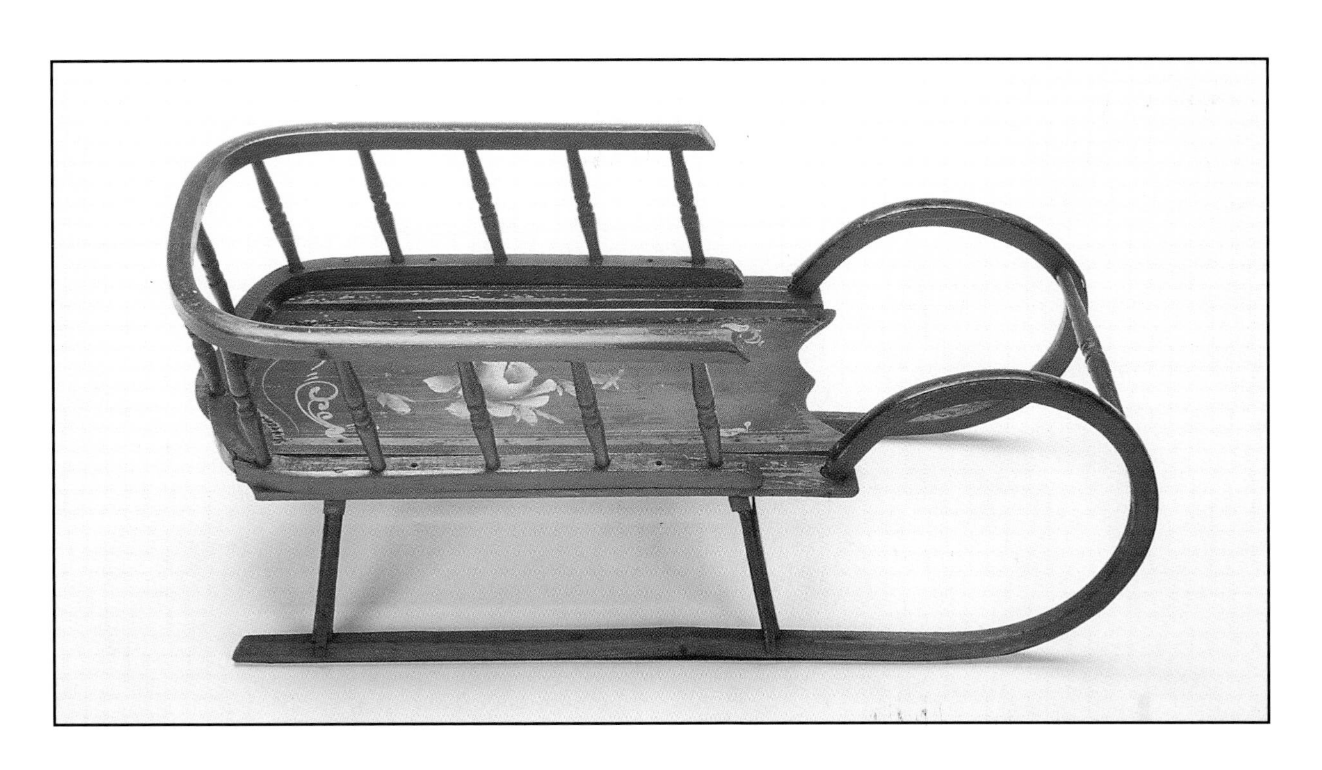

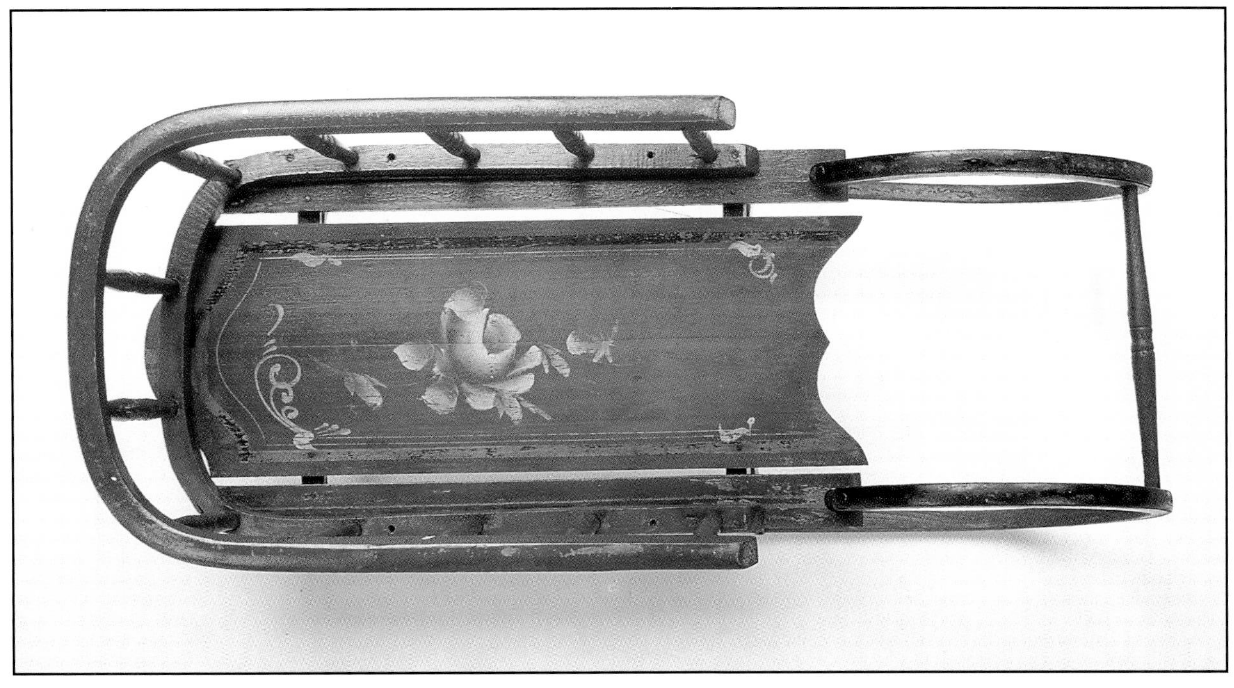

Seat sled with railing. L: 32", W: 12", H: 9". *Courtesy of Lou and Carole Scudillo.* $375-850.

Next Page:
Paris Dasher Push Sleigh. L: 35" w/o handle, W: 16", H: 22" back of seat, 10" runners. *Courtesy of Lou and Carole Scudillo.* $1500-3500.

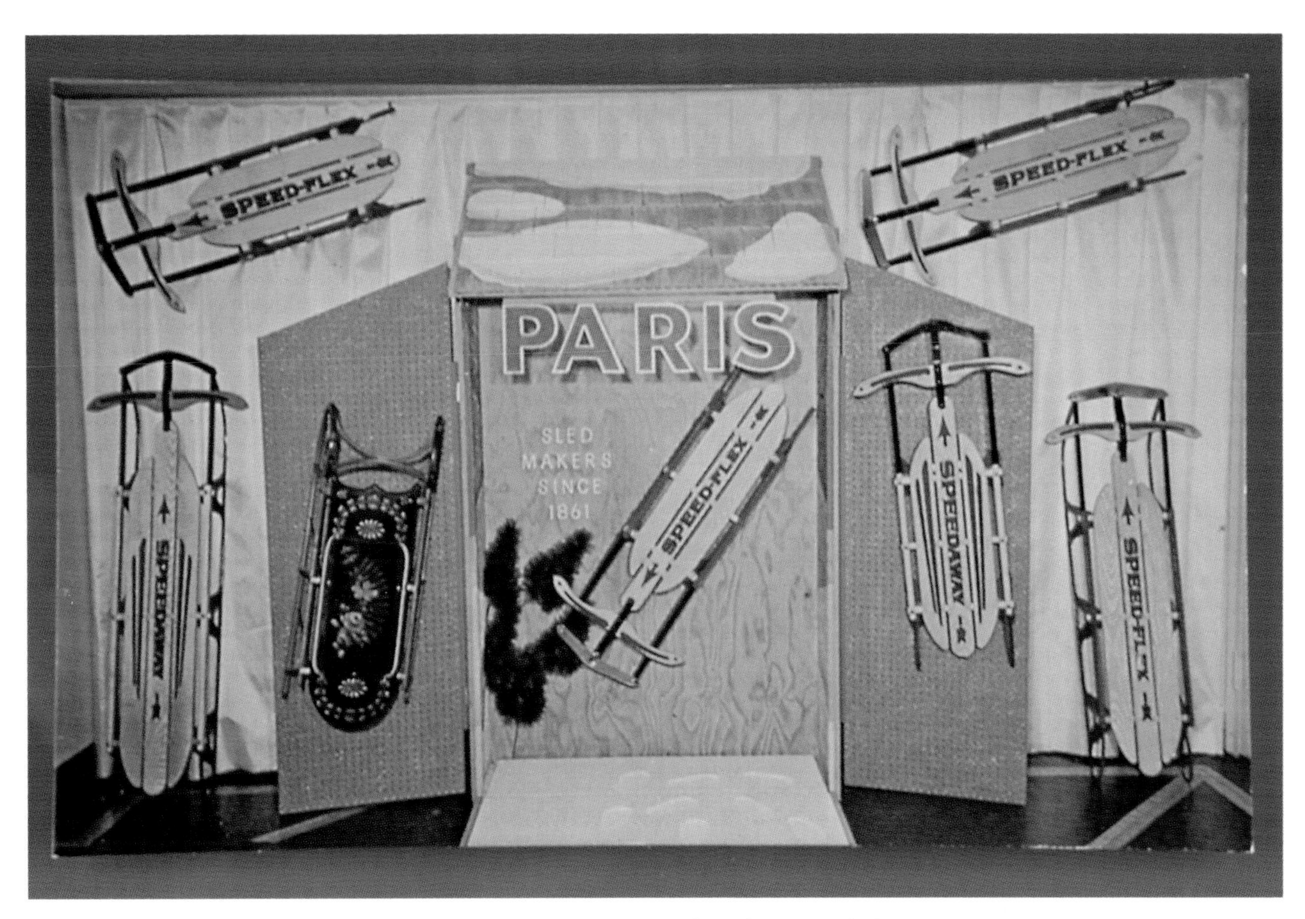

Advertising postcard from Paris Sled Company, S. Paris, Maine. The card is from c. 1950. $25-50.

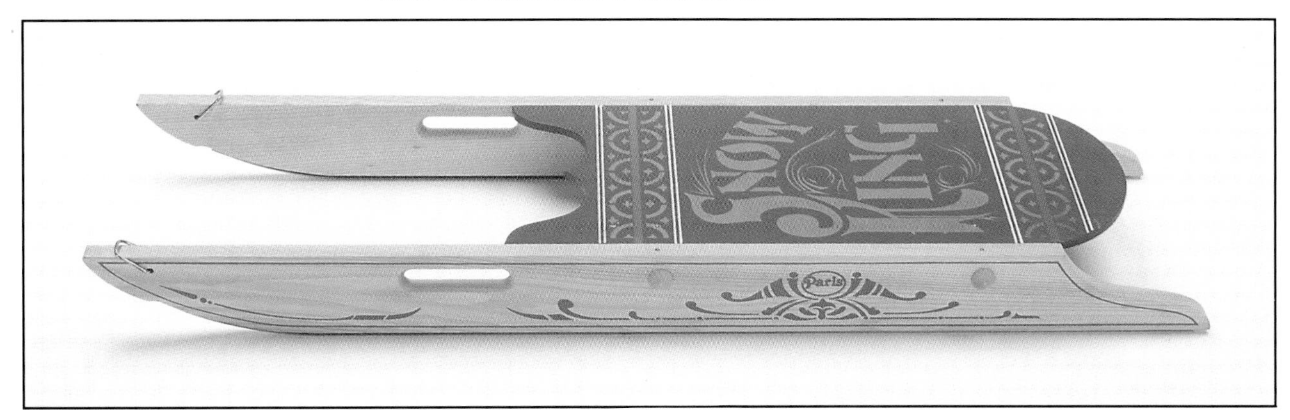

1986. Paris. Reproduction of original Paris Clipper Snow King. Runners different from original. L: 42", W: 12", H: 3.5". $75-125.

You Never in Your Life Saw A Sled Like This One!

It Rolls as Well as Slides!

Little wheels in WHIZZARD runners make it go like sixty—Snow or no snow—it's all the same to a WHIZZARD. Sweeps right through bare spots, past other sleds that have stalled—right to the bottom of the hill.

And a steering wheel or "rudder" in the front saves dragging your feet or twisting the sled. You can steer a WHIZZARD with no effort at all.

You can drag a WHIZZARD on bare sidewalks—no need of going into the street.

A Winter Full of Fun—*AND MORE*
WHIZZARD coasts in the summertime too.

Price $10.00 ($11.00 west of Denver)

If your dealer does not have it, send direct.

THE POLLACK ROLLER RUNNER SLED CO., INC.
Lawyers Bldg., Beacon St., Boston, Mass., U. S. A.

Gentlemen: Please ship immediately one WHIZZARD, the all-season sled. I am enclosing check or money order for $10.00. (If west of Denver, $11.00). ☐Please ship C. O. D.☐(Indicate preference.) You agree to refund my money if I am not thoroughly satisfied, provided I return the sled in good condition within five days.

Name ....

No. .... Street ....

City ....

State ....

1927 advertisement for Pollock's Whizzard sled, with wheels that "make it go like sixty—Snow or no snow."

83

SKIPPY SNO-PLANE
STREAMLINED SLEDS

Here you are, kids—a real streamlined sled. It's ultra-modern in beauty, and it'll go faster than any sled you ever took out on the hill. Remarkably easy to steer. Tested and approved by Professional Drivers on America's most dangerous run (Mt. Van Hovenburg, Lake Placid, N. Y.). Made in 3 sizes. Retail prices, $5.95, $6.95, $7.95.*

● This is how it works—that middle runner gives positive direction and equalized spring suspension which automatically adjusts steering.

RACERS

Skippy now offers a new Streamlined Wagon. It's a beauty. You'll like the new steering system that makes this wagon so easy to handle. Retail price $8.95.*Skippy Racers sold by leading stores everywhere. Skippy Racers, Toledo, Ohio — World's Finest Children's Vehicles.
Prices slightly higher west of the Rockies.

SEE THE NEW SKIPPY CHEVROLET AUTOS WITH FREE FISHER BODY "TURRET TOP" SAFETY HELMETS

BOYS AND GIRLS:
Clip this advertisement!
Take it to the nearest store handling Skippy Racers and receive a Skippy cutout card. You can cut out and build miniature models of the large Skippy Racers.

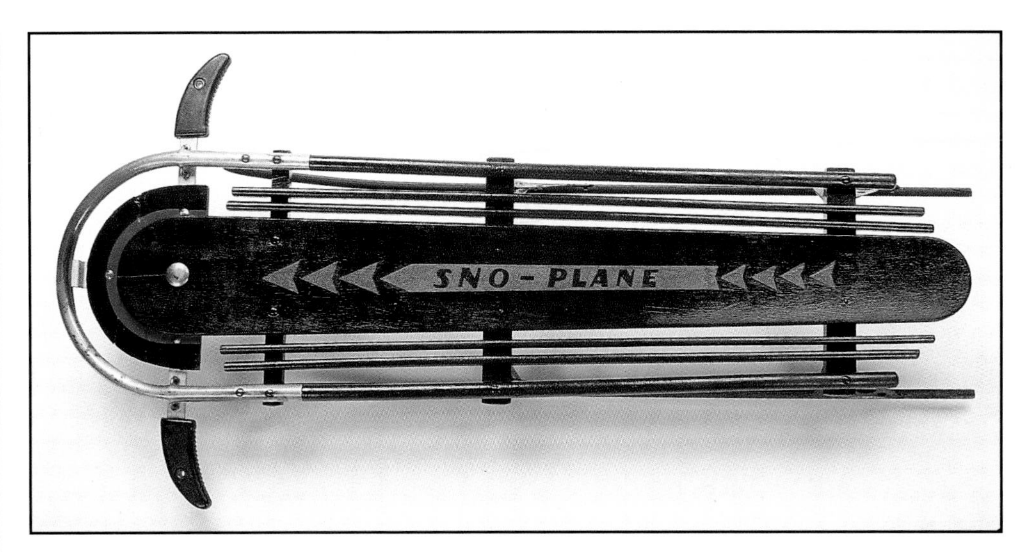

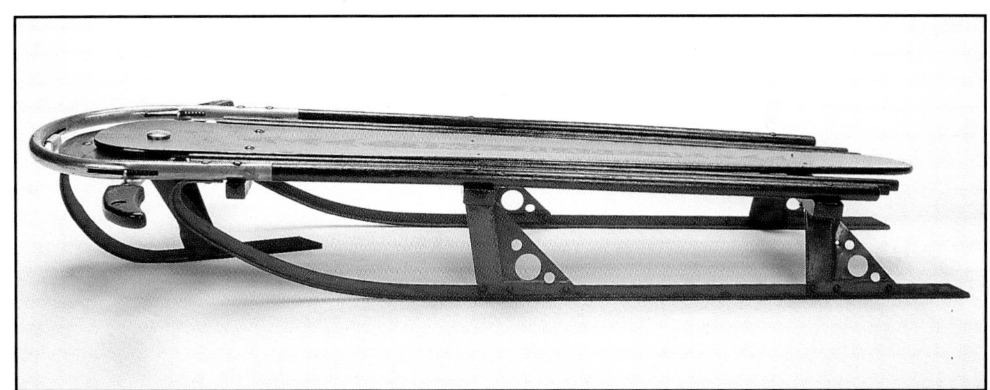

Above:
1933. Sno-Plane (formerly American National). Skippy racers. Unique center runner. L: 53", W: 13", H: 7". $125-200.

Left:
1933 *Saturday Evening Post* advertisement for Skippy's Sno-Plane.

Standard Novelty Works

C.A. Walters, William Wills, and P.F. Duncan established the Standard Novelty Company in 1904 in Duncannon, Pennsylvania. The company manufactured porch swings, gates, furniture, children's wagons, and Lightning Guider Sleds.

Early models included The Racer and Sno-ball, non-steerable sleds, Lightning Guider Sleds, and Bobs. By the 1920s the company was credited with producing more children's sleds than any other American Company.

It is not clear when C.A. Walters sold his fifty shares of stock to William Wills and P.F. Duncan, but in 1968 William Wills and P.F. Duncan sold Standard Novelty Works to the Rosen Family.

The Rosen's continued to manufacture Lightning Guider Sleds introducing new models, the American Racer and Challenger.

In 1984, the Anniversary Sled was released. Production ceased in 1988 and the factory officially closed October, 1990 ending 84 years of sled making.

Today the factory operates as the Old Sled Works, a museum, antique, and craft center.

A happy footnote to a by-gone era!

A Chronology of the Standard Novelty Works

1904: Company Incorporated
1907: Fire destroyed building on Market Street
1915: Lightning Guider bobs introduced
1936: April flood waters threatened sled production. In August the Master Bomber was introduced
1939: Lightning Speedster introduced
1968: Company sold to the Rosen Family
1972: Flood waters halted sled production. Four feet of water in main building, destroying historic documents.
1984: Anniversary Model Released commemorating 80 years of sled production. 414 sleds produced
1988: Last production year
1990: In October the factory officially closes
1991: In April the factory opens as Old Sled Works, a museum, antique, and craft center.
1992: In July the home of Lightning Guider Sleds became an official State Historical Site.

Overhead photographs of The Standard Novelty Works in Duncannon, Pennsylvania. It was in operation making sleds until 1989 when it was converted to an antique mall. $50-100.
Contiued on Next Page.

Overhead photographs of The Standard Novelty Works in Duncannon, Pennsylvania. It was in operation making sleds until 1989 when it was converted to an antique mall. $50-100.

Photograph from the Standard Novelty Works, early 1900s.

Advertising

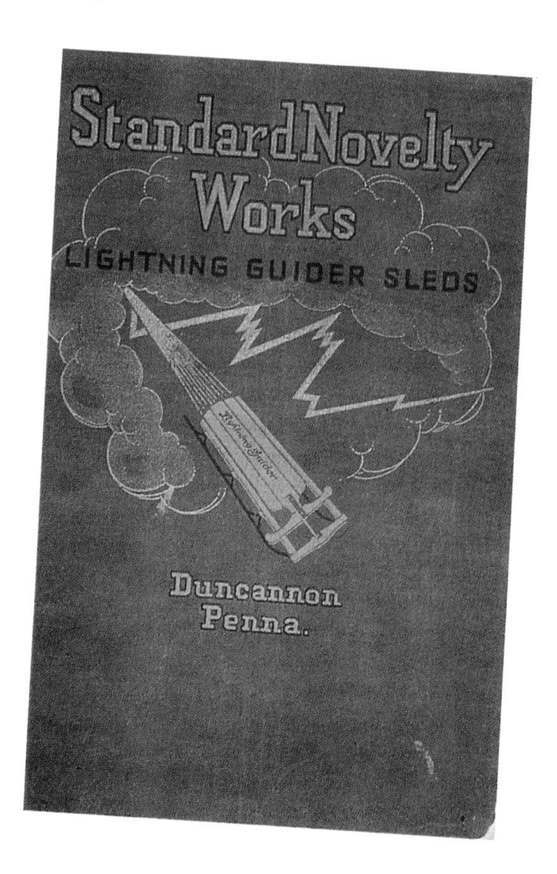

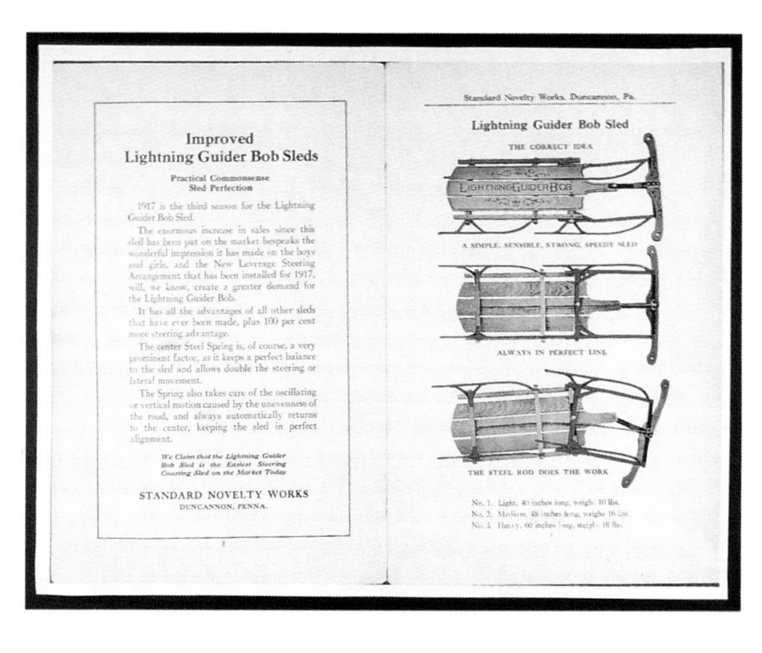

Improved Lightning Guider Bob Sleds

Practical Commonsense Sled Perfection

1917 is the third season for the Lightning Guider Bob Sled.

The enormous increase in sales since this sled has been put on the market bespeaks the wonderful impression it has made on the boys and girls, and the New Leverage Steering Arrangement that has been installed for 1917, will, we know, create a greater demand for the Lightning Guider Bob.

It has all the advantages of all other sleds that have ever been made, plus 100 per cent more steering advantage.

The center Steel Spring is, of course, a very prominent factor, as it keeps a perfect balance to the sled and allows double the steering or lateral movement.

The Spring also takes care of the oscillating or vertical motion caused by the unevenness of the road, and always automatically returns to the center, keeping the sled in perfect alignment.

We Claim that the Lightning Guider Bob Sled is the Easiest Steering Coasting Sled on the Market Today

STANDARD NOVELTY WORKS
DUNCANNON, PENNA.

Standard Novelty Works, Duncannon, Pa.

Lightning Guider Bob Sled

THE CORRECT IDEA

A SIMPLE, SENSIBLE, STRONG, SPEEDY SLED

ALWAYS IN PERFECT LINE

THE STEEL ROD DOES THE WORK

No. 1. Light, 40 inches long, weighs 10 lbs.
No. 2. Medium, 48 inches long, weighs 16 lbs.
No. 3. Heavy, 60 inches long, weighs 18 lbs.

Catalog of the Standard Novelty Works, Duncannon, Pennsylvania, 1917.
Courtesy of Sledworks.

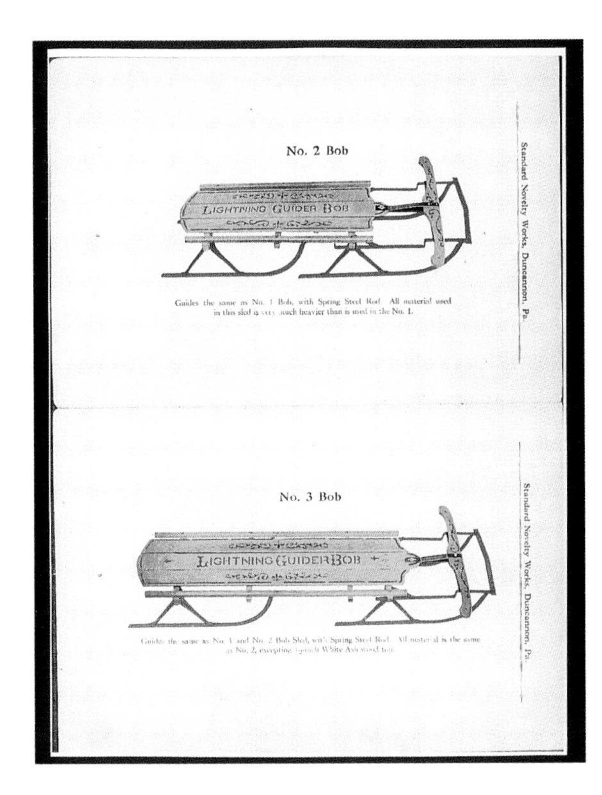
Standard Novelty Works, Duncannon, Pa.

No. 2 Bob

Guides the same as No. 1 Bob, with Spring Steel Rod. All material used in this sled is very much heavier than is used in the No. 1.

Standard Novelty Works, Duncannon, Pa.

No. 3 Bob

Guides the same as No. 1 and No. 2 Bob Sled, with Spring Steel Rod. All material is the same as No. 2, excepting ⅞-inch White Ash wood top.

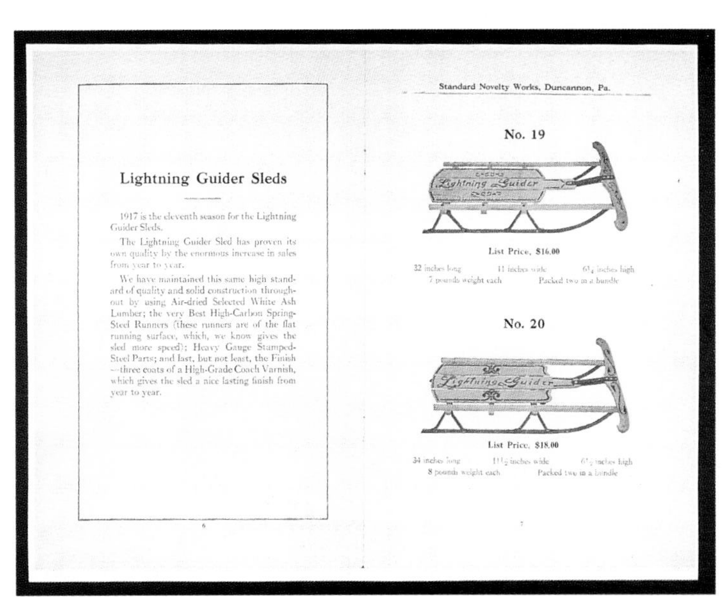
Lightning Guider Sleds

1917 is the eleventh season for the Lightning Guider Sleds.

The Lightning Guider Sled has proven its own quality by the enormous increase in sales from year to year.

We have maintained this same high standard of quality and solid construction throughout by using Air-dried Selected White Ash Lumber; the very Best High-Carbon Spring-Steel Runners (these runners are of the flat running surface, which, we know gives the sled more speed); Heavy Gauge Stamped-Steel Parts; and last, but not least, the Finish—three coats of a High-Grade Coach Varnish, which gives the sled a nice lasting finish from year to year.

6

Standard Novelty Works, Duncannon, Pa.

No. 19

List Price, $16.00

32 inches long 11 inches wide 6¼ inches high
7 pounds weight each Packed two in a bundle

No. 20

List Price, $18.00

34 inches long 11½ inches wide 6½ inches high
8 pounds weight each Packed two in a bundle

7

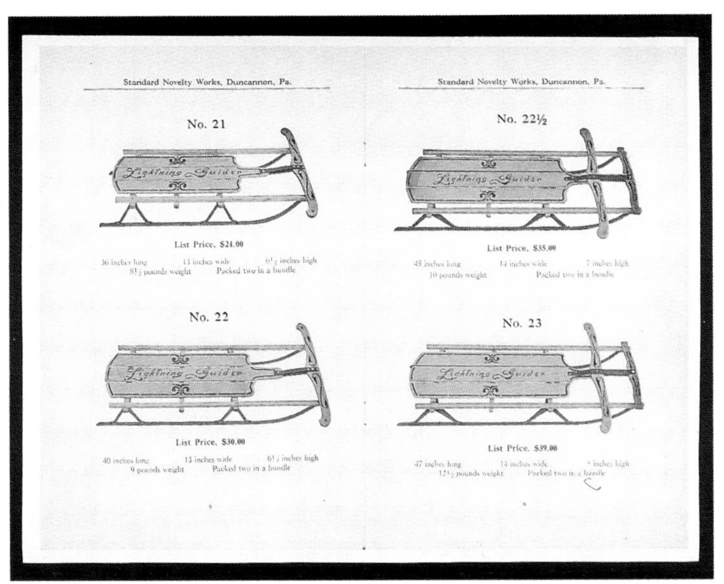
Standard Novelty Works, Duncannon, Pa.

No. 21

List Price, $24.00

36 inches long 13 inches wide 6½ inches high
8½ pounds weight Packed two in a bundle

No. 22

List Price, $30.00

40 inches long 13 inches wide 6½ inches high
9 pounds weight Packed two in a bundle

8

Standard Novelty Works, Duncannon, Pa.

No. 22½

List Price, $35.00

45 inches long 14 inches wide 7 inches high
10 pounds weight Packed two in a bundle

No. 23

List Price, $39.00

47 inches long 14 inches wide 7 inches high
12½ pounds weight Packed two in a bundle

9

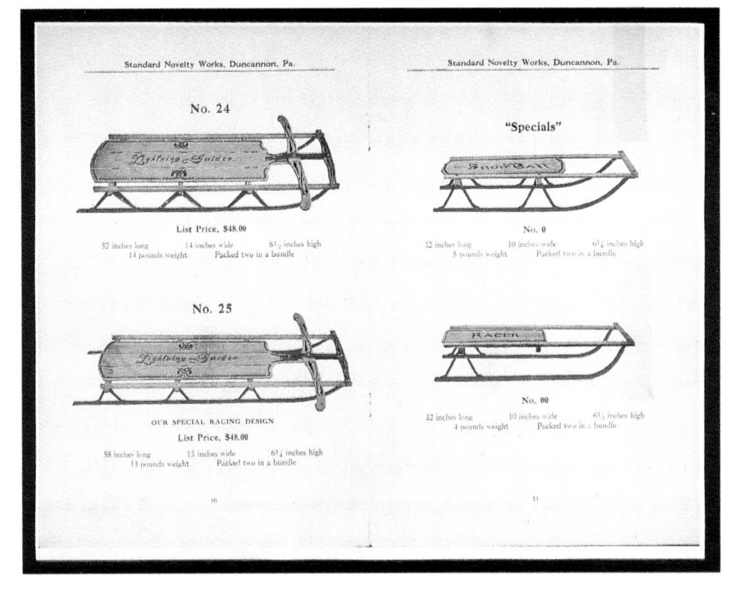
Standard Novelty Works, Duncannon, Pa.

No. 24

List Price, $48.00

52 inches long 14 inches wide 8½ inches high
14 pounds weight Packed two in a bundle

No. 25

OUR SPECIAL RACING DESIGN

List Price, $48.00

58 inches long 13 inches wide 6¼ inches high
13 pounds weight Packed two in a bundle

10

Standard Novelty Works, Duncannon, Pa.

"Specials"

No. 0

32 inches long 10 inches wide 6¼ inches high
5 pounds weight Packed two in a bundle

No. 00

32 inches long 10 inches wide 6½ inches high
4 pounds weight Packed two in a bundle

11

Catalog of the Standard Novelty Works, Duncannon, Pennsylvania, 1917. *Courtesy of Sledworks.*

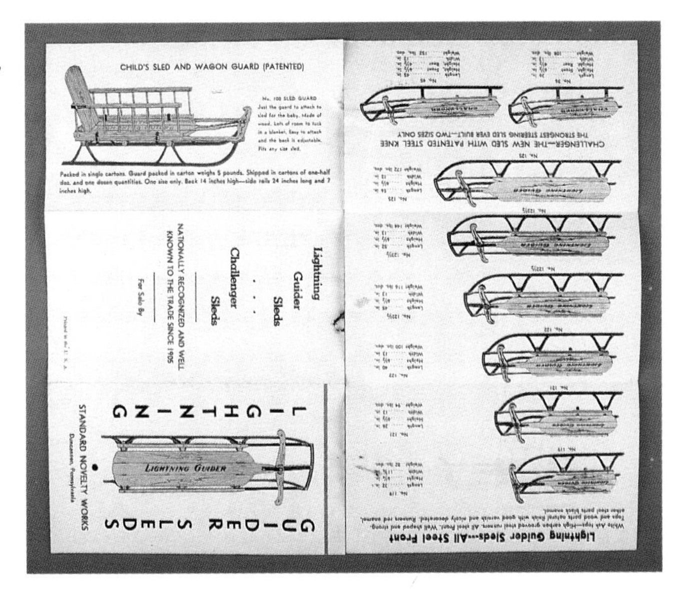
CHILD'S SLED AND WAGON GUARD (PATENTED)

No. 100 SLED GUARD
Just the guard to attach to sled for the baby. Made of wood. Lots of room to tuck in a blanket. Easy to attach and the back is adjustable. Fits any size sled.

Packed in single cartons. Guard packed in carton weighs 5 pounds. Shipped in cartons of one-half doz. and one dozen quantities. One size only. Back 14 inches high—side rails 24 inches long and 7 inches high.

Lightning
Guider
Sleds
. . .
Challenger
Sleds

NATIONALLY RECOGNIZED AND WELL KNOWN TO THE TRADE SINCE 1905

For Sale By

Printed in the U. S. A.

LIGHTNING GUIDER SLEDS

STANDARD NOVELTY WORKS
Duncannon, Pennsylvania

CHALLENGER—THE NEW SLED WITH PATENTED STEEL KNEE
THE STRONGEST STEERING SLED EVER BUILT—TWO SIZES ONLY

Lightning Guider Sleds—All Steel Front

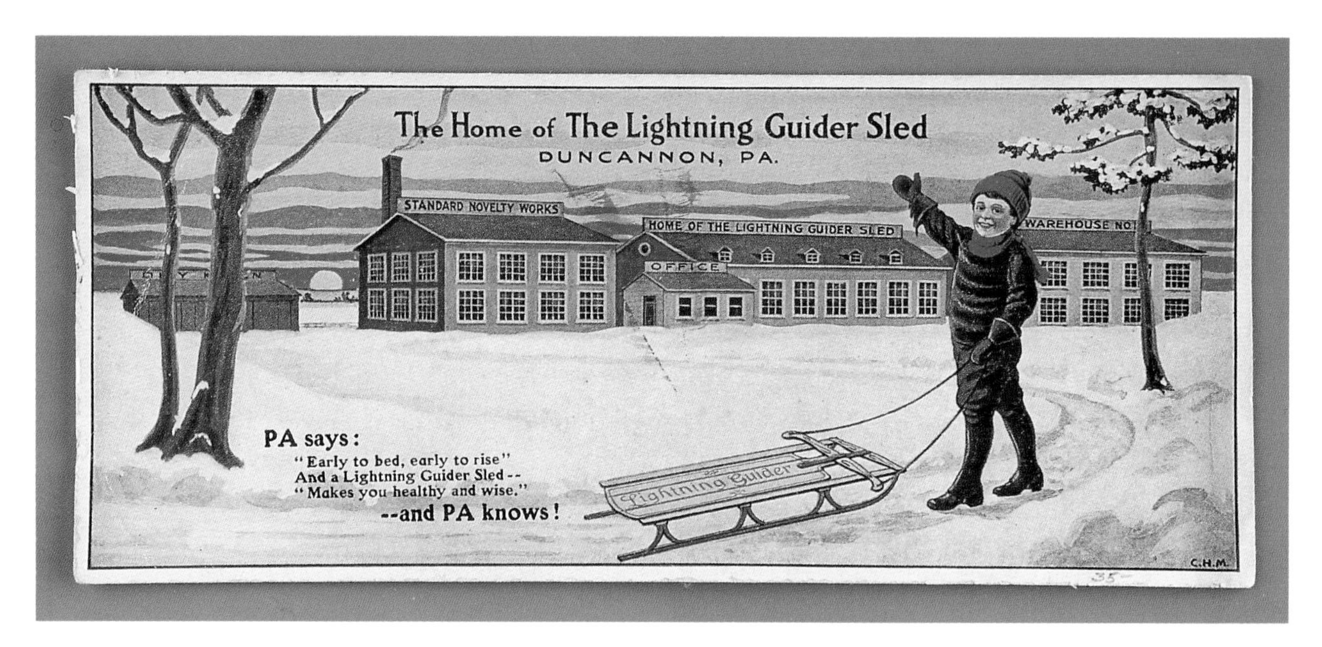

Blotter for the Standard Novelty Works, 1920s. 8.75" x 3.5". $25-50.

Watch fob advertising Standard Novelty Works.
Ca. 1930s. $100-250.

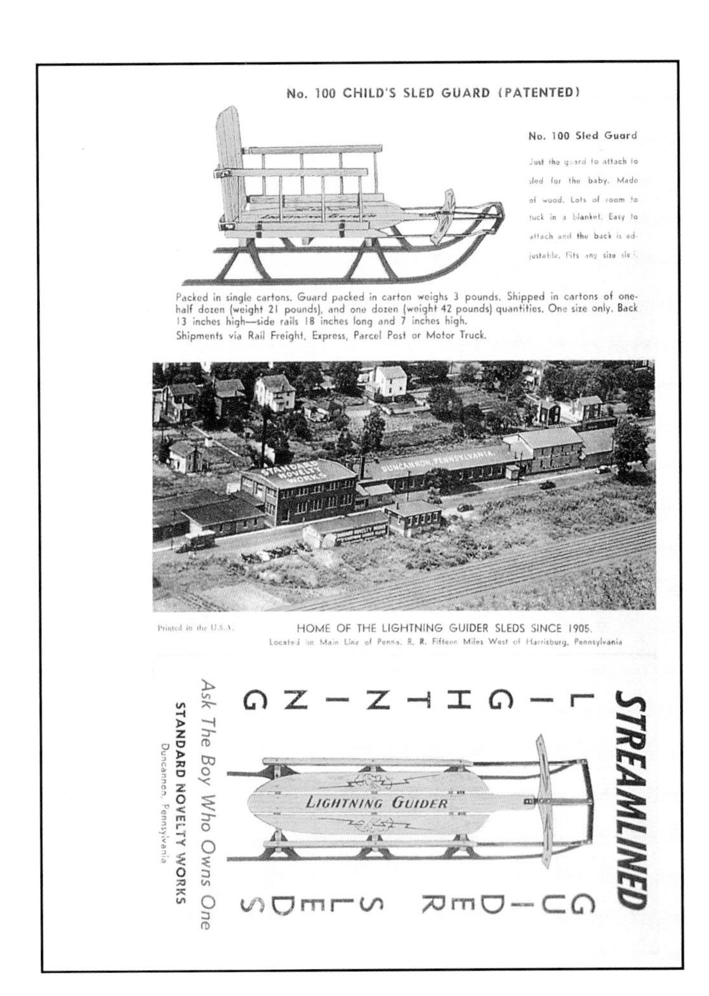

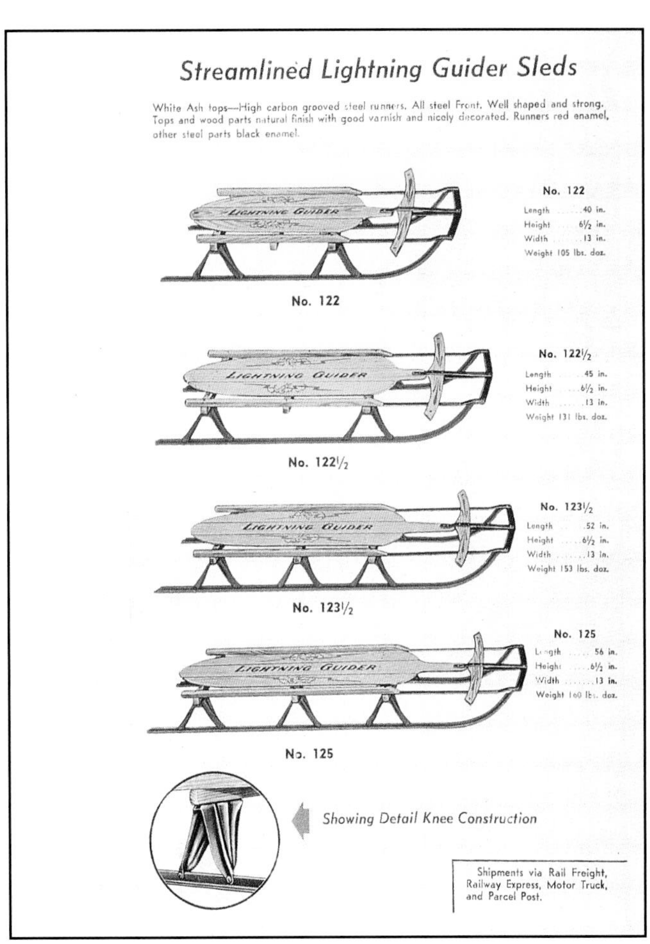

Brochure of Standard Novelty Works, Duncannon, Pennsylvania, featuring the "Streamlined Lightning Guider Sleds, 1950s. *Courtesy of Sledworks.*

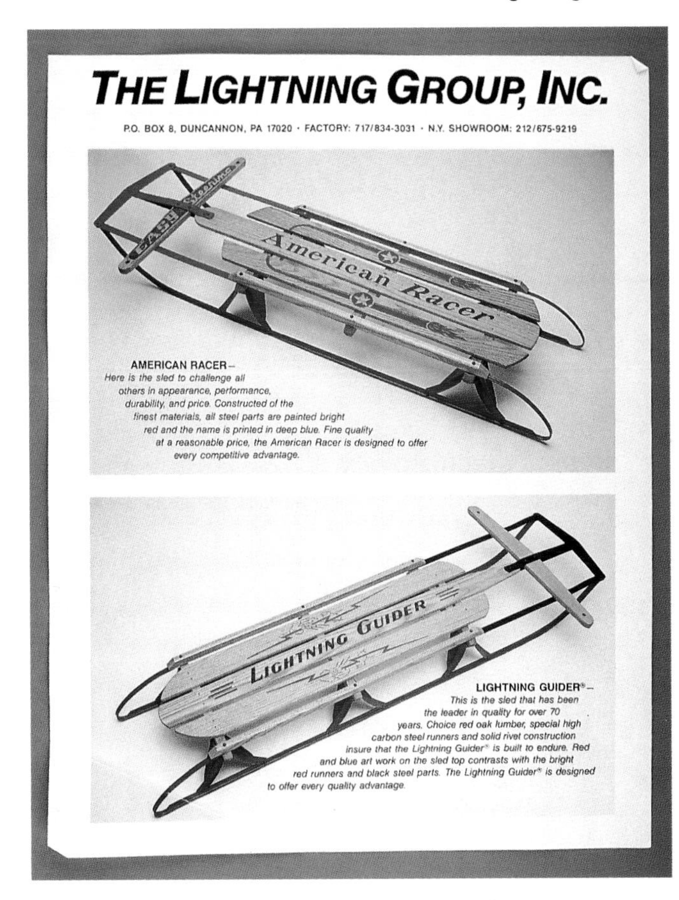

Advertising flier for The Lightning Group Standard Novelty Works, c. 1970. $5-15

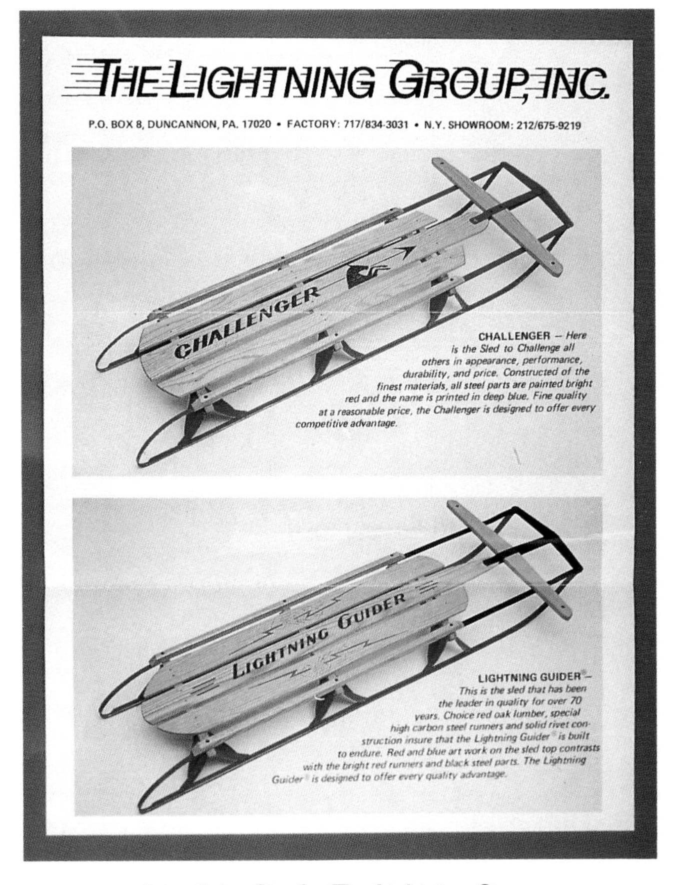

Advertising flier for The Lightning Group Standard Novelty Works, c. 1970. $5-15

Sleds

The sleds pictured here are courtesy of Sled Works, Duncannon, Pennsylvania.

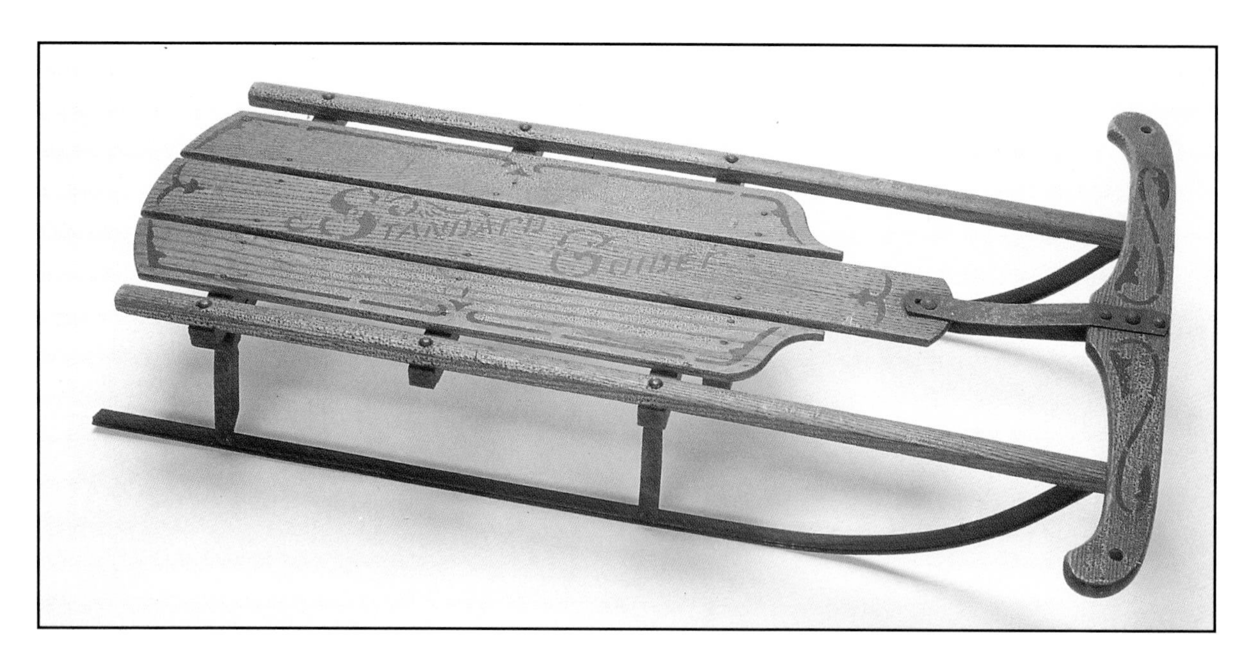

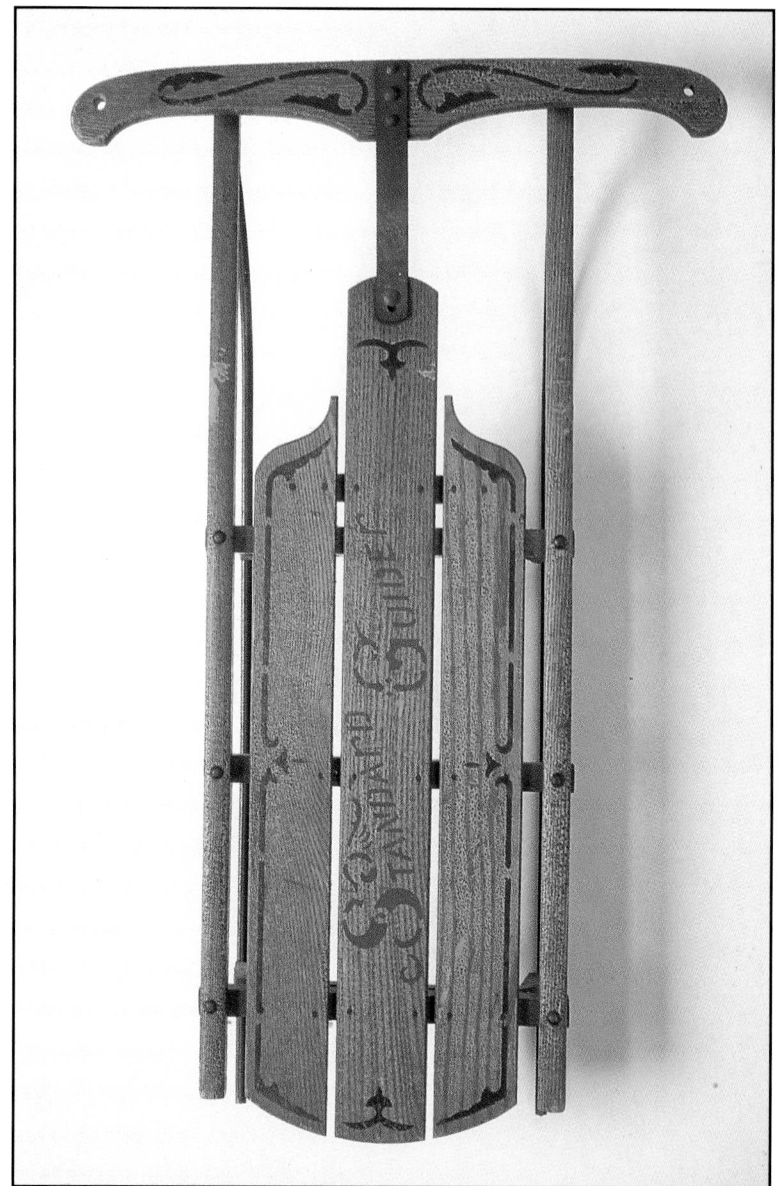

C. 1904. Standard Guider.
L: 35", W: 11.75", H: 6.25".
$50-300.

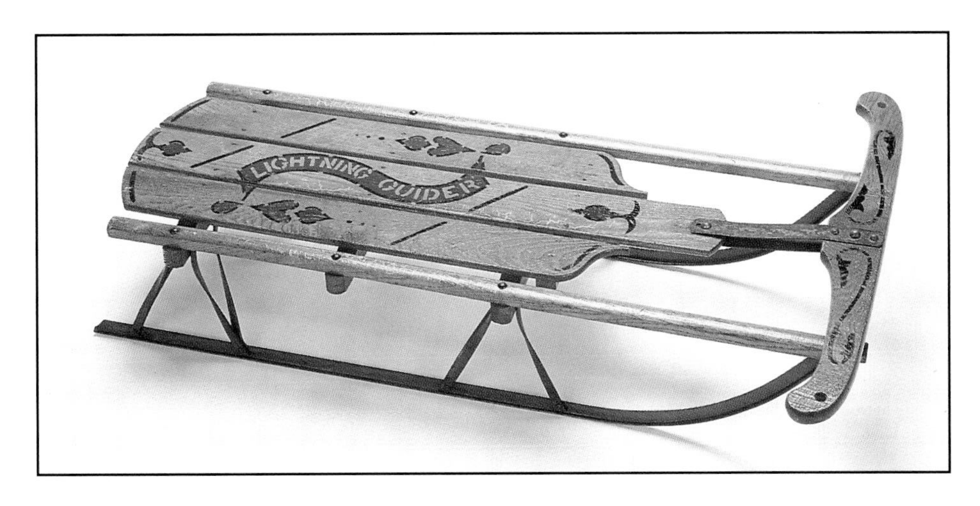

C. 1907. Lightning Guider. L: 34", W: 11.75", H: 6.25". $50-250.

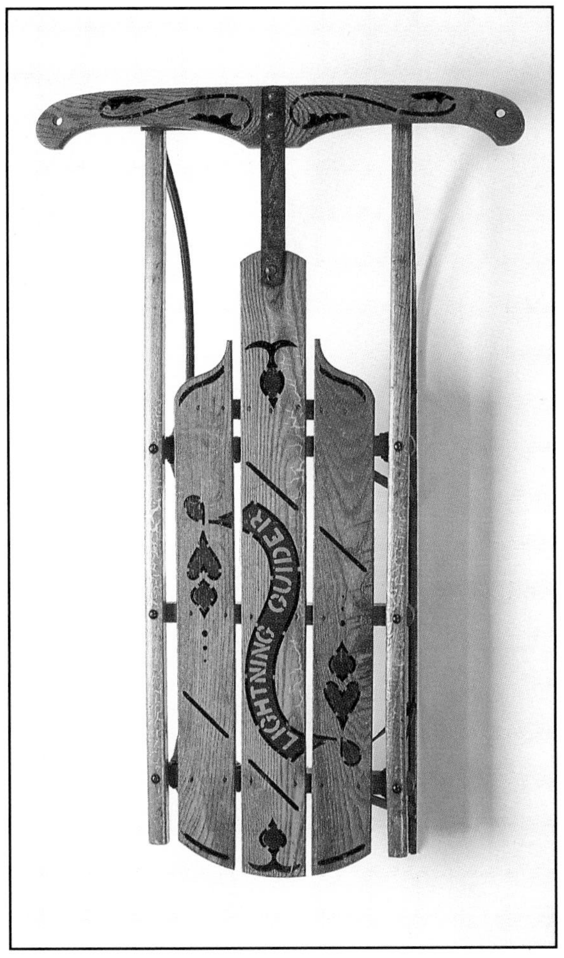

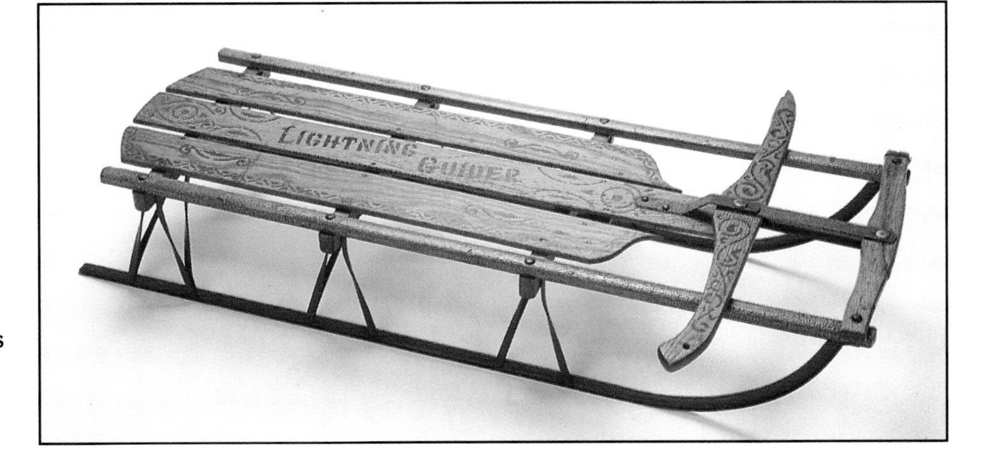

C. 1911. Lightning Guider. The only known example of this rare sled. L: 45", W: 14.25", H: 7.75". $50-450.

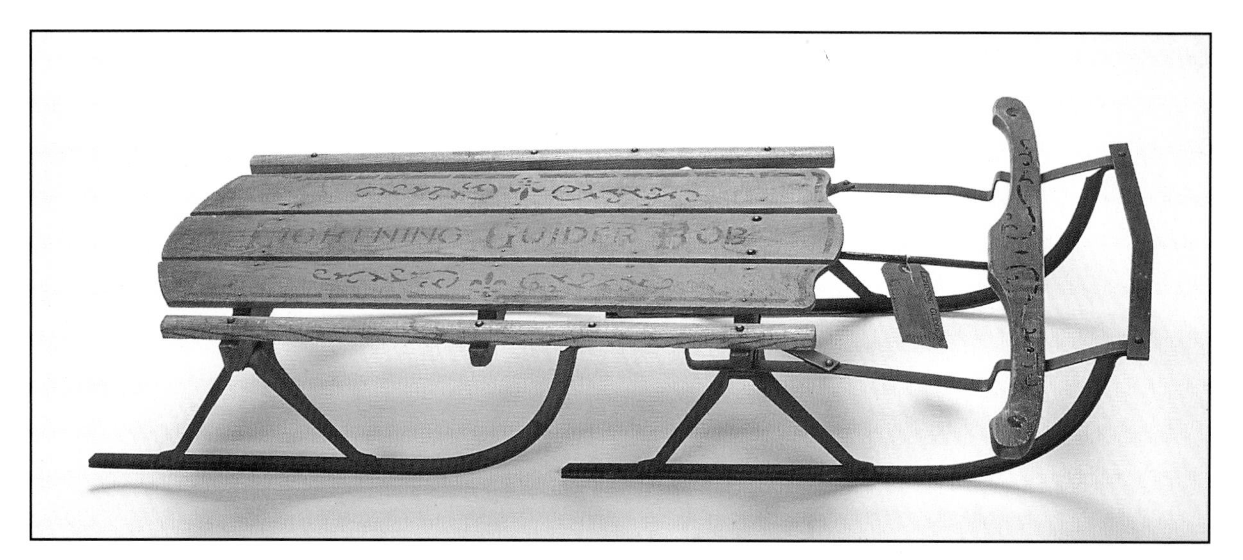

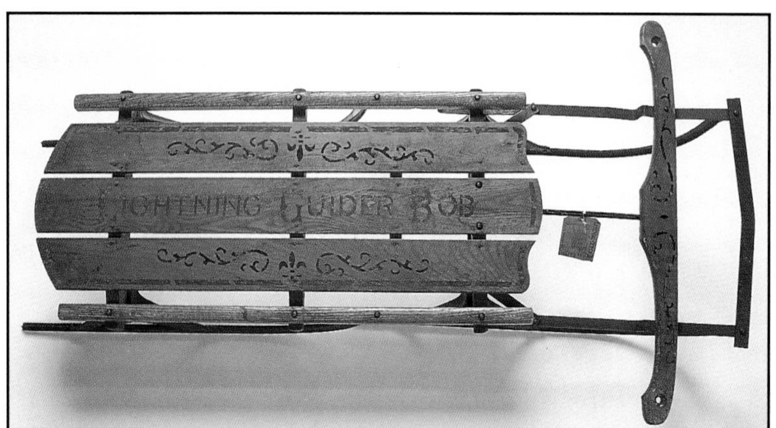

Above: C. 1917. Lightning Guider Bob with original shipping tag. L: 46", W: 14", H: 8.25". $100-900.

Below: C.1927. Lightning Guider, marked 22 1/2 underneath. L: 45", W: 12.75", H: 6.5". $50-150.

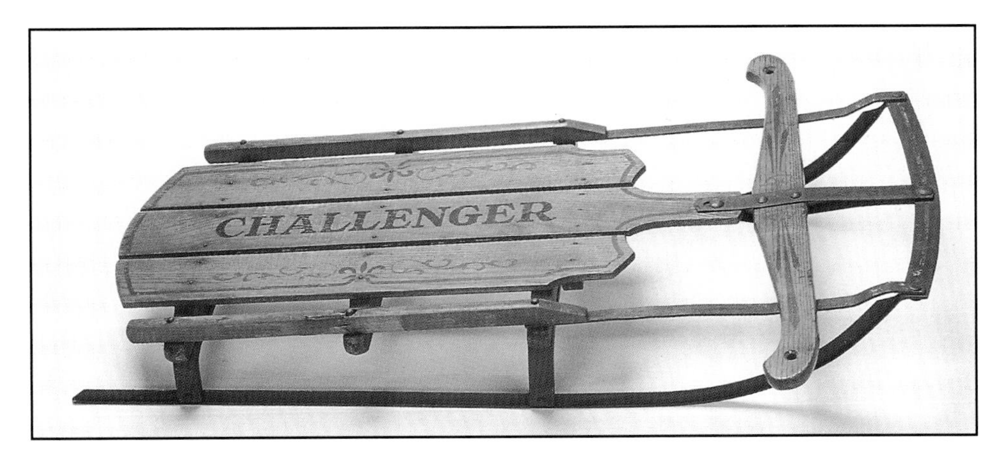

C. 1930s. Challenger.
L: 36", W: 12.5', H: 5".
$25-175.

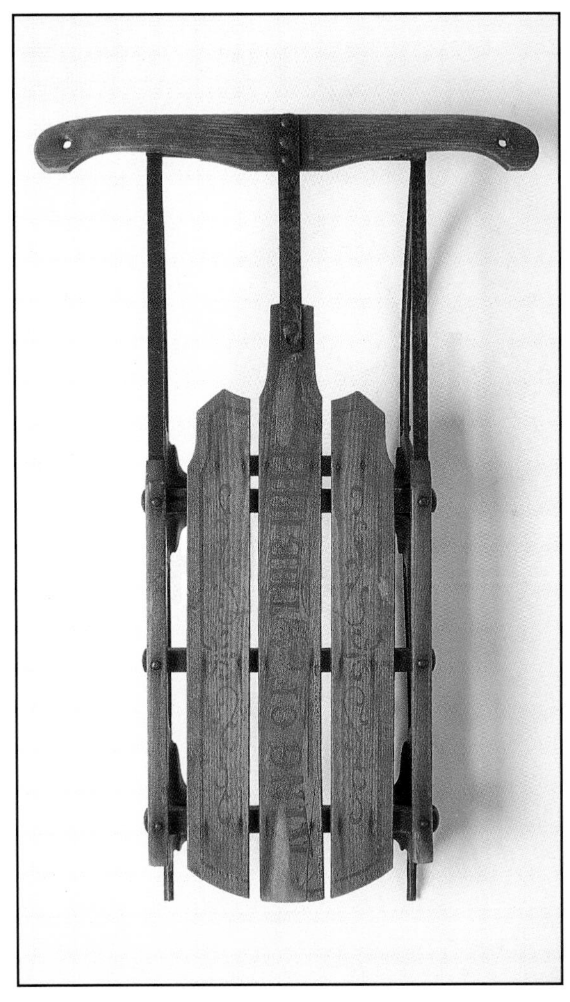

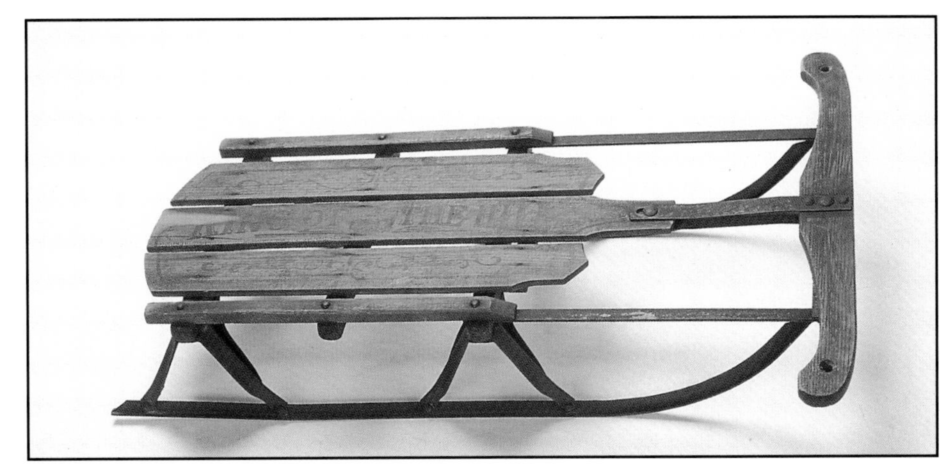

C. 1935. King of the Hill.
L: 32", W: 11.5", H:
6.125". $25-75.

1936. Master Bomber. L: 45", W: 13", H: 6.25". $50-250.

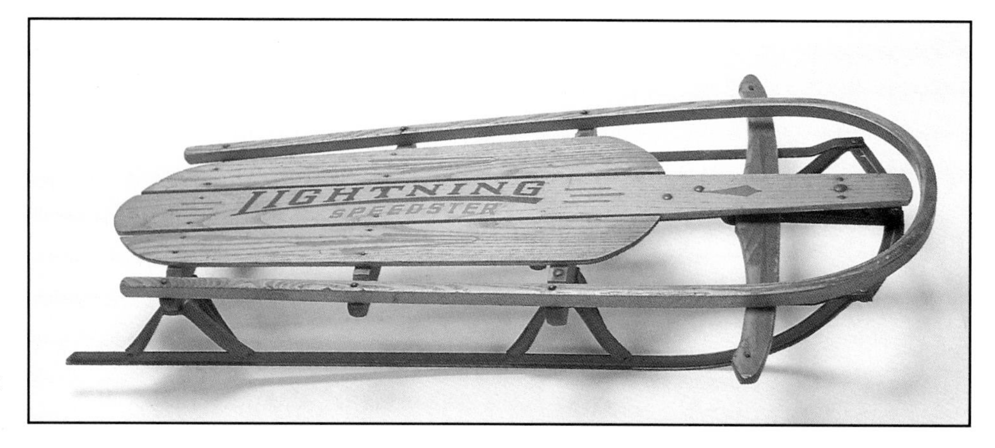

C. 1939. Lightning Speedster. L: 47", W: 14.5", H: 6.5". $50-325.

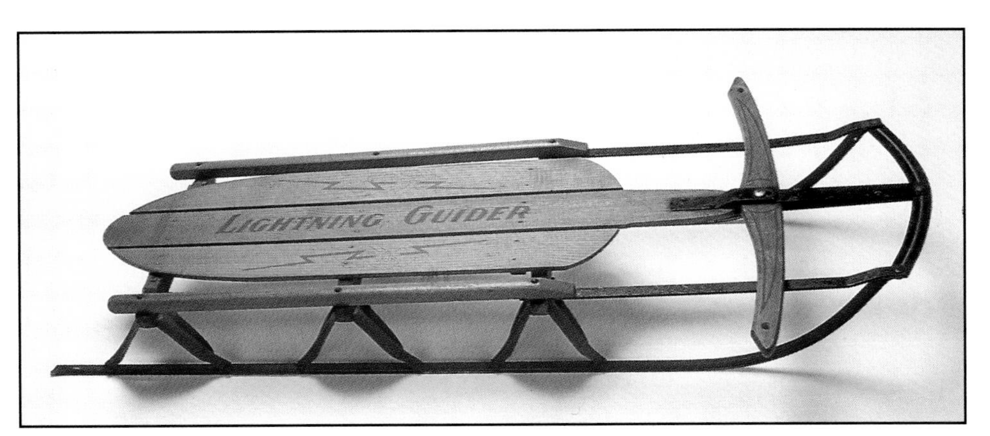

Above: 1950s. Lightning Guider. Marked #123 1/2 underneath. L: 62", W: 12.75", H: 6.25". $25-125.

Below: C. 1975. Lightning Guider. L: 40", W: 13", H: 6". $25-75.

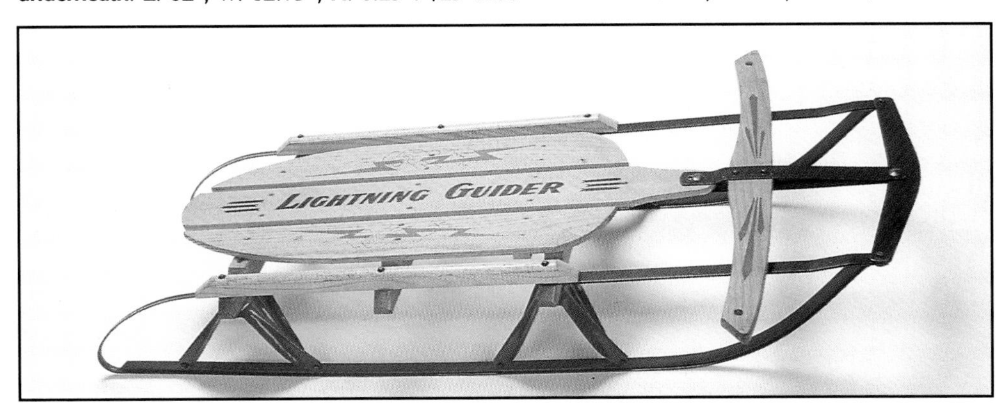

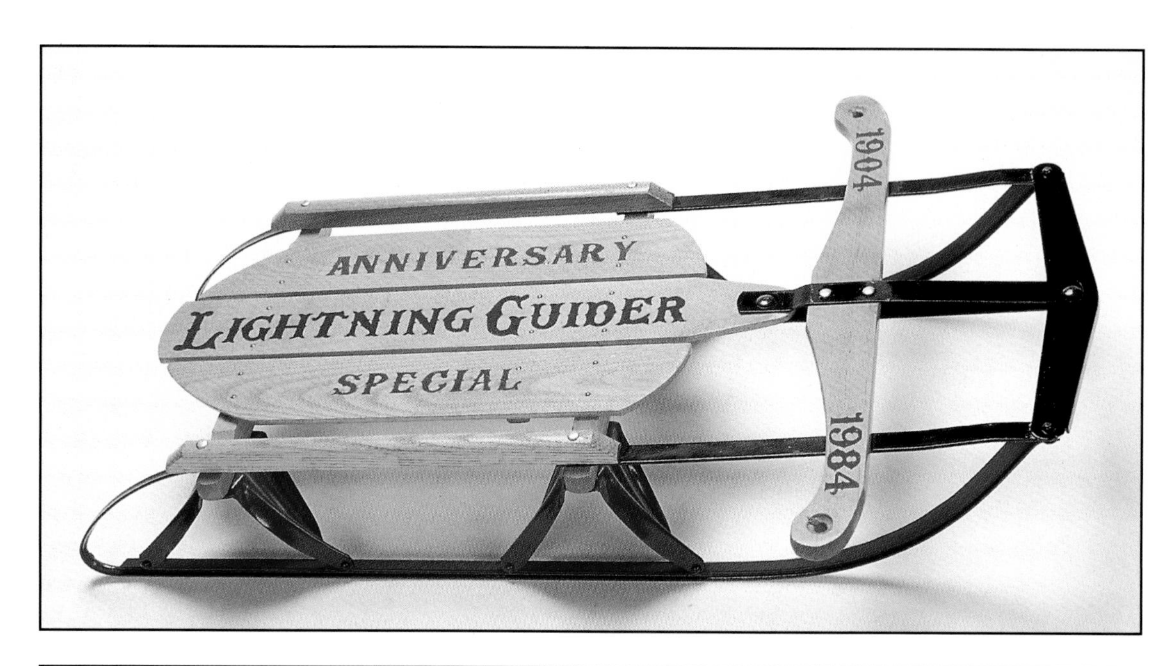

1984. Lightning Guider Anniversary model #34. #1 of 414. L: 34", W: 13", H: 5.75". $25-150.

Adam Wagner founded the company in 1908. Products manufactured in his Cedar Falls, Iowa factory were coaster wagons, steel sleds, and hardware specialties.

Sleds and wagons were referred to by the name Auto Coasters, not to be confused with the Auto Wheel Coaster Company of Tonawanda, New York.

An advertisement in *American Boy,* December, 1916, shows the Wagner Auto Coaster Wagon and Sled.

Very little information is available on the Wagner Auto Coasters. According to the 1978 Early Industries Calendar of Cedar Falls, the company employed 67 people in 1915 and a few years later discontinued the Wagner Steel Sled, because of the advent of a competing model called the "Flexible Flyer®." This new sled had a steering mechanism on the front and reportedly was desired by children of all ages.

Advertisement for Wagner Manufacturing Co. While focusing on wagons, it notes: "We also make a very complete line of sleds—flexible coaster styles and regular types. Dealers everywhere sell both Wagner Wagons and Sleds."

III. Tobogganing

The toboggan, a primitive utility vehicle used by the Indians has been around for a long time. It was not until manufactured toboggans made of strips of wood fitted together and curved up in the front forming a hood were available that thrill seekers were drawn to the sport of coasting.

Tobogganing as a sport had its beginnings on the slopes of Mount Royal, Montreal. Nature's hills and slopes were the original coasting sites. Soon these slopes were replaced with a Russian idea. A high wooden structure called a chute was erected. The design would enable you to pull your toboggan up one side and fly like a rocket down the ice covered chute. Chutes, or runs as they were often called, became very popular in the late 1800s. The most famous chute was none other than the Cresta at St. Moritz, Switzerland, with construction beginning in 1884.

Man's fascination with snow coupled with a competitive spirit and desire for recreation brought about the development of organized clubs to preserve winter sports.

Sporting fever was catching on in the United States and the first toboggan slide in this country opened in 1884 in Saratoga followed by Utica, Albany, Orange, Boston, and New York.

Fashionable private clubs like the Tuxedo Club, Tuxedo, New York and the Ardsley Club at Ardsley-on-Hudson constructed their own toboggan runs, but it was in Europe at St. Moritz that the first organized toboggan club was formed and the fascination for thrill seeking continued.

The Toboggan Club of St. Moritz devised a method of fastening sled-like runners to the toboggan frame making the run faster and more dangerous. They called the new sled a bobsleigh and the Bobsleigh Club of St. Moritz was formed.

Bobsledding became an official sport in 1923. Today bobsledding and luging are official sports headquartered in Lake Placid, New York.

(Reference: *Book of Winter Sports,* McMillan Company, Oct. 1912. Edited by J.C. Dier.)

Illustration for the article: "Tobogganing at Orange."
Harper's Weekly, February 14, 1891.

Star Toboggans

Poster for Star Toboggans in original frame. Johnson Emerson & Co., Burlington, Vermont, "Sole proprietors of the Star Patents and Star * Trademarks on Toboggans hereby warn all infringers that they will be vigorously prosecuted. Copyright ". $500-1500.

114 BURLINGTON CITY DIRECTORY.

Elliott Mary Miss, em C. E. Pease
Ellis Chas. W. em G. A. Hall, h 221 S. Willard
Ellis C. W. mail team, h 158 North
Ellis Fred. J. driver and porter, 174 College, res 221 S. Willard
Elmer Huldah M. Miss, nurse at Home for Destitute Children
Elmore William, em city, bds 154 S. Champlain
EMERSON C. H. & CO. (W. J. Van Patten) manufacturers of cabinets for the trade, boxes, toboggans, etc., office and factory Champlain shops, S. Battery, cor Main. (See page 114)
Emerson Charles H. (C. H. Emerson & Co.) h 18 N. Battery
Emerson Horace, custom boot maker, 17 Center, h do

Advertisement in the Burlington City Directory for Chas. H. Emerson, "Proprietors and Sole Manufacturers of the 'STAR PATENT' the only TOBOGGAN that lessens the frictional surface; 'KING STAR' the only Toboggan without bolts, screws or rivets." 1890

HARPER'S WEEKLY.
JOURNAL OF CIVILIZATION.
NEW YORK, SATURDAY, FEBRUARY 16, 1889.
TEN CENTS A COPY. WITH A SUPPLEMENT.

"The Toboggan Spill." *Harper's Weekly* cover, February 16, 1889. $50-75

132 HARPER'S WEEKLY. [FEBRUARY 13, 1875.

COASTING ON BOSTON COMMON.—[FROM A SKETCH BY E. H. GARRETT.]

COASTING ON BOSTON COMMON.

"Coasting on Boston Common." *Harper's Weekly* cover, February 13, 1875.

Three views of the toboggan slide at Preakness Valley Park in New Jersey. The photos were taken in December, 1939 by Harvey F. Dutcher and show a view from the bottom, the top and a look at the toboggan house. *Courtesy of the Passaic County Historic Society.*

IV. The Finale: Rosebud

In the opening scene of the 1941 classic film "Citizen Kane," Charles Foster Kane, newspaper tycoon, utters his dying word: "Rosebud."

The entire movie plot evolves around the mystery of what the word meant. In the end we find that the billionaire's last thought was of a childhood gift from his mother, the only possession he had that meant more than anything else in the world to him. The final scene shows a miniature sled engulfed in flames...his only true love.

Rosebud was a RKO prop made out of balsa wood. Originally three copies were made from the hardwood prototype. Two of the copies were incinerated during the filming and the third was bought at Sotheby's for $60,500 by Steven Spielberg in 1986, perhaps the largest price ever paid for an American Sled to that point. The hardwood prototype sled was won by Art Bauer of Brooklyn in a movie contest when he was eleven years old. It came up for auction at Christies in Los Angeles in 1996, where it sold for $233,500.

A close up of the design on the deck of Rosebud.
© Southeby's, New York. Used with permission.

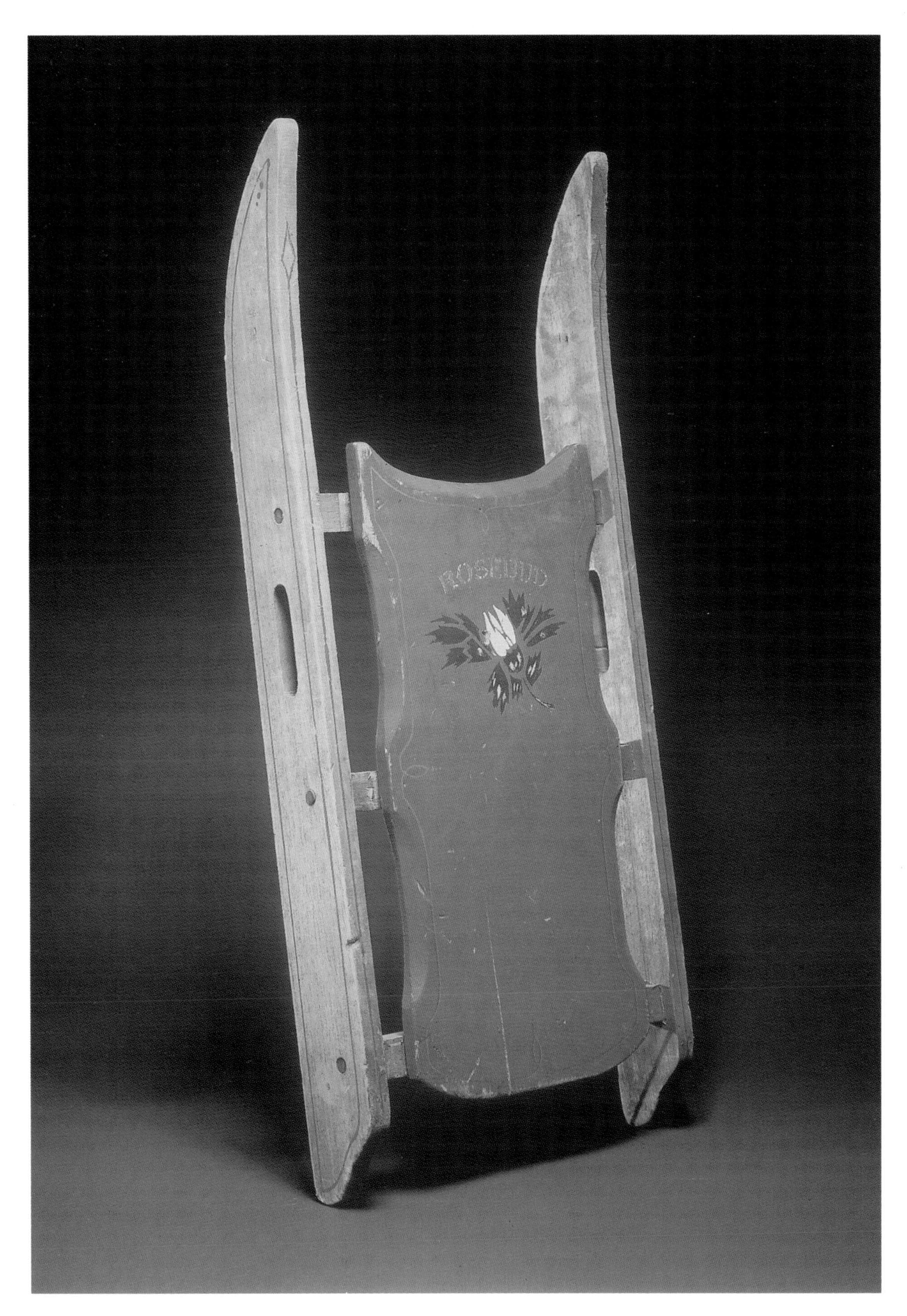

The only survivor of the three original Rosebud sleds used in "Citizen Kane." The others were destroyed in the course of making the movie. © Sotheby's, New York. Used with permission.

References

Greater Harvard Area Historical Society, Harvard, Illinois. Ms. Selma Davidson, Historian (Hunt, Helm, and Ferris)
Mead Public Library, Sheboygan, Wisconsin. (Garton Toy)
Cedar Falls Public Library, Cedar Falls, Iowa. (Wagner Auto Coaster, Wagner MFG)
Historical Society York County, Pennsylvania. Lila Fourhman-Shaull, Assistant Librarian (American Acme Toy Emnis)
Passaic County Freeholders. Edward A. Smyk, Passaic County Historian, Paterson, New Jersey
Kalamazoo Public Library, Kalamazoo, Michigan. Jan Parke, local history
North Tonawanda Public Library, Tonawanda, New York. Janet McKenna, Reference Services and Collection Development Librarian (Buffalo Sled, Auto Wheel Coaster Company)
Fletcher Free Library, Burlington, Vermont. Charles H. Emerson. (Star Toboggan)
Donald Schumacher, Cannonball HNP, Beloit, Wisconsin.
Tom Johnson for tracking down son of inventor of the Icecycle, Contoocook, New Hampshire.
John Everett Bean Jr., Concord, New Hampshire
Historic Society of the Tonawanda, New York. Mr. Willard Dittmar, historian (Autowheel Coaster)